AF769152

1. Nuclear many-body problem

In this chapter we summarize the quantum-mechanical background required for treating the nuclear many-body problem in the low-energy nuclear physics limit.

1.1 Schrödinger equation for nuclear systems

According to first quantum-mechanical principles, the microscopic nuclear theory describes properties of N-particle systems on the basis of the time-dependent Schrödinger equation

$$\hat{H}\,\Psi(t) = i\hbar\,\frac{\partial\Psi(t)}{\partial t}, \tag{1.1}$$

where $\hat{H}$ is the full nuclear Hamiltonian (Schrödinger 1926a,b,c,d). The many-particle wave function $\Psi(t) = \Psi(x_1, x_2, \ldots, x_N; t)$, that is the solution of eqn (1.1), describes the dynamic state of the system which belongs to a complex separable Hilbert space $\mathfrak{L}^N$ (von Neumann 1932).

In the Schrödinger picture, where the operator $\hat{H}$ does not depend explicitly upon the time, eqn (1.1) may be integrated formally to give the solution

$$\Psi(t) = \exp\left(-\frac{i}{\hbar}\,\hat{H}\,t\right)\,\Psi(0), \tag{1.2}$$

where $\Psi(0)$ is the wave function at time $t = 0$ and

$$\exp\left(-\frac{i}{\hbar}\,\hat{H}t\right) = 1 - \left(\frac{i}{\hbar}\,\hat{H}t\right) + \frac{1}{2!}\left(\frac{i}{\hbar}\,\hat{H}t\right)^2 - \frac{1}{3!}\left(\frac{i}{\hbar}\,\hat{H}t\right)^3 + \ldots\,. \tag{1.3}$$

Stationary Schrödinger equation

The solution (1.2) is, in principle, equivalent to the solution of the stationary Schrödinger equation

$$\hat{H}\,\Psi(x_1,x_2,\ldots,x_N) = E\,\Psi(x_1,x_2,\ldots,x_N), \tag{1.4}$$

where $x_i = (\mathbf{r}_i, s_i, \tau_i)$ stands for the spatial $\mathbf{r}_i$, spin s_i, and isospin τ_i coordinates of the i-th nucleon. Indeed, the general solution (1.2) of eqn (1.1) can be represented as a linear combination of its stationary solutions

$$\Psi_n(t) = \exp\left(-\frac{i}{\hbar}E_n t\right)\Psi_n(x_1,x_2,\ldots,x_N), \tag{1.5}$$

where E_n are the eigenvalues and $\Psi_n(x_1,x_2,\ldots,x_N)$ are the associated eigenfunctions of the Hamiltonian $\hat{H}$. We shall therefore be mainly interested in the eigenvalue problem (1.4) and especially in the energy E_0 and the wave function $\Psi_0(x_1,x_2,\ldots,x_N)$ of the ground-state; that is, the state of lowest energy. All other states E_n, Ψ_n with $n > 0$ (and $E_n > E_0$) are called excited states of the system.

Nuclear Hamiltonian

Under the three standard assumptions of low-energy nuclear physics discussed later in Section 1.5, the nuclear Hamiltonian $\hat{H}$ is expressed as a sum of the following two terms:

$$H \equiv \hat{T} + \hat{V} = \sum_{i}^{N} \hat{t}_i + \sum_{i<j}^{N} \hat{\upsilon}_{ij}, \tag{1.6}$$

where the first term $\hat{T}$ is the single-particle kinetic energy operator,

$$\hat{t}_i = \hat{t}(\mathbf{r}_i) = -\frac{\hbar^2}{2m}\nabla^2_{\mathbf{r}_i}, \tag{1.7}$$

and the second term $\hat{V}$ is a sum of two-body forces $\hat{\upsilon}_{ij} = \hat{\upsilon}(x_i,x_j)$.

Symmetry properties

There is a number of symmetry operations under which the nuclear many-body Hamiltonian is invariant, that is, it commutes with the associated symmetry operators $\hat{S}$,

$$[\hat{H}, \hat{S}] = \hat{H}\,\hat{S} - \hat{S}\,\hat{H} = 0, \tag{1.8}$$

and, therefore, the exact wave functions Ψ_n are simultaneous eigenfunctions of $\hat{H}$ and $\hat{S}$. Examples are the three components $\hat{P}_x$, $\hat{P}_y$, and $\hat{P}_z$ of the momentum operator, the particle number operator $\hat{N}$, and the angular momentum operators $\hat{J}^2$ and $\hat{J}_z$. Another important example is the permutation operator upon the particle coordinates (including spin and isospin variables)

$$\hat{P}\,\Psi(x_1,x_2,\ldots,x_N) = C_p\,\Psi(x_1,x_2,\ldots,x_N)\,, \tag{1.9}$$

where $C_p = 1$ for completely symmetric wave functions and $C_p = (-1)^p$ for completely antisymmetric ones, p giving the number of interchanges in $\hat{P}$.

One may verify that the symmetry properties of the many-particle wave function are conserved during the evolution of the system. Suppose that the wave function $\Psi(0)$ at time $t = 0$ is a solution of the Schrödinger equation and, at the same time, it has the symmetry

$$\hat{S}\,\Psi(x_1,x_2,\ldots,x_N;t = 0) = s\,\Psi(x_1,x_2,\ldots,x_N;t = 0), \tag{1.10}$$

where s stands for the eigenvalue of the symmetry operator $\hat{S}$. Then, it follows from eqn (1.2) and eqn (1.8) that the solution $\Psi(t)$ of eqn (1.1) at any moment t has the same symmetry because

$$\begin{aligned}\hat{S}\,\Psi(t) &= \hat{S}\,\exp\left(-\frac{i}{\hbar}\,\hat{H}\,t\right)\,\Psi(0) = \exp\left(-\frac{i}{\hbar}\,\hat{H}\,t\right)\,\hat{S}\,\Psi(0)\\ &= s\,\exp\left(-\frac{i}{\hbar}\,\hat{H}\,t\right)\,\Psi(0) = s\,\Psi(t).\end{aligned} \tag{1.11}$$

The wave functions of a collection of N indistinguishable particles are completely symmetric (the case of boson systems) or antisymmetric (the case of fermion systems) and this property is conserved in time since the Hamiltonian $\hat{H}$ is invariant under the permutation operator $\hat{P}$; that is, $[\hat{H},\hat{P}] = 0$. Therefore, there are two subspaces $\mathfrak{L}^N_B$ and $\mathfrak{L}^N_F$ in the full Hilbert space $\mathfrak{L}^N$ which contain the dynamic states of the boson and fermion systems, respectively (Berezin 1986).

Nuclei are fermion systems so their wave functions must be antisymmetric,

$$\hat{P}\,\Psi(x_1,x_2,\dots,x_N) = (-1)^P\,\Psi(x_1,x_2,\dots,x_N), \tag{1.12}$$

and, therefore, they belong to the Hilbert space $\mathfrak{L}^N_F$. In other words, although it is a solution of eqn (1.4), the function $\Psi \in \mathfrak{L}^N$ will not be the correct nuclear wave function if it does not belong to $\mathfrak{L}^N_F \subset \mathfrak{L}^N$.

Independent-particle model

Let us consider as an example the independent-particle model Hamiltonian

$$\hat{H}_0 = \sum_{i=1}^{N}\left[-\frac{\hbar^2}{2m}\nabla_i^2 + U(x_i)\right], \tag{1.13}$$

where U(x) is a given mean field acting on the nucleon at coordinate **x**. Hamiltonians of the given type are very often used in nuclear theory. For instance, $\hat{H}_0$ may be used as the basis of a perturbation treatment of the difference

$$\hat{H}' = \hat{H} - \hat{H}_0, \tag{1.14}$$

which formally represents the full Hamiltonian into the form

$$\hat{H} = \hat{H}_0 + \hat{H}'. \tag{1.15}$$

The solution of the Schrödinger equation (1.4) with the Hamiltonian (1.13) is a relatively simple matter since the equation

$$\hat{H}_0 \Psi = E\, \Psi \tag{1.16}$$

is separable. We could write for its solution

$$\Psi = \psi_1(x_1)\, \psi_2(x_2) \dots \psi_N(x_N). \tag{1.17}$$

Inserting this in eqn (1.16) and dividing by Ψ gives

$$\sum_{i=1}^{N} \frac{1}{\psi_i} \left[-\frac{\hbar^2}{2m} \nabla_i^2 + U(x_i) \right] \psi_i = E. \tag{1.18}$$

Each term in the sum on the left-hand side depends upon the coordinates of one nucleon only, so that the equation can be separated into the N equations

$$\left[-\frac{\hbar^2}{2m} \nabla_i^2 + U(x_i) \right] \psi_i = \varepsilon_i\, \psi_i, \tag{1.19}$$

with

$$\sum_{i=1}^{N} \varepsilon_i = E. \tag{1.20}$$

Clearly, the ground-state wave function would be just a product wave function (1.17) whose elements are the single-particle wave functions ψ_i corresponding to the lowest energy levels ε_i. However, the function (1.17) is fully symmetric, that is, it belongs to the subspace $\mathfrak{L}_B^N \subset \mathfrak{L}^N$, so that it contradicts eqn (1.12). Therefore, instead of a single product like eqn (1.17) we must take an antisymmetric sum of such products with the particle coordinates permuted in all possible ways. The result is a determinant function

$$\Psi = \frac{1}{(N!)^{1/2}} \begin{vmatrix} \psi_1(x_1) & \psi_2(x_1) & \ldots \psi_N(x_1) \\ \vdots & \vdots & \vdots \\ \psi_1(x_N) & \psi_2(x_N) & \ldots \psi_N(x_N) \end{vmatrix}, \tag{1.21}$$

which belongs to $\mathfrak{L}^N_F \subset \mathfrak{L}^N$ and, therefore, represents a proper wave function of the independent-particle model system considered.

When the term $\hat{H}'$ is included in the Hamiltonian (1.15), so that we have the Schrödinger equation (1.4), this equation can be no longer separated, and the wave function cannot be expressed as a single determinant of one-particle wave functions. However, we should always take into account the antisymmetrization condition (1.12) when describing nuclei and, therefore, to work within the subspace $\mathfrak{L}^N_F \subset \mathfrak{L}^N$. Further, we restrict our consideration to nuclei only, and denote the proper subspace $\mathfrak{L}^N_F \subset \mathfrak{L}^N$ simply by $\mathcal{H}^N \equiv \mathfrak{L}^N_F$.

Slater determinants

In order to employ the Hilbert space $\mathcal{H}^N$ one could construct a proper complete set of N-particle wave functions that must be obviously completely antisymmetric. The eigenfunctions $\Psi_n \in \mathcal{H}^N$ introduced via eqn (1.4) are orthonormalized:

$$\langle \Psi_n | \Psi_m \rangle = \delta_{nm} \tag{1.22}$$

and they form a complete set because the operator $\hat{H}$ is Hermitian. The antisymmetric property (1.12), however, is not explicitly satisfied within the set.

An alternative step is to construct a complete set of explicitly antisymmetric N-particle wave functions in $\mathcal{H}^N$. For this purpose, we start with a complete set of single-particle states $|\alpha\rangle$ which correspond to the single-particle wave functions in coordinate space

$$\varphi_\alpha(x) = \varphi_\alpha(\mathbf{r}, s, \tau) = \langle \mathbf{r}, s, \tau \,|\alpha\rangle, \tag{1.23}$$

where the meaning of the index α is twofold. On one hand, it stands for a complete set of the quantum numbers characterizing the associated nucleon state $|\alpha\rangle$. On the other, α may be considered as a number $\alpha = (1,2,3,\ldots)$ that enumerates each of the states (1.23). We do not specify the actual form of the wave functions (1.23) but only assume that they satisfy the orthogonal and completeness conditions

$$\langle\alpha|\beta\rangle = \int \varphi_\alpha^*(x)\, \varphi_\beta(x)\, dx = \delta_{\alpha\beta} \tag{1.24}$$

and

$$\sum_\alpha \varphi_\alpha^*(x)\, \varphi_\alpha(x') = \delta(x - x'), \tag{1.25}$$

respectively. For continuous quantum numbers as $\mathbf{r}$, the symbol $\delta_{\alpha\alpha'}$ will mean $\delta(\mathbf{r} - \mathbf{r}')$ and the sum $\sum_\alpha$ is to be replaced by $\int d\mathbf{r}$, while $\delta(x - x')$ stands for $\delta(\mathbf{r} - \mathbf{r}')\delta_{ss'}\delta_{\tau\tau'}$. The integration with respect to x implies also a sum over the two values ±1/2 of both the spin and isospin variables, so that

$$\int |\varphi_\alpha(x)|^2\, dx = \sum_{\substack{s=\pm 1/2 \\ \tau=\pm 1/2}} \int |\varphi_\alpha(\mathbf{r}, s, \tau)|^2 d\mathbf{r}\ . \tag{1.26}$$

Using the set $\{\varphi_\alpha(x)\}$ we can construct an orthonormal set of completely antisymmetric N-particle wave functions

$$\Phi_{\alpha_1,\ldots,\alpha_N}(x_1,\ldots,x_N) = \frac{1}{(N!)^{1/2}} \sum_P (-1)^P \hat{P}\{\varphi_{\alpha_1}(x_1),\ldots,\varphi_{\alpha_N}(x_N)\}, \tag{1.27}$$

where the sum runs over all N! different permutations of the numbers $(\alpha_1,\alpha_2,\ldots,\alpha_N)$. This is just a compact way of writing the determinant wave function

$$\Phi_{[\alpha]} \equiv \Phi_{\alpha_1,\ldots,\alpha_N} = \frac{1}{(N!)^{1/2}} \begin{vmatrix} \varphi_1(x_1) & \varphi_2(x_1) & \ldots\varphi_N(x_1) \\ \vdots & \vdots & \vdots \\ \varphi_1(x_N) & \varphi_2(x_N) & \ldots\varphi_N(x_N) \end{vmatrix}, \tag{1.28}$$

the index $[\alpha] = (\alpha_1,\alpha_2,\ldots,\alpha_N)$ denotes the N different integers; that is, the N different functions chosen from the initial set (1.23). The function (1.28) is called the Slater determinant and it is antisymmetric, that is $\Phi_{[\alpha]} \in \mathcal{H}^N$, because interchange of the coordinates of two particles changes two rows in the determinant (1.28) and, thus, its sign.

The Slater determinant (1.28) internally accounts for the Pauli exclusion principle which states that no two fermions in a given assembly can occupy the same quantum state. Indeed, it is obvious from eqn (1.28) that no two single-particle states $\varphi_\alpha(x)$ can be the same in the Slater determinant, because if it were so, two rows would be identical and the determinant would vanish.

We may verify immediately that the Slater determinant (1.28) is normalized to unity and that any two different Slater determinants are orthogonal, see for example (Raimes 1972). It can be also proved (Löwdin 1955) that the Slater determinant orthonormal set $\{\Phi_{[\alpha]}\}$ is a complete set in N variables if the single-particle set $\{\varphi_\alpha(x)\}$ is complete. Therefore, the set $\{\Phi_{[\alpha]}\}$ can be used as a basis in the Hilbert space $\mathcal{H}^N$ and any wave function $\Psi \in \mathcal{H}^N$ can be expanded in terms of this basis,

$$\Psi(x_1,\ldots,x_N) = \sum_{\alpha_1,\ldots,\alpha_N} C_{\alpha_1,\ldots,\alpha_N} \Phi_{\alpha_1,\ldots,\alpha_N}(x_1,\ldots,x_N), \qquad (1.29)$$

with coefficients $C_{[\alpha]}$ being suitably chosen constants. The function (1.29) is antisymmetric, because interchange of the coordinates of two particles changes the sign of every Slater determinant in eqn (1.29).

Occupation numbers

The single-particle states $\{\varphi_\alpha(x)\}$ cannot be associated with any particular particle coordinates $x_1,x_2,\ldots,x_N$ in the wave function (1.29). In any case, however, it is sufficient to list how many quantum states $\varphi_\alpha(x)$ are repeated, which ones, and how often. Such specification can be done in terms of the infinite sequence of integers $n_{\alpha_1}, n_{\alpha_2}, n_{\alpha_3}, \ldots,$ the occupation numbers, where n_α designates the number of times $\varphi_\alpha(x)$ occurs in any one of the Slater determinants $\Phi_{\alpha_1,\alpha_2,\ldots,\alpha_N}(x_1,x_2,\ldots,x_N)$ comprising Ψ.

For a Slater determinant, an occupation number n_α is defined as 0 or 1 according to whether the function $\varphi_\alpha(x)$ does or does not appear in eqn (1.28). If $n_\alpha \geq 2$ the Slater determinant vanishes identically as two or more rows in the determinant are identical. Hence, one Slater determinant may be specified completely by giving the occupation numbers $[n_\alpha] = (n_{\alpha_1}, n_{\alpha_2}, n_{\alpha_3}, \ldots)$,

$$\Phi_{[n_\alpha]}(x_1,x_2,\ldots,x_N) = \Phi_{\alpha_1,\alpha_2,\ldots,\alpha_N}(x_1,x_2,\ldots,x_N), \tag{1.30}$$

where $[n_\alpha]$ is an infinite string of ones and zeros showing us which $\varphi_\alpha(x)$ are ($n_\alpha = 1$) or are not ($n_\alpha = 0$) occupied in eqn (1.28).

1.2 Variational principle

According to the variational principle, the Schrödinger equation (1.4) is equivalent to the variational equation

$$\delta_\Psi\, E[\Psi] = 0, \quad \Psi \in \mathcal{H}^N, \tag{1.31}$$

where

$$E[\Psi] = \frac{\langle\Psi|\,\hat{H}\,|\Psi\rangle}{\langle\Psi|\Psi\rangle} \tag{1.32}$$

is the expectation value of the nuclear Hamiltonian $\hat{H}$ taken between states $\Psi(x_1,x_2,\ldots,x_N)$ belonging to $\mathcal{H}^N$. Therefore, the eigenvalue problem (1.4) within the Hilbert space $\mathcal{H}^N$ may be always derived from the variational principle.

In order to prove this equivalence we consider the variation δE of the energy functional (1.32)

$$\langle\Psi|\Psi\rangle\ \delta E = \delta\langle\Psi|\ \hat{H}\ |\Psi\rangle - E\ \delta\langle\Psi|\Psi\rangle$$

$$= \langle\delta\Psi|(\hat{H} - E)|\Psi\rangle - \langle\Psi|(\hat{H} - E)|\delta\Psi\rangle\ . \tag{1.33}$$

According to eqn (1.31) the following equation is valid for arbitrary $\delta\Psi \in \mathcal{H}^N$:

$$\langle\delta\Psi|(\hat{H} - E)|\Psi\rangle - \langle\Psi|(\hat{H} - E)|\delta\Psi\rangle = 0. \tag{1.34}$$

One can then replace $\delta\Psi$ by $i\delta\Psi \in \mathcal{H}^N$ in eqn (1.34) because the infinitesimal variation $\delta\Psi \in \mathcal{H}^N$ is arbitrary

$$- i\langle\delta\Psi|(\hat{H} - E)|\Psi\rangle - i\langle\Psi|(\hat{H} - E)|\delta\Psi\rangle = 0. \tag{1.35}$$

Together with eqn (1.34), we find

$$\langle\delta\Psi|(\hat{H} - E)|\Psi\rangle = 0 \tag{1.36}$$

and the complex conjugate equation. Finally, eqn (1.36) is equivalent to the eigenvalue problem (1.4) because $\delta\Psi \in \mathcal{H}^N$ is arbitrary (Messiah 1966).

Ground-state variational principle

We are now in a position to define variationally the nuclear ground-state problem, that is, to consider the variational principle leading to the Schrödinger equation

$$\hat{H}\ \Psi_0(x_1,x_2,\ldots,x_N) = E_0\ \Psi_0(x_1,x_2,\ldots,x_N). \tag{1.37}$$

First, we shall demonstrate that the following relation holds for any wave function $\Psi \in \mathcal{H}^N$:

$$E[\Psi] \geq E_0. \tag{1.38}$$

This can easily be done introducing in $\mathcal{H}^N$ the complete orthonormal set of eigenfunctions $\Psi_n \in \mathcal{H}^N$ of the Hamiltonian H,

$$\hat{H}\,\Psi_n(x_1,x_2,\dots,x_N) = E_n\,\Psi_n(x_1,x_2,\dots,x_N). \tag{1.39}$$

Let us develop $\Psi \in \mathcal{H}^N$ in terms of these eigenfunctions; that is,

$$\Psi = \sum_{n=0}^{\infty} a_n\,\Psi_n, \tag{1.40}$$

where $a_n = \langle\Psi|\Psi_n\rangle$ and $n = (0,1,2,\dots)$. This yields

$$E[\Psi] - E_0 = \frac{\langle\Psi|(\hat{H} - E_0)|\Psi\rangle}{\langle\Psi|\Psi\rangle} = \frac{\sum_{n=0}^{\infty}(E_n - E_0)\,|a_n|^2}{\sum_{n=0}^{\infty}|a_n|^2} \geq 0 \tag{1.41}$$

which is precisely eqn (1.38) because $E_n \geq E_0$ and $|a_n|^2 \geq 0$ for all $n = (0,1,2,\dots)$.

When the ground-state energy E_0 is not degenerate, the equal sign in eqn (1.41) is valid if and only if all a_n with $n \neq 0$ vanish, that is, $\Psi \in \mathcal{H}^N$ is proportional to $\Psi_0 \in \mathcal{H}^N$ according to eqn (1.40). In any other cases, that is for all $\Psi \neq \Psi_0$, we will have

$$E[\Psi] > E_0 = E[\Psi_0]. \tag{1.42}$$

The ground-state variational principle can be thus represented into the form

$$E_0 = \min_{\Psi\in\mathcal{H}^N} \frac{\langle\Psi|\,\hat{H}\,|\Psi\rangle}{\langle\Psi|\Psi\rangle}\,. \tag{1.43}$$

In other words, the solution of the Schrödinger equation (1.37) is equivalent to the investigation of all functions $\Psi \in \mathcal{H}^N$ with respect to the energy functional (1.32). Its absolute minimum (1.43) determines the exact energy E_0, while the wave function $\Psi_0 \in \mathcal{H}^N$ coincides with the exact solution of eqn (1.37).

Variational principle for excited states

When analysing the first excited state $\Psi_1 \in \mathcal{H}^N$ (see eqn (1.39) for $n = 1$), one must investigate with respect to $E[\Psi]$ the subspace $\mathcal{H}_1^N \subset \mathcal{H}^N$ orthogonal to $\Psi_0 \in \mathcal{H}^N$, that is, all wave functions (1.40) with $a_0 = 0$. The exact first-excited wave function $\Psi_1 \in \mathcal{H}_1^N$ leads to the minimal expectation value of $\hat{H}$:

$$E_1 \equiv E[\Psi_1] = \min_{\Psi \in \mathcal{H}_1^N} \frac{\langle \Psi | \hat{H} | \Psi \rangle}{\langle \Psi | \Psi \rangle} . \tag{1.44}$$

Therefore, according to the variational principle one must carry out the variation of the functional (1.32) with the supplementary condition $\langle \Psi | \Psi_0 \rangle = 0$. This process can continue and calculate the whole spectrum of $\hat{H}$ by using the variational principle with more and more supplementary conditions.

It should be noted that the energy functional (1.32) does not depend on the norm $\langle \Psi | \Psi \rangle$ of $\Psi \in \mathcal{H}^N$. Therefore, all the above results hold also under any arbitrary conditions on the latter quantity. When considering the ground-state variational principle, for example, one can use the normalization condition

$$\langle \Psi | \Psi \rangle = 1 \tag{1.45}$$

and can define the functional $E[\Psi]$ into the form

$$E[\Psi] = \langle \Psi | \hat{H} | \Psi \rangle . \tag{1.46}$$

In this case, the derivation of the variational equation (1.31) with (1.46) must account for the subsidiary condition (1.45) via the Lagrange multiplier E, the ground-state energy of the system.

Time-dependent variational principle

Before concluding this section, it should be mentioned that the Schrödinger equation (1.1) can also be derived from a time-dependent variational principle (Kerman and Koonin 1976). The

Lagrangian is defined as

$$\mathcal{L}(\Psi,\Psi^*) = \langle\Psi(t)|\ i\hbar \frac{\partial}{\partial t} - \hat{H}\ |\Psi(t)\rangle, \tag{1.47}$$

where $\Psi(t) \in \mathcal{H}^N$ is the general time-dependent many-particle wave function. The action L corresponding to the Lagrangian (1.47) for evolution between times t_1 and t_2 is

$$L = \int_{t_2}^{t_1} \mathcal{L}(\Psi,\Psi^*)\ dt. \tag{1.48}$$

The equation of motion is found by requiring L to be stationary with respect to variations of Ψ and Ψ^* between fixed end points t_1 and t_2, which is just the Schrödinger equation (1.1) and its adjoint.

1.3 Second-quantization representation

The second-quantization representation is an alternative formulation of the usual quantum mechanics which has proved convenient in developments of the nuclear many-body problem. For its introduction, one may first construct the Fock space $\mathcal{H}$ which contains a vacuum (no particles) $|-\rangle = |0,0,0\ldots\rangle$, all the one-particle states, all the antisymmetric two-particle states and so on, that is,

$$\mathcal{H} = \bigoplus_{N\geq 0} \mathcal{H}^N = \left\{\mathcal{H}^0,\ \mathcal{H}^1,\ \mathcal{H}^2\ ,\ldots,\mathcal{H}^N,\ldots\right\}. \tag{1.49}$$

All states belonging to the Fock space $\mathcal{H}$ whose components are equal to zero, except the N-th one, form the subspace $\mathcal{H}^N \subset \mathcal{H}$. The Slater determinants $\Phi_{[n]} \in \mathcal{H}^N$, introduced by eqn (1.30) for $N = 0,1,2,\ldots$, correspond to the basis states $|n_1,n_2\ldots\rangle$ in $\mathcal{H}$. They are characterized by the occupation numbers (occupation number representation), such as

$$\Phi_{[n]}(x_1,x_2,\ldots,x_N) = \langle x_1,x_2,\ldots,x_N|n_1,n_2\ldots\rangle, \tag{1.50}$$

and

$$\langle n'_1,\ldots,n'_\alpha,\ldots|n_1,\ldots,n_\alpha,\ldots\rangle = \delta_{n_1n'_1}\ldots\delta_{n_\alpha n'_\alpha}\ldots \quad . \tag{1.51}$$

Particle creation and annihilation operators

Let us now define the operators a^+_α and a_α which obey the Fermi anticommutation relations

$$\begin{aligned} \left\{a^+_\alpha , a_\beta\right\} &\equiv a^+_\alpha a_\beta + a_\beta a^+_\alpha = \delta_{\alpha\beta}, \\ \left\{a^+_\alpha , a^+_\beta\right\} &= \left\{a_\alpha , a_\beta\right\} = 0 \end{aligned} \tag{1.52}$$

for all α and β. The operator a_α is called the annihilation operator because by definition it annihilates a particle in the state $\varphi_\alpha(x)$ if it is occupied, that is, it lowers the occupation number in the state with number α by one,

$$a_\alpha|n_1,n_2,\ldots,n_\alpha = 1,\ldots\rangle = |n_1,n_2,\ldots,n_\alpha = 0,\ldots\rangle. \tag{1.53}$$

Otherwise, we have $a_\alpha|n_1,n_2,\ldots,n_\alpha = 0,\ldots\rangle = 0$. In particular, the action of a_α on the vacuum state $|-\rangle$ gives

$$a_\alpha\ |-\rangle = 0 \quad \text{for all } \alpha. \tag{1.54}$$

Similarly, the operator a^+_α is a creation operator in the state α because

$$a^+_\alpha|n_1,n_2,\ldots,n_\alpha = 0,\ldots\rangle = |n_1,n_2,\ldots,n_\alpha = 1,\ldots\rangle\ , \tag{1.55}$$

and $a^+_\alpha|n_1,n_2,\ldots,n_\alpha = 1,\ldots\rangle = 0$.

Obviously, the N-particle basis state $|n_1, n_2, \ldots\rangle$ in $\mathcal{H}$ could be expressed as

$$|n_1, n_2, \ldots\rangle = \prod_{\alpha} \left(a_{\alpha}^{+}\right)^{n_\alpha} |-\rangle = a_{\alpha_1}^{+} a_{\alpha_2}^{+} \ldots a_{\alpha_N}^{+} |-\rangle. \tag{1.56}$$

Having thus determined the basis states (1.56) one could represent the arbitrary N-particle wave function (1.29) in the second-quantization representation as

$$|\Psi\rangle = \sum_{\alpha_1, \ldots, \alpha_N} C_{\alpha_1, \ldots, \alpha_N} a_{\alpha_1}^{+} a_{\alpha_2}^{+} \ldots a_{\alpha_N}^{+} |-\rangle . \tag{1.57}$$

Field creation and annihilation operators

We will very often use field creation and annihilation operators

$$\psi^{+}(x) = \sum \varphi_{\alpha}^{*}(x)\, a_{\alpha}^{+}, \tag{1.58}$$

and

$$\psi\,(x) = \sum \varphi_{\alpha}(x)\, a_{\alpha}, \tag{1.59}$$

which create or annihilate a particle at the coordinate $x = (r, s, \tau)$, respectively. From the anticommutators (1.52), one finds

$$\left\{\psi^{+}(x),\ \psi(x')\right\} = \delta(x - x'),$$

and

$$\left\{\psi^{+}(x),\ \psi^{+}(x')\right\} = \left\{\psi(x),\ \psi(x')\right\} = 0. \tag{1.60}$$

By eqn (1.24) and eqn (1.25), the relations (1.58) and (1.59) can be inverted

$$\begin{aligned} a_{\alpha} &= \int dx\ \varphi_{\alpha}^{*}(x)\ \psi(x) \\ a_{\alpha}^{+} &= \int dx\ \varphi_{\alpha}(x)\ \psi^{+}(x). \end{aligned} \tag{1.61}$$

They allow the basis states (1.56) to be represented in terms of the field creation operators,

$$|n_1,n_2,\ldots\rangle = \int dx_1\ldots dx_N \frac{1}{(N!)^{1/2}} \Phi_{[n]}(x_1,\ldots,x_N)\psi^+(x_1)\ldots\psi^+(x_N)|-\rangle, \tag{1.62}$$

where obviously

$$\Phi_{[n]}(x_1,\ldots,x_N) = \frac{1}{(N!)^{1/2}} \langle -|\psi(x_N),\ldots,\psi(x_1)|n_1,n_2,\ldots\rangle. \tag{1.63}$$

By analogy to eqn (1.57) one can represent the arbitrary $\Psi \in \mathcal{H}^N$ in terms of field creation and annihilation operators

$$|\Psi\rangle = \sum_{\alpha_1,\ldots,\alpha_N} C_{\alpha_1,\ldots,\alpha_N}$$

$$\times \int dx_1\ldots dx_N \frac{1}{(N!)^{1/2}} \Phi_{[n]}(x_1,\ldots,x_N)\psi^+(x_1)\ldots\psi^+(x_N)|-\rangle. \tag{1.64}$$

In this way, starting from the vacuum $|-\rangle$, one expresses all N-particle states from $\mathcal{H}^N$ in terms of the operators a^+_α, a_α or the particle field operators $\psi^+(x)$, $\psi(x')$. The same should be done for the operators of the physical quantities acting in $\mathcal{H}^N$.

Operators in second-quantization representations

We consider here the case of one- and two-body operators only. A one-body operator $\hat{F}$ is given as a sum of N operators $\hat{f}_i$ acting on the coordinate of the i-th particle

$$\hat{F} = \sum_{i=1}^{N} \hat{f}_i. \tag{1.65}$$

One can verify that the second-quantization representation of $\hat{F}$ is of the form

$$\hat{F} = \sum_{\alpha,\beta} f_{\alpha\beta}\, a^+_\alpha\, a_\beta, \tag{1.66}$$

where the matrix elements

$$f_{\alpha\beta} = \langle\varphi_\alpha|\ \hat{f}\ |\varphi_\beta\rangle, \tag{1.67}$$

accord to the action

$$\hat{f}_1\varphi_\alpha(x_1) = \sum_\beta f_{\beta\alpha}\ \varphi_\beta(x_1). \tag{1.68}$$

A two-particle operator $\hat{V}$ is given by a sum of operators $\hat{\mathfrak{v}}_{ij}$ which act on the coordinates of the particles i and j,

$$\hat{V} = \sum_{i<j=1}^{N} \hat{\mathfrak{v}}_{ij}. \tag{1.69}$$

In the most general case, $\hat{\mathfrak{v}}$ may be an integral operator with matrix elements

$$\langle\varphi_\alpha\varphi_\beta|\ \hat{\mathfrak{v}}\ |\varphi_{\alpha'}\varphi_{\beta'}\rangle$$
$$=\int dx_1 dx_2 dx_1' dx_2' \varphi_\alpha(x_1)\varphi_\beta(x_2)\hat{\mathfrak{v}}(x_1,x_2,x_1',x_2')\varphi_{\alpha'}(x_1')\varphi_{\beta'}(x_2'). \tag{1.70}$$

In second-quantization representation the operator $\hat{V}$ can be rewritten as

$$\hat{V} = \frac{1}{2}\sum_{\alpha\beta\alpha'\beta'}\langle\varphi_\alpha\varphi_\beta|\ \hat{\mathfrak{v}}\ |\varphi_{\alpha'}\varphi_{\beta'}\rangle\ a_\alpha^+\ a_\beta^+\ a_{\beta'}a_{\alpha'}$$

$$= \frac{1}{4}\sum_{\alpha\beta\alpha'\beta'}\langle\varphi_\alpha\varphi_\beta|\ \hat{\mathfrak{v}}\ |\varphi_{\alpha'}\varphi_{\beta'}\rangle_A\ a_\alpha^+\ a_\beta^+\ a_{\beta'}a_{\alpha'}, \tag{1.71}$$

where

$$\langle\varphi_\alpha\varphi_\beta|\ \hat{\mathfrak{v}}\ |\varphi_{\alpha'}\varphi_{\beta'}\rangle_A = \langle\varphi_\alpha\varphi_\beta|\hat{\mathfrak{v}}|\varphi_{\alpha'}\varphi_{\beta'}\rangle - \langle\varphi_\alpha\varphi_\beta|\hat{\mathfrak{v}}|\varphi_{\beta'}\varphi_{\alpha'}\rangle \tag{1.72}$$

are the antisymmetrized matrix elements of $\hat{\mathfrak{v}}$.

In order to prove eqn (1.66) we consider the action of the operator (1.65) on an arbitrary $\Psi \in \mathcal{H}^N$. By using eqn (1.64) we obtain (Ring and Schuck 1980)

$$\frac{1}{(N!)^{1/2}} \hat{F}\, \Psi$$

$$= \sum_{l=1}^{N} \hat{f}_l \, \langle - | \psi(x_N) \ldots \psi(x_l) \ldots \psi(x_1) | \Psi \rangle$$

$$= \sum_{l=1}^{N} \sum_{\alpha_1, \ldots, \alpha_N} \hat{f}_l \, \varphi_{\alpha_N}(x_N) \ldots \varphi_{\alpha_l}(x_l) \ldots \varphi_{\alpha_1}(x_1) \, \langle - | a_{\alpha_N} \ldots a_{\alpha_1} | \Psi \rangle$$

$$= \sum_{l=1}^{N} \sum_{\alpha} \sum_{\alpha_1, \ldots, \alpha_N} f_{\alpha\alpha_l} \, \varphi_{\alpha_N}(x_N) \ldots \varphi_{\alpha}(x_l) \ldots \varphi_{\alpha_1}(x_1) \, \langle - | a_{\alpha_N} \ldots a_{\alpha_1} | \Psi \rangle . \tag{1.73}$$

On other hand, we find that the action of the operator (1.68) onto the state (1.57) gives the same result:

$$\frac{1}{(N!)^{1/2}} \hat{F}\, |\Psi\rangle = \langle x_1, \ldots, x_N | \sum_{\alpha, \alpha'} f_{\alpha\alpha'} \, a^+_{\alpha} a_{\alpha'} \, |\Psi\rangle$$

$$= \sum_{\alpha\alpha'} \sum_{\alpha_1, \ldots, \alpha_N} f_{\alpha\alpha'} \, \varphi_{\alpha_N}(x_N) \ldots \varphi_{\alpha_1}(x_1) \, \langle - | a_{\alpha_N} \ldots a_{\alpha_1} a^+_{\alpha} a_{\alpha'} | \Psi \rangle$$

$$= \sum_{l=1}^{N} \sum_{\alpha} \sum_{\alpha_1, \ldots, \alpha_N} f_{\alpha\alpha_l} \, \varphi_{\alpha_N}(x_N) \ldots \varphi_{\alpha}(x_l) \ldots \varphi_{\alpha_1}(x_1) \, \langle - | a_{\alpha_N} \ldots a_{\alpha_1} | \Psi \rangle . \tag{1.74}$$

Therefore, we have

$$\sum_{=1}^{N} \hat{f}_l \Psi(x_1, x_2, \ldots, x_N) = \langle x_1, \ldots, x_N | \sum_{\alpha, \alpha'} f_{\alpha\alpha'} \, a^+_{\alpha} a_{\alpha'} \, |\Psi\rangle, \tag{1.75}$$

which shows that the forms (1.65) and (1.68) of the operator $\hat{F}$ are identical due to the arbitrariness of $\Psi \in \mathcal{H}^N$. By complete analogy of eqn (1.75) one can identify both forms (1.69) and (1.71) of the two-body operator $\hat{V}$.

One can also express the operators $\hat{F}$ and $\hat{V}$ in terms of the field operators (1.58) and (1.59) by inserting eqn (1.61) into eqn (1.66) and eqn (1.71), respectively. One obtains

$$\hat{F} = \int dx\ \psi^+(x)\ \hat{f}(x)\ \psi(x) \tag{1.76}$$

and

$$\hat{V} = \frac{1}{2} \int dx\ dx'\ \psi^+(x)\psi^+(x')\ \hat{\omega}(x,x')\ \psi(x')\psi(x). \tag{1.77}$$

Concluding, we obtain the following two different expressions for the nuclear Hamiltonian (1.6) in second-quantization representation

$$\hat{H} = \sum_{\alpha,\beta} \langle\varphi_\alpha|\ \hat{t}\ |\varphi_\beta\rangle\ a_\alpha^+\ a_\beta + \frac{1}{2} \sum_{\alpha,\beta,\gamma,\delta} \langle\varphi_\alpha\varphi_\beta|\hat{\omega}|\varphi_\gamma\varphi_\delta\rangle\ a_\alpha^+\ a_\beta^+\ a_\delta\ a_\gamma \tag{1.78}$$

and

$$\hat{H} = \int dx\ \psi^+(x)\hat{t}(x)\psi(x) + \frac{1}{2}\int dx\ dx'\ \psi^+(x)\psi^+(x')\hat{\omega}(x,x')\psi(x')\psi(x). \tag{1.79}$$

1.4 Many-body problem in terms of density matrices

In any case, the N-particle wave function contains a great deal of information which is almost wholly irrelevant in considering many of the most important physical properties. Thus, it is very advantageous to introduce the so-called density matrices defined from the many-body wave function by integrating out much of the redundant information. The density matrix formulation of the many-body problem enables us to focus attention directly on physically important quantities, thus making more transparent the investigation of nuclear phenomena. In addition, the formulation of the many-body problem in terms of density matrices is more general because it allows us also to describe some mixed states which have no single wave function representation (von Neumann 1932).

Density matrix operators

One may first define the most general density matrix operator of the order n (n = 1,2,3,...,N) in terms of the field creation and annihilation operators

$$\hat{\rho}^{(n)}(x_1,x_2,\ldots,x_n;x'_1,x'_2,\ldots,x'_n) = \psi^+(x'_n)\ldots\psi^+(x'_1)\psi(x_1)\ldots\psi(x_n). \tag{1.80}$$

Among the operators (1.80) we shall concentrate attention on the one- and two-body density matrix operators

$$\hat{\rho}(x_1,x'_1) \equiv \hat{\rho}^{(1)}(x_1,x'_1) = \psi^+(x'_1)\ \psi(x_1) \tag{1.81}$$

and

$$\hat{\Gamma}(x_1,x_2;x'_1,x'_2) \equiv \hat{\rho}^{(2)}(x_1,x_2;x'_1,x'_2) = \psi^+(x'_2)\psi^+(x'_1)\psi(x_1)\psi(x_2), \tag{1.82}$$

respectively. They can also be represented as

$$\hat{\rho} \equiv \hat{\rho}(x_1,x'_1) = \sum_{\alpha,\beta} a^+_\alpha\ a_\beta\ \varphi^*_\alpha(x'_1)\ \varphi_\beta(x'_1) \tag{1.83}$$

and

$$\hat{\Gamma} \equiv \hat{\Gamma}(x_1,x_2;x'_1,x'_2) = \sum_{\beta\gamma\delta} a^+_\alpha a^+_\beta a_\gamma a_\delta \varphi^*_\alpha(x'_2)\varphi^*_\beta(x'_1)\varphi_\gamma(x_1)\varphi_\delta(x_2). \tag{1.84}$$

One-body density matrix

Density matrices are called the expectation values of the operators $\hat{\rho}^{(n)}$ taken on the N-particle states $\Psi \in \mathcal{H}^N$. The one-particle density matrix may be expressed in the following equivalent representations

$$\rho(x_1,x_1') = N \int \Psi^*(x_1',x_2,\ldots,x_N)\, \Psi(x_1,x_2,\ldots,x_N)\, dx_2\ldots dx_N$$

$$= \langle\Psi|\psi^+(x_1')\psi(x_1)|\Psi\rangle = \sum_{\alpha,\beta} \rho_{\beta\alpha}\, \varphi_\alpha^*(x_1')\, \varphi_\beta(x_1), \tag{1.85}$$

where

$$\rho_{\beta\alpha} = \langle\Psi|a_\alpha^+\, a_\beta|\Psi\rangle \tag{1.86}$$

are the matrix elements of the one-body operator (1.81) in the arbitrary basis (1.23).

It is obvious that the operators (1.80) are Hermitian and consequently the matrix (1.86) can be diagonalized by an unitary transformation $\hat{D}$ of the single-particle basis; that is,

$$(\hat{D}^+\, \hat{\rho}\, \hat{D})_{\alpha\beta} = \rho_\alpha\, \delta_{\alpha\beta}\,, \tag{1.87}$$

where

$$\rho_\alpha = \langle\Psi|c_\alpha^+\, c_\alpha|\Psi\rangle\,, \quad 0 \le \rho_\alpha \le 1 \tag{1.88}$$

and $c_\alpha^+ = \sum_\beta D_{\beta\alpha}\, a_\beta^+$. The basis $\bar{\varphi}_\alpha = \sum_\alpha D_{\beta\alpha}\, \varphi_\beta$ in which the one-body density matrix (1.85) is diagonal is called the basis of natural single-particle orbitals (Löwdin 1955), whereby the associated natural occupation numbers (1.88) express the probability that the level $\bar{\varphi}_\alpha$ is occupied in the wave function $\Psi \in \mathcal{H}^N$. In natural occupation numbers representation the exact one-body density matrix (1.85) obviously has the simplest form

$$\rho(x_1,x_1') = \sum_\alpha \rho_\alpha\, \bar{\varphi}_\alpha^*(x_1')\, \bar{\varphi}_\alpha(x_1). \tag{1.89}$$

The importance of the one-body density matrix is due to the fact that it allows the expectation value of any single-particle operator to be calculated without making reference to the complete many-particle wave function $\Psi \in \mathcal{H}^N$. For example, we have from eqn (1.66) and eqn (1.85) that

$$\langle\Psi|\hat{F}|\Psi\rangle = \sum_{\alpha,\beta} f_{\alpha\beta}\, \rho_{\beta\alpha} = \mathrm{tr}(\hat{f}.\hat{\rho}). \tag{1.90}$$

The right-hand side of eqn (1.90) is convenient because it does not depend on the particular form of the single-particle basis in which the matrix elements (1.67) are calculated. In particular, the kinetic energy part of the functional (1.46) can be represented by the following equivalent forms:

$$\begin{aligned} T[\Psi] &= \langle\Psi|\hat{T}|\Psi\rangle = \langle\Psi|\sum_{i=1}^{N} \hat{t}(\mathbf{r}_i)|\Psi\rangle \\ &= \frac{\hbar^2}{2m}\int dx\, dx'\ \delta(\mathbf{r}-\mathbf{r}')\ \nabla\, .\, \nabla'\ \rho(x,x') \\ &= \sum_{\alpha,\beta} t_{\alpha\beta}\, \rho_{\beta\alpha} = \sum_{\alpha} \langle\bar{\varphi}_\alpha|\hat{t}|\bar{\varphi}_\alpha\rangle\, \rho_\alpha = \mathrm{tr}(\hat{t}\ .\ \hat{\rho}). \end{aligned} \tag{1.91}$$

Another important example of eqn (1.90) is the local density distribution whose operator is of the form

$$\hat{\rho}(\mathbf{r}) = \sum_{i=1}^{N} \delta(\mathbf{r}_i - \mathbf{r}), \tag{1.92}$$

where $\mathbf{r}_i$ is the space operator of the i-th particle and the radius vector $\mathbf{r}$ is a parameter. The expectation value of $\hat{\rho}(\mathbf{r})$ with respect to $\Psi \in \mathcal{H}^N$ defines the local density distribution of the particles

$$\rho(\mathbf{r}) = N\int \Psi^*(x_1,x_2,\ldots,x_N)\left[\sum_{i}^{N} \delta(\mathbf{r}_i - \mathbf{r})\right]\Psi(x_1,x_2,\ldots,x_N)\ dx_1\ldots dx_N$$

$$= \sum_{s,\tau} \langle\Psi|\psi^{+}(\mathbf{r}, s, \tau)\, \psi(\mathbf{r}, s, \tau)|\Psi\rangle$$

$$= \sum_{s,\tau} \rho(\mathbf{r}, s, \tau; \mathbf{r}, s, \tau). \tag{1.93}$$

Having in mind eqn (1.93), it is useful to define the spinless first-order density matrix

$$\rho(\mathbf{r}, \mathbf{r}') = \sum_{s,\tau} \rho(\mathbf{r}, s, \tau; \mathbf{r}', s, \tau) \tag{1.94}$$

whose diagonal element $\rho(\mathbf{r}) = \rho(\mathbf{r},\mathbf{r}' \equiv \mathbf{r})$ gives just the local density (1.93).

By using eqn (1.94), the full one-body density matrix $\rho(x,x')$ can be exactly represented into the form (Bohr and Mottelson 1969)

$$\rho(x,x') = \frac{1}{4}\rho(\mathbf{r},\mathbf{r}')\, I + \frac{1}{2}\sum_{k} (\rho_{\sigma}(\mathbf{r},\mathbf{r}'))_{k}\, \sigma_{k} + \frac{1}{2}\sum_{\ell} (\rho_{\tau}(\mathbf{r},\mathbf{r}'))_{\ell}\, \tau_{\ell}$$
$$+ \sum_{k,\ell} (\rho_{\sigma\tau}(\mathbf{r},\mathbf{r}'))_{k\ell}\, \sigma_{k}\, \tau_{\ell}\,, \tag{1.95}$$

where σ_k (τ_k), $(k = 1,2,3)$ are the Pauli matrices which, together with the identity 2 x 2 matrix I, form a complete set in the spin (isospin) 2 x 2 matrix space. We have introduced in eqn (1.95) also the one-body spin and isospin density matrices

$$\rho_{\sigma}(\mathbf{r},\mathbf{r}') = \sum_{\tau} \rho(\mathbf{r}, s, \tau; \mathbf{r}', s', \tau) \tag{1.96}$$

and

$$\rho_{\tau}(\mathbf{r},\mathbf{r}') = \sum_{s} \rho(\mathbf{r}, s, \tau; \mathbf{r}', s, \tau'). \tag{1.97}$$

Their components are respectively

$$(\rho_\sigma(\mathbf{r}, \mathbf{r}'))_k = \sum_{s,s'} \rho_\sigma(\mathbf{r}, s; \mathbf{r}', s')\,(\sigma_k)_{ss'} = \mathrm{tr}(\hat{\rho}\,.\,\sigma_k) \qquad (1.98)$$

and

$$(\rho_\tau(\mathbf{r}, \mathbf{r}'))_k = \sum_{\tau,\tau'} \rho_\tau(\mathbf{r}, \tau; \mathbf{r}', \tau')\,(\tau_k)_{\tau\tau'} = \mathrm{tr}(\hat{\rho}.\tau_k), \qquad (1.99)$$

and by analogy

$$(\rho_{\sigma\tau}(\mathbf{r}, \mathbf{r}'))_{k\ell} = \mathrm{tr}(\hat{\rho}.\sigma_k.\tau_\ell). \qquad (1.100)$$

Two-body density matrix

The two-body density matrix is defined from eqn (1.84) as

$$\Gamma(x_1,x_2;x_1',x_2') = \langle\Psi|\psi^+(x_2')\psi^+(x_1')\psi(x_1)\psi(x_2)|\Psi\rangle$$

$$= \tfrac{1}{2}N(N-1)\int\Psi^*(x_1',x_2',\ldots,x_N)\Psi(x_1,x_2,\ldots,x_N)dx_3\ldots dx_N$$

$$= \sum_{\alpha\beta\gamma\delta} \Gamma_{\alpha\beta\gamma\delta}\;\varphi_\alpha^*(x_2')\varphi_\beta^*(x_1')\varphi_\delta(x_1)\varphi_\gamma(x_2), \qquad (1.101)$$

where its matrix elements in the basis (1.23) are

$$\Gamma_{\alpha\beta\gamma\delta} = \langle\Psi|a_\alpha^+a_\beta^+\,a_\delta a_\gamma|\Psi\rangle\;. \qquad (1.102)$$

The symmetry properties $\Gamma_{\alpha\beta\gamma\delta} = -\Gamma_{\beta\alpha\gamma\delta} = -\Gamma_{\alpha\beta\delta\gamma} = \Gamma_{\beta\alpha\delta\gamma}$ immediately result from the anticommutation relations (1.52). By analogy to eqn (1.94) one can consider the two-body density matrix depending on the radius-vector coordinates only; that is,

$$\Gamma(\mathbf{r}_1,\mathbf{r}_2;\mathbf{r}_1',\mathbf{r}_2') = \sum_{s_1 s_2 \tau_1 \tau_2} \Gamma(\mathbf{r}_1,s_1,\tau_1,\mathbf{r}_2,s_2,\tau_2;\mathbf{r}_1',s_1',\tau_1',\mathbf{r}_2',s_2',\tau_2'). \qquad (1.103)$$

Its diagonal matrix elements characterize the correlation properties of the system; for example, $\Gamma(\mathbf{r}_1,\mathbf{r}_2;\mathbf{r}_1,\mathbf{r}_2)$ is in fact equal to the probability of finding simultaneously one nucleon in position $\mathbf{r}_1$ and another nucleon in $\mathbf{r}_2$ (multiplied by the number of the different pairs in the system, $N(N-1)/2$).

The two-particle density matrix (1.101) allows one to calculate expectation values of two-body operators on the state $\Psi \in \mathcal{H}^N$. For example, one can express the potential energy of the system as

$$V[\Psi]= \langle\Psi|\hat{V}|\Psi\rangle = \langle\Psi|\sum_{i<j}^{N}\hat{\omega}_{ij}|\Psi\rangle$$

$$=\int dx_1dx_2dx'_1dx'_2\;\hat{\omega}(x_1,x_2;x'_1,x'_2)\Gamma(x_1,x_2;x'_1,x'_2) = \mathrm{tr}(\hat{\omega}\,.\,\hat{\Gamma}), \quad (1.104)$$

where

$$\hat{\omega}(x_1,x_2;x'_1,x'_2) = \delta(\mathbf{r}_1-\mathbf{r}'_1)\,\delta(\mathbf{r}_2-\mathbf{r}'_2)\,\hat{\omega}(x_1,x'_2).$$

The combination of eqn (1.91) and eqn (1.104) shows that the energy functional (1.46) can be exactly represented in terms of one- and two-body density matrices when the Hamiltonian is of the form of eqn (1.6); that is,

$$E[\Psi] = \langle\Psi|\int dx\;\psi^+(x)\hat{t}(x)\psi(x) + \frac{1}{2}\int dx\;dx'\,\psi^+(x)\psi^+(x')\hat{\omega}(x,x')\psi(x')\psi(x)|\Psi\rangle$$

$$= \frac{\hbar^2}{2m}\int dx\;dx'\;\delta(\mathbf{r}-\mathbf{r}')\;\nabla.\nabla'\rho(x,x')$$

$$+\int dx_1dx_2dx'_1dx'_2\;\hat{\omega}(x_1,x_2;x'_1,x'_2)\;\hat{\Gamma}(x_1,x_2;x'_1,x'_2)$$

$$= \mathrm{tr}(\hat{t}\,.\,\hat{\rho}) + \mathrm{tr}(\hat{\omega}\,.\,\hat{\Gamma}). \quad (1.105)$$

Finally, the time-dependent Schrödinger equation (1.1) can be represented in terms of density matrices as an infinite system of equations of motion (Bertsch 1978)

$$i\hbar \frac{\partial}{\partial t} \rho^{(n)}(r_1,\ldots,r_n;r'_1,\ldots,r'_n;t)$$

$$-\frac{\hbar^2}{2m}\sum_{i=1}^{n} (\nabla^2-\nabla'^2)\,\rho^{(n)} + \sum_{i<j}^{n}\int dr\left[\hat{\omega}(r_i,r_j)-\hat{\omega}(r'_i,r'_j)\right]\rho^{(n)}$$

$$+\sum_{i=1}^{n}\int dr\left[\hat{\omega}(r_i,r) - \hat{\omega}(r'_i,r)\right]\rho^{(n+1)}(r_1,..,r_n,r;r'_1,..,r'_n,r;t). \quad (1.106)$$

The equation for each density matrix $\rho^{(n)}$ involves the density matrix for one additional particle $\rho^{(n+1)}$. The goal of the theory is to truncate this set of equations at some lower order, by making a suitable approximation in terms of lower order density matrices (see Section 8.1).

1.5 Nuclear forces problem

The Schrödinger equation (1.1) or (1.4) immediately shows the first general difficulty that occurs within the nuclear many-body problem. There exists no derivation of the Hamiltonian $\hat{H}$ from first principles. This is true even with the following rather drastic assumptions made when writing the Hamiltonian in the form (1.6). Actually one assumes that: (i) the relativistic effects are negligible and the nucleus consists only of nonrelativistic nucleons; (ii) the dynamic mesonic degrees of freedom are ignored and the interaction between the nucleons is represented by a potential; (iii) the three-body and higher forces are neglected assuming that only two-body forces are important. Of course, the bare nucleon-nucleon interaction may be obtained within the modern theory of gauge fields for quarks and gluons, but attempts in this direction are only presently under way (Liberman 1977, De Tar 1978a,b, Matveev and Sorba 1978, Aerts, Mulders, and de Swart 1980).

Bare nucleon-nucleon interaction

Quite successful description of bare nucleon-nucleon forces is obtained within theories starting out from an effective Lagrangian for interacting mesons and nucleons (Cottingham *et al.* 1973, Brown and Jackson 1976, Durso *et al.* 1977, Larsson, Leander, and Ragnarsson 1978, Vinh Mau 1979).

Using the pion-nucleon coupling constant, which is known from the experiment, the bare nucleon force is obtained without free parameters for particle distances greater than 0.8 fm. For example, a phase shift analysis of the nucleon-nucleon scattering data shows (Bryan 1967) that the one-pion exchange potential (OPEP) is well able to reproduce the nuclear force properly at large distances ($r \geq 2$ fm). For smaller distances (the medium-range part of the force) one must also introduce the two-pion exchange and the ρ- and ω-meson exchange. Only the short-range part from zero to 0.8 fm still has to be fitted by a phenomenological ansatz containing six parameters in each isospin channel.

Entirely phenomenological potentials consist of combinations of central, tensor, spin-orbit, and higher terms, and more or less arbitrary radial functions containing up to 50 parameters, which are fitted to the experimental scattering phase shifts and the deuteron data. At large distances they go over into the OPEP-potential, whereas at short distances they have an extremely repulsive core with a core distance $r_c \simeq 0.4$ fm. Several authors such as, for example, Hamada and Johnston (1962), Lassila *et al.* (1962), use a hard-core ($\mathcal{V}_{12} = \infty$ for $r \equiv |\mathbf{r}_1 - \mathbf{r}_2| \leq r_c$), while others use a very repulsive core which goes to infinity only for $r = 0$, the so-called soft core potentials (Reid 1968).

Effective nucleon-nucleon forces

The bare nucleon-nucleon interactions are ill behaved in view of their applications to nuclei. On one hand, they show strong repulsion at short distances (hard or soft core) and cannot be treated straightforwardly by the usual many-body techniques. On the other hand, the nucleons within the nucleus do not feel the bare nucleon-nucleon interaction. They interact with one another, but in the presence of many other nucleons. This permits us to introduce an effective interaction from the bare nucleon-nucleon force.

Much work has been done to derive the effective nucleon-nucleon interaction and a great deal has been learned about its

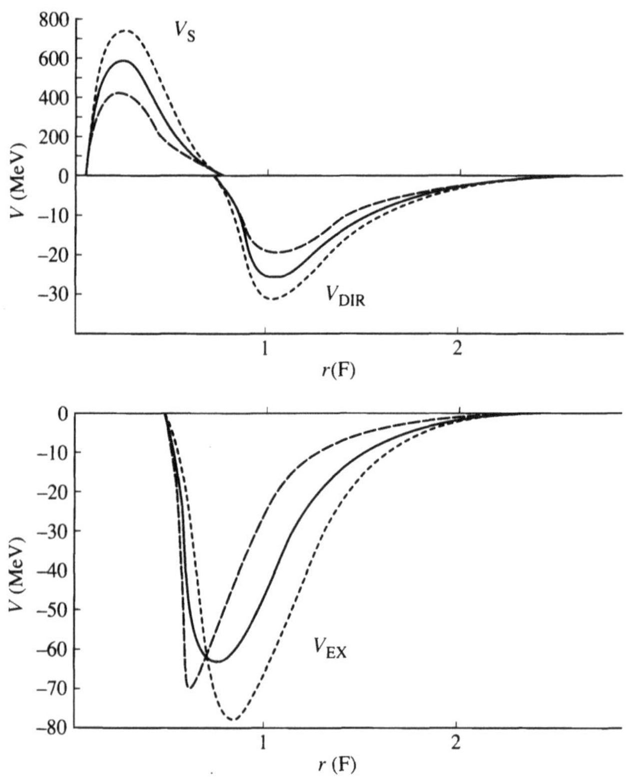

Figure 1.1. Density-dependent effective forces at $k_F = 1.4\ \mathrm{fm}^{-1}$ derived by Negele (1970). The solid lines, short dashes and long dashes correspond to the average, unlike and like interaction, respectively. V_S, V_{DIR} and V_{EX} stand for the short-range direct, long-range direct and long-range exchange interactions, respectively.

properties such as, for example, its rather well behaved character at short distances (see Figure 1.1). It lies, however, outside the scope of the book to discuss the range of validity and the success of the concept for the effective forces in nuclei. In fact, the bare and effective nucleon-nucleon forces represent a separate discipline involving complicated and rather expensive techniques. For example, the effective forces can be derived microscopically in terms of the so-called Brückner G-matrix and using the Bethe-Goldstone equation (Brückner 1955, Day 1967 and references therein). Bethe (1971) has shown how difficult it is in principle to calculate the Brückner G-matrix and to get qualitative agreement with experiment.

Instead, the use of phenomenological effective forces, which contain a certain number of parameters adjusted to reproduce the experimental data, has been adopted from the early days of nuclear physics. Presently there exists an enormous number of different phenomenological effective interactions that have been applied to various problems in nuclear physics. Among them, the Skyrme-type force is nowadays one of the widely used effective forces in nuclear theory.

Effective forces of Skyrme

In 1956 Skyrme proposed an effective interaction with a three-body term (Skyrme 1956, 1957, 1959)

$$\hat{V} = \sum_{i<j} \hat{\omega}(i,j) + \sum_{i<j<k} \hat{\omega}(i,j,k), \tag{1.107}$$

where

$$\hat{\omega}(1,2) = t_0(1 + x_0 P_\sigma)\delta(\mathbf{r}_1 - \mathbf{r}_2) + \frac{1}{2}t_1[\delta(\mathbf{r}_1 - \mathbf{r}_2)\mathbf{k}'^2 + \mathbf{k}^2\delta(\mathbf{r}_1 - \mathbf{r}_2)]$$

$$+ t_2\mathbf{k}'.\delta(\mathbf{r}_1 - \mathbf{r}_2)\mathbf{k} + iW_0(\sigma_1 + \sigma_2).\mathbf{k}' \times \delta(\mathbf{r}_1 - \mathbf{r}_2)\mathbf{k}, \tag{1.108}$$

and

$$\hat{\omega}(1\ 2\ 3) = t_3\ \delta(\mathbf{r}_1 - \mathbf{r}_2)\ \delta(\mathbf{r}_2 - \mathbf{r}_3). \tag{1.109}$$

Here $\mathbf{k} = (1/\hbar)\mathbf{p} = (1/2i)(\nabla_1 - \nabla_2)$ is the operator of the relative momentum acting on the right, whereas $\mathbf{k}' = -(1/2i)(\nabla_1 - \nabla_2)$ acts on the left; $P_\sigma = (1 - \sigma_1 . \sigma_2)/2$ is the spin-exchange operator and σ_k are the Pauli spin matrices. There are six phenomenologically adjusted constants t_0, t_1, t_2, t_3, x_0, and W_0.

The zero-range force (1.108) whose radial dependence is described by δ-functions is a simple but appropriate ansatz because the range of the nuclear forces is rather short, which is true also for the range of the effective G-matrix. Furthermore, an effective finite range is additionally simulated by the momentum dependence in eqn (1.108). To see this one can transform a function $V(\mathbf{r})$ of the relative distance $\mathbf{r} = (\mathbf{r}_1 - \mathbf{r}_2)$ into momentum space

$$\langle \mathbf{p} | V | \mathbf{p}' \rangle = (2\pi\hbar)^{-3} \int \exp[- \frac{i}{\hbar}(\mathbf{p} - \mathbf{p}') \, . \, \mathbf{r}] \, V(\mathbf{r}) \, d\mathbf{r}. \qquad (1.110)$$

Obviously, a δ-function for $V(\mathbf{r})$ will give a constant and any finite range will be represented as a p-dependence in momentum space. The simplest rotational invariant part

$$(2\pi\hbar)^2 \langle \mathbf{p} | V | \mathbf{p}' \rangle = V_0 + V_1 \, \mathbf{p}'^2 + V_1 \, \mathbf{p}^2 + V_2 \, \mathbf{p} \, . \, \mathbf{p}', \qquad (1.111)$$

corresponds in the coordinate space to the momentum dependent operator

$$V(\mathbf{r}) = V_0 \delta(\mathbf{r}) + V_1 (\mathbf{p}'^2 \delta(\mathbf{r}) + \delta(\mathbf{r}) \mathbf{p}^2) + V_2 \mathbf{p} \, . \, \delta(\mathbf{r}) \mathbf{p}'. \qquad (1.112)$$

Equation (1.108) is of the form of eqn (1.112). Therefore, the term containing the parameter t_0 describes a pure δ-force with a spin-exchange, while the terms proportional to t_1 and t_2 simulate an effective finite range. The fourth term in eqn (1.108) represents a two-body spin-orbit interaction which can be obtained from a normal spin-orbit term in the short-range limit (Bell and Skyrme 1956).

The density dependence of the effective interaction, also known from the G-matrix calculations, is simulated by the three-body term (1.109). Indeed, for spin-saturated even-even nuclei, this term turns out to be equivalent to a density-dependent two-body term

$$\hat{v}_\rho(1,2) = \frac{1}{6}\, t_3(1 + P_\sigma)\; \delta(\mathbf{r}_1 - \mathbf{r}_2)\; \rho((\mathbf{r}_1 + \mathbf{r}_2)/2). \tag{1.113}$$

Usually, such two-body terms are simply regarded as a phenomenological representation of the density dependence of the microscopic effective interaction.

Skyrme-type forces currently in use

The renewed interest in Skyrme forces is due to the works of Vautherin and Brink (1970, 1972). They have reproduced binding energies and radii over the whole periodic table with a set of eight parameters by adopting eqn (1.108) and eqn (1.109) within the Hartree-Fock framework. Negele and Vautherin (1972) have given the connection between this Skyrme-type forces and the more fundamental G-matrix.

The present versions of the Skyrme force can be generalized by the following Skyrme-type ansatz

$$\begin{aligned}\hat{v}(1,2) = {} & t_0(1 + x_0P_\sigma)\; \delta(\mathbf{r}_1 - \mathbf{r}_2) + \frac{1}{2}\, t_1(1 + x_1P_\sigma)\; (\mathbf{k}'^2\delta(\mathbf{r}_1 - \mathbf{r}_2) \\ & + \delta(\mathbf{r}_1 - \mathbf{r}_2)\mathbf{k}^2) + t_2(1 + x_2P_\sigma)\mathbf{k}'.\; \delta(\mathbf{r}_1 - \mathbf{r}_2)\mathbf{k} \\ & + \frac{1}{6}\, t_3(1 + x_3P_\sigma)\rho^\alpha\delta(\mathbf{r}_1 - \mathbf{r}_2) + iW_0(\sigma_1 + \sigma_2).\mathbf{k}' \times \delta(\mathbf{r}_1 - \mathbf{r}_2)\mathbf{k}\;,\end{aligned} \tag{1.114}$$

where the forces are written in their density dependent form (Van Giai and Sagawa 1981). The expression (1.114) differs from the original Skyrme force mainly by the new density dependence ($\cong\rho^\alpha$ with $\alpha \le 1$) which is strongly correlated to the nuclear incompressibility as it will be discussed later. Several sets of all ten parameters resulting from different fits to experimental data are listed in Table 1.1. Their ability to reproduce nuclear properties is discussed later in Section 1.7.

Table 1.1. Different sets of parameters of Skyrme-type effective forces currently in use.

Force	t_0	t_1	t_2	t_3	W_0	x_0	x_1	x_2	x_3	α
SI	-1057.30	235.90	-100.00	14463.5	120	0.560	0.0	0.0	1.000	1
SII	-1169.90	585.60	-27.10	9331.1	105	0.340	0.0	0.0	1.000	1
SIII	-1128.75	395.00	-95.00	14000.0	120	0.450	0.0	0.0	1.000	1
SIV	-1205.60	765.00	35.00	5000.0	150	0.050	0.0	0.0	1.000	1
SV	-1248.29	970.56	107.22	0.0	150	-0.170	0.0	0.0	1.000	1
SVI	-1101.81	271.67	-138.33	17000.0	115	0.583	0.0	0.0	1.000	1
SkE2	-1299.30	802.41	-67.89	19558.96	120	0.270	0.0	0.0	1.000	1
Ska	-1602.78	570.88	-67.70	8000.0	125	-0.020	0.0	0.0	-0.286	1/3
Skb	-1602.78	570.88	-67.70	8000.0	125	-0.165	0.0	0.0	-0.286	1/3
T	-1788.90	301.50	502.50	12764.0	130	0.353	-2.5	-1.7	0.475	1/3
T1	-1794.00	298.00	-298.00	12812.0	110	0.154	-0.5	-0.5	0.089	1/3
SkM	-2645.00	385.00	-120.00	15595.0	130	0.090	0.0	0.0	0.000	1/6
SkM*	-2645.00	410.00	-135.00	15595.0	130	0.090	0.0	0.0	0.000	1/6

The values are taken as follows: SI and SII from Vautherin and Brink (1972); SIII though SVI from Beiner *et al.* (1975); SkE2 from Waroquier, Heyde and Wenes (1983) where there is an additional force parameter t_4 = -15808.79 MeV fm^8; T from Tondeur (1983); T1 from Tondeur *et al.* (1984); Ska and Skb from Köhler (1976); SkM from Krivine, Treiner, and Bohigas (1980); SkM* from Bartel *et al.* (1982). The units are MeV fm^3 for t_1, MeV fm^5 for t_1, t_2, W_0, and MeV $fm^{3+3\alpha}$ for t_3.

1.6 Variational method

The second difficulty that occurs within the nuclear many-body theory even when the Hamiltonian has a quite definite form is the usual many-body problem. The Schrödinger equation admits exact solutions only in rare cases corresponding to simple quantum-mechanical systems. Therefore, one needs some approximate methods in order to treat realistic situations. The variational method is one of the best and most widely used methods for approximate determination of energies and wave functions of many-particle systems.

According to the variational principle considered in Section 1.2, one can find all eigenvalues E_n and eigenfunctions Ψ_n of the nuclear Hamiltonian $\hat{H}$ by calculating the energy functional $E[\Psi]$, eqn (1.32), for all wave functions $\Psi \in \mathcal{H}^N$ and further searching for its stationary points $E_n = E[\Psi_n]$. Of course, this program is not possible to be carried out in practice. One is able, however, to investigate the functional (1.32) within a restricted set, the so-called trial variational set $\mathcal{H}^N_{tr} \subset \mathcal{H}^N$, of mathematically simple (often denoted as trial) many-particle wave functions $\Phi \in \mathcal{H}^N_{tr} \subset \mathcal{H}^N$.

Ground-state variational method

The ground-state variational method consists of finding a local minimum of the energy functional (1.32) defined on a given trial class $\mathcal{H}^N_{tr} \subset \mathcal{H}^N$,

$$E[\Phi] = \frac{\langle\Phi|\ \hat{H}\ |\Phi\rangle}{\langle\Phi|\Phi\rangle}\ . \tag{1.115}$$

This can be done by solving the variational equation

$$\delta_\Phi\ E[\Phi] = 0, \qquad \Phi \in \mathcal{H}^N_{tr} \subset \mathcal{H}^N, \tag{1.116}$$

which gives us an approximation to the Schrödinger equation (1.37). As soon as the exact wave function $\Psi_0 \in \mathcal{H}^N$ is not in the trial set $\mathcal{H}^N_{tr} \subset \mathcal{H}^N$, the minimal solution $\Phi_0 \in \mathcal{H}^N_{tr}$ of eqn (1.116) is no longer the exact eigenfunction of $\hat{H}$, but only an approximation. The associated approximate ground-state energy

$$E_{\Phi_0} = \min_{\Phi\in\mathcal{H}^N_{tr}} E[\Phi] = E[\Phi_0] \tag{1.117}$$

satisfies the inequality

$$E[\Phi] \geq E_{\Phi_0} \geq E_0 = E[\Psi_0], \tag{1.118}$$

which is a consequence of the general relation (1.38). Therefore, the approximate ground-state energy (1.117) is always higher than or equal to the exact one E_0.

Obviously the accuracy of the variational method depends strongly on the set of trial wave functions used. The trial wave function must be sufficiently simple for actual calculations to be made and, at the same time, $\mathcal{H}^N_{tr}$ must allow Φ to vary in a sufficiently large or well chosen region for the obtained solution of eqn (1.116) to be near to the exact one, $\Psi_0 \in \mathcal{H}^N$.

It may be demonstrated that the variational method gives us an approximation E_{Φ_0} to the exact energy E_0 which is better than the approximation of the associated trial function $\Phi_0 \in \mathcal{H}^N_{tr}$ to the exact one $\Psi_0 \in \mathcal{H}^N$. For this purpose, one can introduce the difference $\delta\Phi$ between Φ_0 and Ψ_0 and represent the solution of eqn (1.116) in the form

$$\Phi_0 = \Psi_0 + \delta\Phi. \tag{1.119}$$

Inserting eqn (1.119) into eqn (1.115) and using eqn (1.37) and the equality

$$\langle\Psi_0|\Psi_0\rangle + \langle\Psi_0|\delta\Phi\rangle + \langle\delta\Phi|\Psi_0\rangle + \langle\delta\Phi|\delta\Phi\rangle = \langle\Phi_0|\Phi_0\rangle, \tag{1.120}$$

one obtains

$$E_{\Phi_0} = E_0 + \frac{\langle\delta\Phi|(\hat{H}-E_0)|\delta\Phi\rangle}{\langle\Phi_0|\Phi_0\rangle}. \tag{1.121}$$

This equation clearly shows that the error in energy $\delta E = E_{\Phi_0} - E_0$

obtained after performing the variational method is of second order with respect to the error in the wave function $\delta\Phi = \Phi_0 - \Psi_0$.

A nice feature of the variational method is that its results can be systematically improved by enlarging the trial variational set used. For example, if the set $\mathcal{H}^N_{tr} \subset \mathcal{H}^N$ is a subset of another trial set, the larger one leads to an energy value lower or at least equal to E_{Φ_0} because it contains all wave functions, $\Phi \in \mathcal{H}^N_{tr} \subset \mathcal{H}^N$, of the first set. This improvement, of course, concerns the results with respect to the ground-state energy only.

Generally speaking, there is no unique criterion to decide which of two trial sets is the better one because usually there are no correlations between the accuracy in determining the energy on the one hand and that of the other characteristics of the system on the other. One usually hopes that out of two trial wave functions, the one for which the associated energy is closest to E_0 is better. But other criteria, such as the maximal overlap of Φ_0 with the exact function Ψ_0 or the magnitude of the dispersion $\langle\Phi_0|(H - E_{\Phi_0})^2|\Phi_0\rangle$ are also used within the variational method (Epstein 1974).

In actual calculations we have to make sure that the extremum following from eqn (1.116) exists and that it corresponds to a minimum of the functional (1.115). This can be verified by calculating the second variational derivative $\delta^{(2)}E[\Phi]$ of the functional $E[\Phi]$. The solution of eqn (1.116) should lead to the minimal condition

$$\delta^{(2)}E[\Phi_0] > 0. \tag{1.122}$$

Usually this condition is very expensive to verify in practice. In many cases, its validity is checked numerically, but there are also situations when the second variation may be explicitly outlined. We will come back to this point considering some particular variational techniques.

Ritz variational method

So far, we have assumed that the extremal condition (1.116) does exist within $\mathcal{H}^N_{tr} \subset \mathcal{H}^N$. In the general case, however, this assumption is not clear a priori and one has to be very careful in applying the variational method. An important exception is the linear or the so-called Ritz variational method (Epstein 1974) in which the trial set $\mathcal{H}^N_{tr}$ forms a linear variational subspace of the full Hilbert space $\mathcal{H}^N$. In this case, one can introduce a projection operator $\hat{R}$ by the equation

$$\hat{R}\Psi = \begin{cases} \Psi \text{ for } \Psi\in\mathcal{H}^N_{tr} \\ 0 \text{ for } \Psi\notin\mathcal{H}^N_{tr} \end{cases}, \tag{1.123}$$

where Ψ is any function from $\mathcal{H}^N$. The operator $\hat{R}$ is idempotent ($\hat{R}^2 = \hat{R}$) and the projected Hamiltonian

$$\hat{H}_R = \hat{R}\,\hat{H}\,\hat{R}^+ \tag{1.124}$$

is a linear and Hermitian operator acting in $\mathcal{H}^N_{tr} \subset \mathcal{H}^N$. Therefore, the variational equation (1.116) is equivalent to the eigenvalue problem

$$\hat{H}_R\,\Phi_0 = E_{\Phi_0}\Phi_0, \tag{1.125}$$

which should be easier to be solved in comparison with the exact Schrödinger equation, since $\mathcal{H}^N_{tr}$ is a restricted subspace of $\mathcal{H}^N$.

Variational method for excited states

The variational method can be applied also to excited nuclear states, that is, for an approximate solution of the Schrödinger equation (1.39) with $n \geq 1$. For a variational determination of the approximation Φ_n to the exact excited state Ψ_n of eqn (1.39), one has to solve the variational equation (1.116) obtained with the supplementary conditions

$$\langle\Phi_n|\Phi_m\rangle = \delta_{nm} \qquad (m = 0,1,2,\ldots,n), \tag{1.126}$$

which are coupled to the problem via Lagrange multipliers. The method quickly becomes too complicated for higher excited states. In some cases, however, the supplementary conditions (1.126) are easily satisfied (because of symmetry properties, for instance) and the variational method becomes an effective tool for investigating nuclear excited states as well.

1.7 Hartree-Fock theory

The Hartree-Fock theory (Hartree 1927, Fock 1930) is an excellent example demonstrating the usefulness of the many-body variational method. It will prove convenient for future reference to consider this theory in the contact of a large number of many-body techniques presently available.

Hartree-Fock approximation

From the variational point of view, the Hartree-Fock (HF) theory operates with a restricted variational space $\mathcal{H}^N_{HF} \subset \mathcal{H}^N$ which contains as trial wave functions all single Slater determinants of the type of eqn (1.28). According to the variational method, the best function will be this Slater determinant $\Phi_0 \in \mathcal{H}^N_{HF}$ which minimizes the energy functional (1.115) defined within $\mathcal{H}^N_{HF} \subset \mathcal{H}^N$.

In order to obtain the variational equation (1.116) within the trial space $\mathcal{H}^N_{HF}$, one should first clarify how to define a small variation $\delta\Phi \in \mathcal{H}^N_{HF}$ since the set $\mathcal{H}^N_{HF}$ is not a linear subspace of $\mathcal{H}^N$; a sum of two Slater determinants is not, in general, a Slater determinant again (Löwdin 1955, Foldy 1962). For this purpose, one may introduce the proper ground-state Slater determinant $\Phi_0 \in \mathcal{H}^N_{HF}$ in the second-quantization representation (see eqn (1.56))

$$|\Phi_0\rangle = \prod_{l=1}^{N} a^+_l \, |-\rangle. \tag{1.127}$$

Thouless (1960) proves that any other Slater determinant $\Phi \in \mathcal{H}^N_{HF}$, which is not orthogonal to (1.127) can be written in the form

$$|\Phi\rangle = \exp\left(\sum_{l,m} C_{lm}\, a^+_m a_l \right) |\Phi_0\rangle, \tag{1.128}$$

where C_{lm} are arbitrary coefficients. In the following we characterize the occupied levels in eqn (1.127) by letters i,j (hole states) and the empty levels by m,n (particle states); if no distinction is made we use letters $\alpha,\beta,\gamma,\delta$. For example, the operator $a^+_m a_l$ in eqn (1.128) creates a particle-hole state $|\Phi^l_m\rangle = a^+_m a_l |\Phi_0\rangle$ with respect to the Slater determinant (1.127).

Expanding the exponential (1.128) to first order in the coefficients C_{lm} and taking $\delta\Phi$ to be the difference between Φ and Φ_0, one obtains the small variation

$$|\delta\Phi\rangle = \sum_{l,m} C_{lm}\, a^+_m a_l\, |\Phi_0\rangle. \tag{1.129}$$

It is obviously orthogonal to Φ_0; that is,

$$\langle\Phi_0|\delta\Phi\rangle = \sum_{l,m} C_{lm}\, \langle\Phi_0|\, a^+_m a_l\, |\Phi_0\rangle = 0, \tag{1.130}$$

because $\langle\Phi_0|a^+_m$ is zero, since m is not occupied state in Φ_0.

By using eqn (1.130), the stationary condition (1.116) in $\mathcal{H}^N_{HF}$ can be rewritten in the form

$$\delta\langle\Phi_0|\hat{H}|\Phi_0\rangle = \langle\delta\Phi|\hat{H}|\Phi_0\rangle + \langle\Phi_0|\hat{H}|\delta\Phi\rangle$$

$$= \sum_{l,m} C^*_{ml}\, \langle\Phi_0|a^+_l a_m \hat{H}|\Phi_0\rangle + \sum_{l,m} C_{ml}\, \langle\Phi_0|\hat{H} a^+_m a_l|\Phi_0\rangle = 0, \tag{1.131}$$

where we restrict ourselves within Slater determinants normalized to unity according to the comment at the end of Section 1.2. The first term in eqn (1.131) is just the complex conjugate of the second term and, therefore, each term identically can be equated to zero. Furthermore, C_{ml} are arbitrary coefficients and thus we obtain the main equation of the HF theory

$$\langle\Phi^{i}_{m}|\hat{H}|\Phi_0\rangle \equiv \langle\Phi_0|a^{+}_{i}a_{m}\hat{H}|\Phi_0\rangle = 0 \quad \text{for all i and m.} \tag{1.132}$$

This equation implies that all matrix elements of the nuclear Hamiltonian between the HF ground state Φ_0 and one-particle one-hole states built on Φ_0 are zero. The anticommutation relations (1.52) allow eqn (1.132) to be expressed in the form (Eisenberg and Greiner 1972)

$$\langle\varphi_m|\hat{t}|\varphi_i\rangle + \sum_{j=1}^{N} \langle\varphi_m\varphi_j|\hat{v}|\varphi_i\varphi_j\rangle_A = 0, \tag{1.133}$$

where the antisymmetrized matrix elements (1.72) are used.

The so-called HF single-particle Hamiltonian $\hat{H}_{HF}$ can be defined by its matrix elements between any two single-particle states

$$\langle\varphi_\alpha|\hat{H}_{HF}|\varphi_\beta\rangle = \langle\varphi_\alpha|\hat{t}|\varphi_\beta\rangle + \sum_{j=1}^{N} \langle\varphi_\alpha\varphi_j|\hat{v}|\varphi_j\varphi_\beta\rangle_A \,. \tag{1.134}$$

This Hamiltonian is Hermitian when the original one $\hat{H}$ is Hermitian. Obviously $\hat{H}_{HF}$ is also invariant up to unitary transformation among the occupied levels or among the empty ones. One can use this freedom in order to define the HF single-particle wave functions as a solution of the Hermitian eigenvalue problem following from eqn (1.133)

$$\langle\varphi_\alpha|\hat{H}_{HF}|\varphi_\beta\rangle \equiv \langle\varphi_\alpha|\hat{T}|\varphi_\beta\rangle + \sum_{j=1}^{N} \langle\varphi_\alpha\varphi_j|\hat{v}|\varphi_j\varphi_\beta\rangle_A = \varepsilon_\alpha\delta_{\alpha\beta}, \tag{1.135}$$

where the eigenvalues ε_α can be further identified with the single-particle energies of the nucleons (Koopmans 1933).

To give a better interpretation of the HF equations (1.135) one can analyse them in coordinate representation. It is useful to write first $\hat{H}_{HF}$ as a sum of a kinetic energy operator $\hat{T}$ and a one-body potential energy operator $\hat{U}_{HF}$ whose matrix elements are given by the second term on the right of eqn (1.134):

$$\langle\varphi_\alpha|\hat{U}_{HF}|\varphi_\beta\rangle = \sum_{j=1}^{N} \langle\varphi_\alpha\varphi_j|\hat{\omega}|\varphi_j\varphi_\beta\rangle_A. \tag{1.136}$$

The HF potential (1.136) is an one-body field and averages over all two-body interactions. In coordinate representation, it consists of the so-called local or Hartree potential

$$U_H(x) = \sum_{j=1}^{N} \int dx' \; \varphi_j^*(x')\hat{\omega}(x,x')\varphi_j(x'), \tag{1.137}$$

and the nonlocal or exchange potential

$$U_{EX}(x,x') = \sum_{j=1}^{N} \varphi_j^*(x)\hat{\omega}(x,x')\varphi_j(x'). \tag{1.138}$$

The latter appears due to the Pauli principle and the antisymmetrization of the matrix elements entering eqn (1.136).

Using eqns (1.136), (1.137) and (1.138), we find the HF equations (1.135) in the coordinate space

$$\left\{-\frac{\hbar^2}{2m}\Delta + U_H(x)\right\}\varphi_\alpha(x) + \int dx' \; U_{EX}(x,x')\varphi_\alpha(x') = \varepsilon_\alpha\varphi_\alpha(x). \tag{1.139}$$

This is a nonlinear equation because the potentials U_H and U_{EX} depend on its solution; that is, on the single-particle states $\varphi_\alpha(x)$ satisfying eqn (1.139). It is usually solved by the so-called HF self-consistent iterative procedure. As a first step, one starts with a set of phenomenological independent-particle model wave functions and calculates the potentials (1.137) and (1.138). From eqn (1.139) one then gets new single-particle wave functions for the first step and so forth until convergence is achieved, that is, the self-consistent HF potential stays unchanged in two consecutive steps.

At the conclusion of the self-consistent HF iterations one gets the single-particle energies ε_α and wave functions $\varphi_\alpha(x)$. From these, one can calculate the total energy of the HF theory

$$E_{HF} = \langle\Phi_0|\hat{H}|\Phi_0\rangle = \sum_{i=1}^{N} \langle\varphi_i|\hat{t}|\varphi_i\rangle + \frac{1}{2}\sum_{i,j=1}^{N} \langle\varphi_i\varphi_j|\hat{v}|\varphi_i\varphi_j\rangle_A$$

$$= \sum_{i=1}^{N} \varepsilon_i - \frac{1}{2}\sum_{i,j=1}^{N} \langle\varphi_i\varphi_j|\hat{v}|\varphi_i\varphi_j\rangle_A = \sum_{i=1}^{N} \varepsilon_i + \frac{1}{2}\sum_{i=1}^{N} \langle\varphi_i|\hat{t}|\varphi_i\rangle$$

$$= \sum_{i=1}^{N} \varepsilon_i - \frac{1}{2}\sum_{i=1}^{N} \langle\varphi_i|U_{HF}|\varphi_i\rangle, \tag{1.140}$$

which is not the sum of the single-particle energies (compare with eqn (1.20)), but differs from it by 1/2 the sum over the interaction energies.

In order to verify whether the energy (1.140) is the absolute minimum of the HF energy functional we should check the second variation of this functional as stated in eqn (1.122). For this purpose one must evaluate the normalization $\langle\Phi|\Phi\rangle$ and the Hamiltonian expectation value $\langle\Phi|\hat{H}|\Phi\rangle$ up to second order in the coefficients C_{mi} from eqn (1.128). Bearing in mind that no terms of the first order in C_{mi} are retained, since these are removed by the stationary HF condition (1.132), one obtains (Thouless 1960)

$$\delta^{(2)}E[\Phi_0] = \varepsilon_m - \varepsilon_i - \langle\varphi_m\varphi_i|\hat{v}|\varphi_m\varphi_i\rangle_A > 0. \tag{1.141}$$

Therefore, the HF energy E_{HF} will be a true minimum only if all particle-hole excitation energies

$$E_m^i = \langle\Phi_0|a_i^+a_m\,\hat{H}\,a_m^+a_i|\phi_0\rangle = E_{HF} + \varepsilon_m - \varepsilon_i - \langle\varphi_m\varphi_i|\hat{v}|\varphi_m\varphi_i\rangle_A \tag{1.142}$$

are positive with respect to the energy (1.140). One should look at the condition (1.141) as a supplementary requirement for a satisfactory HF solution, although it is very physical condition and one should expect it to arise.

Density-dependent effective forces

Density-dependent effective forces applied to the HF theory are not only convenient because they solve the hard core problem (the two-body matrix elements entering the HF potential (1.136) all become infinite for a hard core force) but they also lead to a consistent re-summation of certain higher order terms of the full many-body problem.

The basic idea of the density-dependent HF theory is that the many-body aspects of the nucleus can to a good approximation be taken into account by replacing the density-matrix dependence of the nuclear G-matrix by dependence only on the local density ρ. One then proceeds to extract the form of this dependence $G(\rho)$ of G on ρ from the detailed Brückner HF calculations (see the review article by Svenne (1979)). In addition, a phenomenological adjustment of the resulting effective interaction $\omega \equiv G(\rho)$ may also be performed in order to get good values of some observables and thus to take into account, in some average sense, the contributions of higher-order effects and possible deficiencies in the underlying two-body forces (Bethe 1971).

The density-dependent effective interaction cannot simply be used in the variational HF theory. Now, because of its ρ-dependence, an extra term involving $\partial\hat{\omega}/\partial\rho$ appears in the potential (1.136) which can be rewritten formally as

$$\langle\varphi_\alpha|\hat{U}_{HF}|\varphi_\beta\rangle = \sum_{j=1}^{N}\langle\varphi_\alpha\varphi_j|\hat{\omega}|\varphi_j\varphi_\beta\rangle_A + \sum_{i,j=1}^{N}\langle\varphi_i\varphi_j|\frac{\partial\hat{\omega}}{\partial\rho}\Big(|\varphi_\beta\rangle\langle\varphi_\alpha|\Big)_R|\varphi_i\varphi_j\rangle_A. \tag{1.143}$$

Here the subscript R indicates that the orbitals in the brackets are to be evaluated at the centre of mass of the two particles

$\mathbf{R} = (\mathbf{r}_1 + \mathbf{r}_2)/2$. For density-independent forces eqn (1.143) just reduces to eqn (1.136). For density-dependent forces, however, the additional term ($\simeq \partial\hat{\omega}/\partial\rho$), usually called the rearrangement or saturation potential (see extensive discussions of its particular forms by Hodgson (1975)), plays an important role in the density-dependent HF calculations.

The solution of the HF equations with effective interaction obtained from more or less realistic attempts at extraction from the self-consistent G-matrix is a successive but very complicated task (Negele 1970, Negele and Vautherin 1972, 1975). An alterative approach may replace the density-dependent effective interaction by a purely phenomenological one. A spectacularly successful force is that of the Skyrme-type interaction considered earlier in Section 1.5.

Skyrme-Hartree-Fock theory

We start with the Skyrme-type force given in eqn (1.114) and calculate the expectation value of the corresponding Hamiltonian with respect to the Slater determinants $\Phi \in \mathcal{H}^{N}_{HF}$. For a time-reversal invariant system the energy is given as

$$E[\Phi]= \int d\mathbf{r}\ H(\mathbf{r}) = \int d\mathbf{r}\ H(\rho(\mathbf{r}),\tau(\mathbf{r}),\mathbf{J}(\mathbf{r})), \tag{1.144}$$

where the energy density $H(\mathbf{r})$ is an algebraic function of the nucleon density

$$\rho(\mathbf{r}) = \sum_{i,s,\tau} |\varphi_i(\mathbf{r},s,\tau)|^2, \tag{1.145}$$

the kinetic energy density

$$\tau(\mathbf{r}) = \sum_{i,s,\tau} |\nabla\ \varphi_i(\mathbf{r},s,\tau)|^2, \tag{1.146}$$

and the so-called spin-orbit or spin-density

$$\mathbf{J}(\mathbf{r}) = (-i) \sum_{l,s,s'\tau} \varphi_l^*(\mathbf{r},s,\tau) \left[\nabla \varphi_l(\mathbf{r},s',\tau) \times \sigma_{ss'}\right]. \tag{1.147}$$

A lengthy but straightforward calculation (Vautherin and Brink (1972)) gives (Tondeur *et al.* (1984))

$$\begin{aligned} H(\mathbf{r}) = {} & \frac{\hbar^2}{2m}\tau + \frac{1}{2} t_0\left[(1 + \frac{1}{2} x_0) \rho^2 - (x_0 + \frac{1}{2}) (\rho_n^2 + \rho_p^2)\right] \\ & + \frac{1}{4} t_1 \left[(1 + \frac{1}{2}x_1)\left(\rho\tau + \frac{3}{4}(\nabla\rho)^2\right) - (x_1 + \frac{1}{2})\left(\rho_n\tau_n + \rho_p\tau_p + \frac{3}{4}(\nabla\rho_n)^2 + \frac{3}{4}(\nabla\rho_p)^2\right)\right] \\ & + \frac{1}{4} t_2 \left[(1 + \frac{1}{2}x_2)\left(\rho\tau - \frac{1}{4}(\nabla\rho)^2\right) + (x_2 + \frac{1}{2})\left(\rho_n\tau_n + \rho_p\tau_p - \frac{1}{4}(\nabla\rho_n)^2 - \frac{1}{4}(\nabla\rho_p)^2\right)\right] \\ & + \frac{1}{2} W_0 \, (\mathbf{J}.\nabla\rho + \mathbf{J}_n.\nabla\rho_n + \mathbf{J}_p.\nabla\rho_p) - \frac{1}{16}\left[t_1\left(x_1 \mathbf{J}^2 - (\mathbf{J}_n^2 + \mathbf{J}_p^2)\right) - t_2\left(x_2 \mathbf{J}^2 + (\mathbf{J}_n^2 + \mathbf{J}_p^2)\right)\right] \\ & + \frac{1}{12} t_3 \left[(1 + \frac{1}{2}x_3)\rho^2 - (x_3 + \frac{1}{2})(\rho_n^2 + \rho_p^2)\right] \rho^\alpha + H_{\mathrm{Coul}}(\mathbf{r}), \end{aligned} \tag{1.148}$$

where the index n (p) refers to neutron (proton) densities and restricts the summations to neutron (proton) occupied states in formulae (1.145), (1.146) and (1.147); the unlabelled densities being implicitly summed over q (= n or p). As usual, one can add to eqn (1.148) the Coulomb energy density (Beiner *et al.* 1975)

$$H_{\mathrm{Coul}}(\mathbf{r}) \equiv H_{\mathrm{CD}} + H_{\mathrm{CEX}} = \frac{e^2}{2}\int d\mathbf{r}' \, \frac{\rho_p(\mathbf{r}')}{|\mathbf{r} - \mathbf{r}'|} - \frac{3}{4} e^2 \left(\frac{3}{\pi}\right)^{1/3} \rho_p^{4/3}. \tag{1.149}$$

According to the concept of the density-dependent HF theory we have to vary the functional (1.144) with respect to the single-particle wave functions $\varphi_i(x)$ under the condition that they are normalized to unity. We use Lagrange multipliers ε_i for these subsidiary conditions and find

$$\delta E=\sum_{i=1}^{N}\int dr\delta\varphi_i^*(x)\left[-\nabla\frac{\hbar^2}{2m_{q_i}^*(x)}.\nabla+U_{q_i}(x)-iW_{q_i}(x).(\nabla x\sigma)-\varepsilon_i\right]\varphi_i(x) = 0. \tag{1.150}$$

Due to the arbitrariness of the infinitesimal variations $\delta\varphi_i^*(x)$ we obtain from eqn (1.150) the Skyrme-HF self-consistent equations

$$\left\{-\nabla\frac{\hbar^2}{2m_{q_i}^*(x)}\,.\,\nabla + U_{q_i}(x) - i\,\mathbf{W}_{q_i}(x)\,.\,(\nabla x\sigma)\right\}\varphi_i(x) = \varepsilon_i\varphi_i(x). \tag{1.151}$$

The exchange potential (1.138) is now local and it is included in the HF potential

$$U_q(r) = \frac{\delta H(\mathbf{r})}{\delta\rho_q(\mathbf{r})} = \frac{\partial H(\mathbf{r})}{\partial\rho_q(\mathbf{r})} - \nabla\,.\,\frac{\partial H(\mathbf{r})}{\partial\nabla\rho_q(\mathbf{r})} + \Delta\,\frac{\partial H(\mathbf{r})}{\partial\Delta\rho_q(\mathbf{r})}$$

$$= t_0(1 + \tfrac{1}{2}x_0)\rho - t_0(x_0 + \tfrac{1}{2})\rho_q + \tfrac{1}{6}\,t_3\left[(1 + \tfrac{1}{2}x_3)\rho - (x_3 + \tfrac{1}{2})\rho_q\right]\rho^{\alpha}$$

$$+ \tfrac{\alpha}{12}\,t_3\left[(1 + \tfrac{1}{2}x_3)\rho^2 - (x_3 + \tfrac{1}{2})(\rho_n^2 + \rho_p^2)\right]\rho^{(\alpha-1)} - \tfrac{1}{2}\,W_0\,(\nabla\,.\,\mathbf{J} + \nabla\,.\,\mathbf{J}_q)$$

$$+ \tfrac{1}{4}\left[t_1(1 + \tfrac{1}{2}x_1) + t_2(1 + \tfrac{1}{2}x_2)\right]\tau - \tfrac{1}{4}\left[t_1(\tfrac{1}{2} + x_1) - t_2(\tfrac{1}{2} + x_2)\right]\tau_q$$

$$- \tfrac{1}{8}\left[3t_1(1 + \tfrac{1}{2}x_1) - t_2(1 + \tfrac{1}{2}x_2)\right]\Delta\rho + \tfrac{1}{8}\left[3t_1(\tfrac{1}{2} + x_1) + t_2(\tfrac{1}{2} + x_2)\right]\Delta\rho_q$$

$$+ \left[e^2\int dr\,\frac{\rho_p(\mathbf{r}')}{|\mathbf{r} - \mathbf{r}'|} - e^2\left(\frac{3}{\pi}\right)^{1/3}\rho_p^{1/3}\right]\delta_{pq}. \tag{1.152}$$

Therefore, eqn (1.151) is a pure differential single-particle Schrödinger equation with a Hamiltonian (apart from the integral introduced by the Coulomb forces)

$$\hat{H}_{HF} \equiv \hat{H}_{HF}(\rho,\tau,\mathbf{J}) = -\nabla \frac{\hbar^2}{2m^*(x)} \cdot \nabla + U + (\nabla W \times \hat{\mathbf{p}}) \cdot \sigma, \qquad (1.153)$$

and effective mass

$$\frac{\hbar^2}{2m_q^*(x)} = \frac{\partial H(\mathbf{r})}{\partial \tau_q(\mathbf{r})} = \frac{\hbar^2}{2m} + \frac{1}{4}\left[t_1(1 + \frac{1}{2}x_1) + t_2(1 + \frac{1}{2}x_2)\right]\rho$$

$$- \frac{1}{4}\left[t_1(\frac{1}{2} + x_1) - t_2(\frac{1}{2} + x_2)\right]\rho_q, \qquad (1.154)$$

and spin-orbit potential

$$\mathbf{W}_q = \frac{\partial H(\mathbf{r})}{\partial \mathbf{J}_q(\mathbf{r})} = -\frac{1}{8}(x_1t_1 + x_2t_2)\mathbf{J} + \frac{1}{8}(t_1 - t_2)\mathbf{J}_q + \frac{1}{2}W_0(\nabla\rho + \nabla\rho_q). \quad (1.155)$$

Actual calculations

Numerous calculations have been carried out numerically by solving eqn (1.151) in coordinate space with very fast codes. Vautherin and Brink (1972) originally solved eqn (1.151) for spherical closed shell nuclei and introduced the sets SI and SII (see the parameters of the Skyrme-type effective forces listed in Table 1.1). Later, many other spherical nuclei were investigated (Beiner *et al.* 1975) and other sets of parameters were determined (SIII–SVI).

It soon becomes apparent that the original forces SI and SII as well as Skyrme–Orce forces SIII–SVI could be further improved, and that a more detailed agreement with experiment, including dynamical properties, was possible. New forces Ska and Skb were

Table 1.2. Results for binding energies per nucleon E/N (in MeV) and root mean square charge radii r_c (in fm) from the density-dependent Hartree-Fock calculations.

^{16}O		^{40}Ca		^{48}Ca		^{90}Zr		^{208}Pb	
E/N	r_c	E/N	r_c	E/N	r_c	E/N	r_c	E/N	r_c
Negele (1970)									
-7.98	2.73	-8.55	3.49	-8.67	3.48	-8.71	4.23	-7.87	5.50
Campi and Sprung (1972)									
-7.68	2.75	-8.33	3.49	-8.40	3.51	-8.63	4.27	-7.87	5.45
Beiner *et al.* (1975), SIII-force									
-8.01	2.69	-8.54	3.48	-8.71	3.53	-8.72	4.32	-7.87	5.57
Bartel *et al.* (1982), SkM*-force									
-7.98	2.81	-8.53	3.52	-8.75	3.54	-8.72	4.30	-7.87	5.51
Tondeur (1983), T-force									
-7.87	2.75	-8.5	3.49	-8.67	3.49	-8.71	4.27	-7.87	5.49
Experiment									
-7.98	2.73	-8.55	3.49	-8.67	3.48	-8.71	4.27	-7.87	5.50

proposed by Köhler (1976). They have a lower compression modulus possible when $\alpha \neq 1$ is allowed. Good predictions for the giant monopole resonances were also obtained after introducing the set SkM (Krivine, Treiner, and Bohigas 1980). Its minor modification, the set SkM* introduced by Bartel *et al.* (1982), predicts also the fission barriers for heavy nuclei.

As an illustration of the results obtained in this way we list in Table 1.2 binding energies per nucleon E/N and root mean square charge radii r_c as calculated by the most sophisticated HF calculations with effective interaction derived from the bare nucleon-nucleon force and also with some Skyrme-type effective

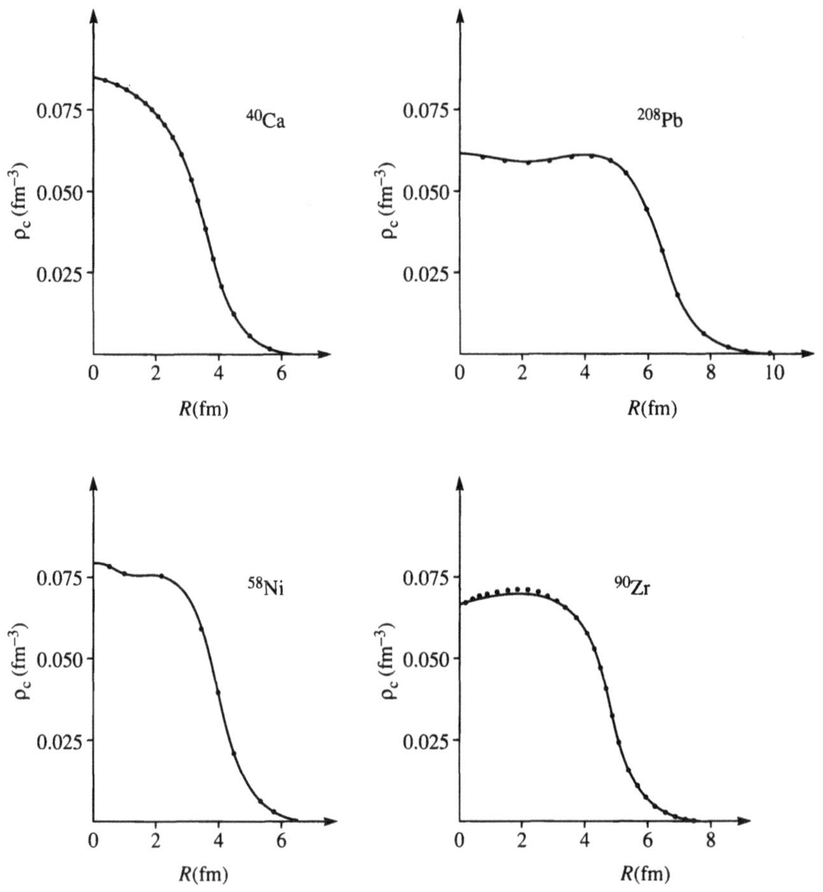

Figure 1.2. Skyrme-Hartree-Fock charge distributions (full lines) compared with experimental results (dots) by Tondeur (1983).

forces. Skyrme-HF densities of four spherical nuclei obtained with the set T (Table 1.1) are compared with experimental charge distributions in Figure 1.2. Other Skyrme-HF results are outlined throughout the book.

In conclusion, it seems now that it might be possible to obtain simultaneously a fair agreement with experiments for nuclear masses, densities, radii, giant resonances and fission barriers, although none of the above forces fully reaches this goal. Many works (Tondeur *et al.* 1984, Friedrich and Reinhard 1986) represent very interesting steps towards a good multipurpose force, which however still has to be built.

2. Foundation of density functional theory

2.1. Introduction

The aim of density functional theory is to reduce the many-body quantum mechanical problem formulated in terms of N-particle wave functions $\Psi(x_1, x_2, \ldots, x_N) \in \mathcal{H}^N$, or more generally, in terms of N-particle density matrices $\rho^{(N)}(x_1, \ldots, x_N; x'_1, \ldots, x'_N)$ to the one-particle level where the natural mathematical object is the local density distribution $\rho(\mathbf{r})$. This alternative formulation of the many-body problem, employing the local density $\rho(\mathbf{r})$ rather than the wave function $\Psi \in \mathcal{H}^N$, is confirmed by the Hohenberg-Kohn theorem (Hohenberg and Kohn 1964).

The local density formulation of the many-body problem is a conceptually new and very attractive idea for the following reasons:

- There is a considerable economy in dealing with the local density, rather than the wave function $\Psi \in \mathcal{H}^N$, since the former is a function in the three-dimensional vector space R^3 independent of the mass number N.

- Being a function in R^3, $\rho(\mathbf{r})$ can be easily visualized, unlike the wave function $\Psi(x_1, x_2, \ldots, x_N)$ defined in multidimensional configuration space.

- The local density $\rho(\mathbf{r})$ has a transparent physical meaning, while no physical significance can be attributed directly to $\Psi \in \mathcal{H}^N$.

- Unlike the wave function $\Psi \in \mathcal{H}^N$, $\rho(\mathbf{r})$ is an observable that can be measured by experimental methods such as diffraction and scattering.

2.2 Hohenberg-Kohn theorem

The Hohenberg-Kohn theorem states that the nondegenerate ground-state energy of a system of identical spinless fermions is a unique functional of the local density $\rho(\mathbf{r})$. This functional $E[\rho]$ attains its minimum value, the ground-state of the system, when the density has its correct ground-state value.

In proving their theorem Hohenberg and Kohn (1964) considered a collection of an arbitrary number of electrons enclosed in a large box, moving under the influence of an external potential $\upsilon(\mathbf{r})$ and the mutual Coulomb repulsion. We outline their proof below for an N-particle nuclear system with Hamiltonian

$$\hat{H}_{\upsilon} = \hat{H} + \hat{V}, \tag{2.1}$$

where $\hat{H}$ is given by eqn (1.79) and the term $\hat{V}$ in the second-quantization representation is of the form

$$\hat{V} = \int dx\, \upsilon(\mathbf{r})\, \psi^{+}(x)\, \psi(x). \tag{2.2}$$

That is, we suppose that the system is under the influence of an external field which is derivable from a local potential $\upsilon(\mathbf{r})$.

Proof of the Hohenberg-Kohn theorem

Let us consider the nondegenerate ground-state solution $\Psi_{\upsilon} \in \mathcal{H}^N$ of the Schrödinger equation

$$\hat{H}_{\upsilon}\, \Psi_{\upsilon}(x_1,x_2,\ldots,x_N) = E_{\upsilon}\, \Psi_{\upsilon}(x_1,x_2,\ldots,x_N), \tag{2.3}$$

with Hamiltonian (2.1). Obviously, $\Psi_{\upsilon} \in \mathcal{H}^N$ is a functional of $\upsilon(\mathbf{r})$ because $\upsilon(\mathbf{r})$ specifies the Hamiltonian (2.1) and, therefore, its eigenfunction $\Psi_{\upsilon} \in \mathcal{H}^N$. It is clear that the local density associated with $\Psi_{\upsilon} \in \mathcal{H}^N$

$$\rho(\mathbf{r}) = \sum_{\sigma,\tau} \int |\Psi_{\upsilon}(x,x_2,\ldots,x_N)|^2\, dx_2\ldots dx_N, \tag{2.4}$$

is also a functional of the external field $v(\mathbf{r})$. One can prove that conversely $v(\mathbf{r})$ is a unique functional of $\rho(\mathbf{r})$, apart from a trivial additive constant. The proof proceeds by *reductio ad absurdum* (Hohenberg and Kohn 1964).

Assume that another potential $v'(\mathbf{r})$, leading to the ground-state wave function $\Psi_{v'}$ and the energy $E_{v'}$, gives rise to the same density $\rho'(\mathbf{r}) \equiv \rho(\mathbf{r})$. Clearly, Ψ_v cannot be equal to $\Psi_{v'}$ (unless $v'(\mathbf{r}) - v(\mathbf{r}) = \text{const}$) since they both, Ψ_v and $\Psi_{v'}$, satisfy different Schrödinger equations. Therefore, due to condition (1.42), one has

$$E_{v'} < \langle\Psi_v|\hat{H}_{v'}|\Psi_v\rangle = \langle\Psi_v|\hat{H}_v + \hat{V}' - \hat{V}|\Psi_v\rangle = E_v + \langle\Psi_v|\hat{V}' - \hat{V}|\Psi_v\rangle, \quad (2.5)$$

so that

$$E_{v'} < E_v + \int d\mathbf{r}\,(v'(\mathbf{r}) - v(\mathbf{r}))\,\rho(\mathbf{r}). \quad (2.6)$$

Exactly in the same way, interchanging primed and unprimed quantities, one finds that

$$E_v < E_{v'} - \int d\mathbf{r}\,(v'(\mathbf{r}) - v(\mathbf{r}))\,\rho(\mathbf{r}). \quad (2.7)$$

After adding eqn (2.6) and eqn (2.7), one finds the inconsistency $E_v + E_{v'} < E_v + E_{v'}$. Hence, $\rho(\mathbf{r})$ and $\rho'(\mathbf{r})$ must be different and, therefore, $v(\mathbf{r})$ is a unique functional of $\rho(\mathbf{r})$.

A by-product of the above statement is that the full many-particle wave function $\Psi_v \in \mathcal{H}^N$ is a functional of $\rho(\mathbf{r})$ as well. Indeed, the complete Hamiltonian (2.1) contains the external potential $\hat{V}$ which is specified in principle if the ground-state density is known. Consequently, the ground-state local density specifies not only the Hamiltonian $\hat{H}_v$ but also its eigenfunction Ψ_v and, therefore, the expectation values of all operators of interest. In particular, one may define the following universal functional of $\rho(\mathbf{r})$

$$F[\rho] = \langle \Psi_{v} | \hat{H} | \Psi_{v} \rangle, \tag{2.8}$$

which is valid for any number of particles N and for any external field $v(\mathbf{r})$. One may also define the energy functional

$$E[\rho] = \langle \Psi_{v} | \hat{H}_{v} | \Psi_{v} \rangle = F[\rho] + \int d\mathbf{r}\, v(\mathbf{r})\, \rho(\mathbf{r}), \tag{2.9}$$

which clearly, for the correct density $\rho(\mathbf{r})$, equals the ground-state energy E_{v} entering eqn (2.3).

One can further prove that $E[\rho]$ assumes its minimal value for the correct $\rho(\mathbf{r})$ if the admissible functions are restricted by the normalization condition

$$N[\rho] \equiv \int d\mathbf{r}\, \rho(\mathbf{r}) = N. \tag{2.10}$$

The desired result is a consequence of the general ground-state variational principle considered in Section 1.2.

Suppose that $\Psi_{v} \in \mathcal{H}^{N}$ is the correct ground-state wave function and that $\Psi_{v'} \in \mathcal{H}^{N}$ is a different trial wave function obtained as a solution of the Schrödinger equation (2.3) with a different potential $v'(\mathbf{r})$. The local density associated with $\Psi_{v'} \in \mathcal{H}^{N}$ is $\rho'(\mathbf{r})$. Then, it follows from eqns (2.8) and (2.9) that

$$E[\rho'] = F[\rho'] + \int d\mathbf{r}\, v(\mathbf{r})\rho'(\mathbf{r}) = \langle \Psi_{v'} | \hat{H}_{v} | \Psi_{v'} \rangle$$

$$\geq \langle \Psi_{v} | \hat{H}_{v} | \Psi_{v} \rangle = F[\rho] + \int d\mathbf{r}\, v(\mathbf{r})\, \rho(\mathbf{r}) = E[\rho]. \tag{2.11}$$

Thus, the minimal property of the unique functional (2.9) is established relatively to all densities $\rho'(\mathbf{r})$ associated with some external potentials $v'(\mathbf{r})$. The expression 'unique functional' means that $E[\rho]$ depends neither on the particular number of particles, except through the density $\rho(\mathbf{r})$, nor on the external potential $v(\mathbf{r})$, except through the term $\int d\mathbf{r}\, v(\mathbf{r})\rho(\mathbf{r})$.

Euler–Lagrange equation for the density

The minimum property of the energy $E[\rho]$ as a functional of the local density $\rho(\mathbf{r})$ means that the correct density $\rho(\mathbf{r})$ can be variationally determined by minimizing $E[\rho]$ with respect to $\rho(\mathbf{r})$, that is, by using the stationary condition

$$\delta\ [E[\rho] - \mu(N[\rho] - N)] = 0. \tag{2.12}$$

Here μ is a Lagrange multiplier accounting for the normalization condition (2.10). This variational equation can be regarded also as telling us that the Lagrange multiplier μ can be interpreted as the chemical potential of the system $\mu = (\partial E/\partial N)$.

Performing the variation in eqn (2.12) one obtains the Euler–Lagrange equation for the density $\rho(\mathbf{r})$

$$\frac{\delta E[\rho]}{\delta\rho(\mathbf{r})} - \mu = 0, \tag{2.13}$$

which is the main equation of the density functional theory. Its first term is the variational derivative of the functional $E[\rho]$ with respect to the density $\rho(\mathbf{r})$. This term varies from point to point through the system unless the density is the true density of the system. In this case the value of $\delta E[\rho]/\delta\rho$ is the same everywhere and is equal to the value of the chemical potential μ.

υ-representability problem

The Euler–Lagrange equation (2.13) demonstrates a significant restriction of the Hohenberg–Kohn functional (2.9) because it appears to be defined only for those $\rho(\mathbf{r})$ which can be obtained by a solution of the Schrödinger equation containing some external potential $\upsilon(\mathbf{r})$. In order to clarify this problem let us consider the real three-dimensional vector space

$$R^3 \equiv \{\ \mathbf{r} = (r, \mathbf{r}_0);\ r \in R^1,\ \mathbf{r}_0 \in \Omega^1\ \}, \tag{2.14}$$

where R^1 and Ω^1 are spaces in which the length $r \equiv |\mathbf{r}|$ of the vector $\mathbf{r} \in R^3$ and its direction $\mathbf{r}_0 \equiv \mathbf{r}/r$ are defined,respectively.

One can define everywhere in R^3 the set of all functions $\rho(\mathbf{r})$ which satisfy the following conditions:

(i) nonnegativity: $\rho(\mathbf{r}) \geq 0$ everywhere on R^3;

(ii) normalization: $\int d\mathbf{r}\, \rho(\mathbf{r}) \leq \infty$, i.e., $\rho^{1/2}(\mathbf{r}) \in L^2(R^3)$;

(iii) $\rho(\mathbf{r})$ is continuously differentiable function in R^3.

One can also isolate the subset D_N of these $\rho(\mathbf{r})$ which satisfy conditions (i)-(iii) and for which eqn (2.10) is valid for an integer number N. Elements of D_N are further interpreted as local densities of N-particle systems.

A density $\rho(\mathbf{r}) \in D_N$ is said to be N-representable if it is derivable from a wave function $\Psi \in \mathcal{H}^N$ via eqn (1.93). In addition, if this $\Psi \in \mathcal{H}^N$ can be realized as a nondegenerate ground state of some N-particle Hamiltonian $\hat{H}_{\upsilon}$ characterized by some external field $\upsilon(\mathbf{r})$ via eq. (2.1), then the corresponding $\rho(\mathbf{r}) \in D_N$ is said to be υ-representable. Therefore, the υ-representable densities form a subset $D_N^{\upsilon} \subset D_N$ of the N-representable densities, that is, the υ-representability imposes stronger conditions on the densities in comparison with the N-representability.

Although the N-representability is an extremely difficult and perhaps intractable problem with respect to high-order density matrices (Coleman 1963), it seems to have been fully solved with respect to the local densities (Gilbert 1975, Harriman 1981, Lieb 1982, 1983). According to Gilbert (1975), the three conditions (i)- (iii) given above are the necessary and sufficient criteria local densities $\rho(\mathbf{r})$ to be N-representable. Harriman (1981) gives an explicit construction of N-particle wave functions related to a density $\rho(\mathbf{r}) \in D_N$. The necessary and sufficient criteria for $\rho(\mathbf{r})$ to be υ-representable, i.e. for $\rho(\mathbf{r})$ to belong to D_N^{υ}, may be found in the works of Lieb (1982, 1983), although many authors develop different approaches for defining the υ-representability condition (see the review by Ludeña and Kryachko 1989).

As already mentioned, the Hohenberg-Kohn functional $E[\rho]$, as it stands, is defined only for υ-representable local densities $\rho(\mathbf{r}) \in D_N^{\upsilon} \subset D_N$. At the same time, the Euler-Lagrange equation (2.13) assumes variations with respect to all possible local density distributions $\rho(\mathbf{r}) \in D_N$. However, it cannot be proved (Hohenberg and Kohn 1964) whether an arbitrary positive $\rho(\mathbf{r})$ which satisfies the condition (2.10) can be realized by some external field $\upsilon(\mathbf{r})$.

In other words, the υ-representability problem within the density functional theory consists in understanding that the universal functional $E[\rho]$ following from the construction of Hohenberg and Kohn (1964) is defined for the exact local density and for those $\rho(\mathbf{r}) \in D_N^{\upsilon} \subset D_N$, but it remains completely undefined for any other trial densities $\rho(\mathbf{r}) \subset \tau D_N$.

Levy-Lieb approach

In order to solve the υ-representability problem, Levy (1979, 1982) extended the domain $F[\rho]$ by defining a new universal functional

$$Q[\rho] = \min \langle \Psi_\rho | \hat{H} | \Psi_\rho \rangle = \langle \Psi_\rho^{\min} | \hat{H} | \Psi_\rho^{\min} \rangle, \tag{2.15}$$

which exists for any N-representable $\rho(\mathbf{r})$. The functional $Q[\rho]$ searches all wave functions $\Psi_\rho \in \mathcal{H}^N$ yielding a given local density $\rho(\mathbf{r}) \subset D_N$ and provides a minimal expectation value of $\hat{H}$ among them. Obviously $Q[\rho]$ becomes identical with $F[\rho]$ for any $\rho(\mathbf{r}) \subset D_N^{\upsilon}$, that is, for any $\rho(\mathbf{r})$ which is υ-representable. Lieb (1982, 1983) has shown that $\Psi_\rho^{\min} \in \mathcal{H}^N$ always exists.

Obviously, there exist many wave functions $\Psi_\rho \in \mathcal{H}^N$ yielding the same local density $\rho(\mathbf{r})$ and, therefore, the one-to-one mapping between the density and the wave function no longer exists in the Levy-Lieb approach. Valone (1980a, 1980b) has solved this problem by providing an ensemble-search generalization of eqn (2.15).

The Hohenberg-Kohn theorem seems to be restricted also due to the assumption that the ground state of the system is to be nondegenerate. This assumption, however, is not a strong restriction as mentioned by Kohn (1966). As pointed out by Parr *et al.* (1978) only a minor modification of the theorem is required in order to remove this restriction. Thus, the Hohenberg-Kohn theorem has also been found to be valid for degenerate ground states (Katriel, Appellof, and Davidson 1981, English and English 1984a,b).

Concluding, we observe that many efforts have been devoted since 1964 towards the strict mathematical formulation and foundation of the density functional theory. To this end, there is a large number of important developments summarized in the review works by Gombas (1948), Rajagopal (1980), Ghosh and Deb (1982), Lundqvist and March (1983), Parr (1983), Callaway and March (1984), Dahl and Avery (1984), Dreizler and Providencia (1985), Erdahl and Smith (1987), March and Deb (1987), Ludeña and Kryachko (1989), Parr and Yang (1989).

2.3 Extensions of the Hohenberg-Kohn theorem

Many extensions going beyond the ground-state problem have been provided after the Hohenberg-Kohn theorem has proved to be mathematically correct and physically significant. We summarize the main extensions of the original theorem in order to stimulate their possible application to nuclear many-body problems which is not as yet a well developed area.

Extension to finite temperatures

Mermin (1965) has proved an underlining extension of the Hohenberg-Kohn theorem to finite temperatures presently known as Hohenberg-Kohn-Mermin theorem. Its essential result is that the grand-potential of a N-particle system at finite temperature T and chemical potential μ is a functional of the local density at that temperature. In order to define this functional Mermin (1965) introduces the expression

$$\Omega[\rho] = F[\rho] + \int d\mathbf{r} \; (v(\mathbf{r}) - \mu) \; \rho(\mathbf{r}), \tag{2.16}$$

which corresponds to the structure of the grand-potential

$$\Omega = \mathrm{tr} \; \{\hat{\rho} \; (\hat{H}_{v} - \mu\hat{N} + \beta^{-1} \ln \hat{\rho})\}. \tag{2.17}$$

Here $\beta^{-1} = T$ is the temperature (measured in MeV), $\hat{H}_{v}$ is the Hamiltonian (2.1), $\hat{N}$ is the particle number operator and $\hat{\rho}$ is the density matrix operator which is used to calculate thermal averages of physical quantities (the apparatus of statistical physics is considered in more detail in Chapter 10).

It is possible to repeat the arguments of the previous Section in order to show that the complete density matrix $\hat{\rho}$ is a functional of the local density $\rho(\mathbf{r})$, as are all averages based on the grand canonical ensemble. In particular, one may use the local density representation of $\hat{\rho}$ in eqn (2.17) in order to define the functional (2.16). As before, the proof is by *reductio ad absurdum*. Then Mermin (1965) proved (see also Kohn and Vashishta 1983) that eqn (2.16) is a minimum and equal to the grand potential when $\rho(\mathbf{r})$ is the equilibrium density of the system.

The Hohenberg-Kohn-Mermin theorem is again too restrictive because one can define $\rho(\mathbf{r})$ for a wider class of grand canonical ensembles (v-representability problem). In this context, there is a Levy-Lieb theorem formulated by Rajagopal (1985), whereby the resulting temperature-dependent local density functionals are constructed without any reference to external fields. We shall return to the Hohenberg-Kohn-Mermin theorem in Section 10.1 again.

Relativistic generalization

Starting from a relativistic point of view Rajagopal and Callaway (1973) consider a system of an arbitrary number of electrons in a large box with volume V under the influence of a four-vector potential $A_\mu^{EXT}(x) = [v(\mathbf{r},t), \mathbf{A}_{EXT}(\mathbf{r},t)]$. The variable x is used to denote coordinates and time. The operators in all equations that follow must be normal ordered although we have no explicit notation for this. The Hamiltonian used contains four parts

$$\hat{H} = \hat{H}_0 + \hat{H}_C + \hat{H}_I + \hat{H}_{EXT}. \tag{2.18}$$

The operator $\hat{H}_0$ describes noninteracting electromagnetic and Dirac fields

$$\hat{H}_0 = \hat{H}_{em} + \int d\mathbf{r}\ \psi^+(x)\hat{h}(x)\psi(x), \tag{2.19}$$

where $\hat{H}_{em}$ describes a free radiation field in the radiation gauge so that the Coulomb interaction between electrons $\hat{H}_C$ appears explicitly in eqn (2.18). The Dirac Hamiltonian $\hat{h}(x)$ entering eqn (2.19) is of the form

$$\hat{h}(x) = (i\hbar\gamma^\mu\partial_\mu - mc), \tag{2.20}$$

and $\psi^+(x)$, $\psi(x)$ are creation and annihilation operators of the free Dirac field.

The interaction between matter and the transverse portion of the radiation field is contained in the term

$$\hat{H}_I = \frac{e}{c}\int d\mathbf{r}\ \mathbf{j}(x)\ \mathbf{A}(x), \tag{2.21}$$

where the components of the current operator are

$$j^\mu(x) = c\ \psi^+(x)\ \gamma^0\gamma^\mu\ \psi(x), \tag{2.22}$$

with γ^μ ($\mu = 0,1,2,3$) being the Dirac matrices. The last term in eqn (2.18)

$$\hat{H}_{EXT} = -\frac{e}{c}\int d\mathbf{r}\ j_\mu(x)\ A^\mu_{EXT}(x), \tag{2.23}$$

represents the interaction between the current and the external electromagnetic field $A^{\mu}_{EXT}(x)$.

The relativistic Hamiltonian (2.18) is not bounded from below and therefore it does not have a ground state. Therefore, in the conventional way, one has to suppose that all negative energy states are filled, and one has to define the ground-state in this sense. In this ground-state the four-current density is an ordinary function of the position

$$J_{\mu}(x) = \langle j_{\mu}(x) \rangle, \tag{2.24}$$

where $\langle ... \rangle$ stands for the ground-state expectation value. The four components (2.24) are not all independent since they satisfy the continuity equation

$$\partial_{\mu} J^{\mu}(x) = 0, \tag{2.25}$$

which results from the gauge invariance of the Hamiltonian given by eqn (2.18).

Rajagopal (1978) has proved step-by-step the following statements: (1) The external (classical) field $A^{\mu}_{EXT}(x)$ is a unique functional of $J_{\mu}(x)$; (2) The nondegenerate ground-state energy per volume V is a functional of $J_{\mu}(x)$; that is,

$$\langle \hat{H} \rangle / V \equiv E[J_{\mu}] = F[J_{\mu}] - \frac{e}{c} \int dr\ J_{\mu}(x)\ A^{\mu}(x), \tag{2.26}$$

where

$$F[J_{\mu}] = \langle \hat{H}_0 + \hat{H}_C + \hat{H}_I \rangle \tag{2.27}$$

is a universal functional of $J_{\mu}(x)$ independent on $A^{\mu}_{EXT}(x)$; and (3) The correct $J_{\mu}(x)$ minimizes eqn (2.26) subject the condition (2.25) and the constant total number of electrons $\int dr\ J^0(x) = eN$.

In a similar way, Rajagopal (1980) has proved a relativistic extension of the Hohenberg-Kohn-Mermin theorem which states that in a grand canonical ensemble with fixed temperature T and chemical potential μ, the components of the four-current density

$$J^{\mu}(x) = \mathrm{tr}\,(\hat{\rho}\; j_{\mu}(x)) \tag{2.28}$$

are determined by a static external vector potential $A_{\mu}^{EXT}(x)$. The correct divergenceless $J_{\mu}(x)$ minimizes the grand-potential for given $A_{\mu}^{EXT}(x)$.

Spin generalization

The extension of the Hohenberg-Kohn theorem to include spin has been made by different authors (Stoddart and March 1971, Pant and Rajagopal 1972, Barth and Hedin 1972). This leads to spin-density functional theory: the ground-state energy is a unique functional $E[\rho,\mathbf{s}]$ of the local density $\rho(\mathbf{r})$ and the spin-density vector $\mathbf{s}(\mathbf{r})$. The functional $E[\rho,\mathbf{s}]$ attains its minimum value when the quantities $\rho(\mathbf{r})$ and $\mathbf{s}(\mathbf{r})$ are correct. Rajagopal and Callaway (1973) derived the functional $E[\rho,\mathbf{s}]$ quite naturally by taking the nonrelativistic limit of the relativistic approach outlined above.

Extensions to excited states

The difficulty in the variational treatment of the excited states has been mentioned in Section 1.2. Any excited state must be orthogonal to the lower states. Therefore, all lower states must be known in order to determine the particular excited state.

An exception noted earlier arises in the case when the orthogonality problem has a trivial solution by reason of symmetry. Gunnarson and Lundqvist (1976) have shown that if this is the case, then a procedure like that of Hohenberg and Kohn (1964) may be used to develop a density functional theory for the lowest energy of each symmetry. One can find a one-to-one correspondence between the local density and the lowest state of given symmetry. Barth (1979) developed a density functional theory of mixed symmetry in order to improve the Gunnarson-Lundqvist approach for higher excited states. Theophilou (1987) has shown that a rigorous formulation of the Von Barth theory is possible.

Mikolas and Tomasek (1977) deduce that the total energy of a particular state (ground or excited) is a functional of the local density of the same state only. In the general case, however, this statement cannot be proved and it is probably false (Callaway and March 1984).

A more realistic attack on the excited-state problem has been made by Theophilou (1979, 1987), who has extended the Hohenberg-Kohn theorem to excited states in the general case. The density, however, is not the density corresponding to a single excited state, but represents a sum of all lowest-energy eigenstate densities. The same composite structure holds also for the energy and all other measured quantities.

An interesting construction has been used by Valone and Capitani (1981). According to McDonald (1979), if λ is a given real parameter then the minimum of the functional

$$F(\Psi;\lambda) = \langle\Psi|(\hat{H}_{\lambda\vartheta} - \lambda I)^2|\Psi\rangle \geq 0, \tag{2.29}$$

occurs for the eigenstate Ψ_n of $\hat{H}_{\lambda\vartheta}$ whose eigenvalue $E_{\lambda\vartheta}$ is closest to λ. Valone and Capitani (1981) introduce the above functional $F(\Psi;\lambda)$ in density functional theory, also defining similar functionals for some even powers of $(\hat{H}_{\lambda\vartheta} - \lambda I)$. The advantage is that $F(\Psi;\lambda)$ is a single-state functional, whereas Theophilou's is defined on many groups of mutually orthogonal states.

The above examples show that excited-state density functional theory is still an open field and many of its problems have not yet been investigated.

Time-dependent processes

The development of a general time-dependent density functional theory is also still incomplete. The main difficulty is due to the lack of a suitable time-dependent variational principle corresponding to an energy minimum. There has been considerable success in quantum fluid dynamics which is interlinked to density functional theory (Deb and Ghosh 1987). As for the excited states, however, the time-dependent processes are the main drawback of contemporary density functional theory.

Momentum space

It is interesting to consider whether there is a form of the Hohenberg-Kohn theorem in momentum space where the natural ingredient is the particle momentum distribution

$$n(\mathbf{k}) = \int d\mathbf{r}\, d\mathbf{r}' \; \rho(\mathbf{r},\mathbf{r}') \exp\,[-\mathbf{k}.(\mathbf{r} - \mathbf{r}')]. \qquad (2.30)$$

The local density $\rho(\mathbf{r})$ and the momentum distribution $n(\mathbf{k})$ are not simple Fourier transforms of each other. The former is the diagonal element of the one-body density matrix (1.94), while the nondiagonal elements of $\rho(\mathbf{r},\mathbf{r}')$ appear in eqn (2.30). Therefore, there is no trivial transcription of the Hohenberg-Kohn theorem into momentum space.

Nevertheless, there is a corresponding density functional theory in momentum space given by Henderson (1981). He observes the existence of an energy functional of $n(\mathbf{k})$, but it is not a universal functional since it requires the external potential (nonlocal operator in the momentum space) to be known. Thus, some of the power of the Hohenberg-Kohn theorem is lost because a specific functional of $n(\mathbf{k})$ must be studied for each physical system. Some aspects of the density functional theory in coordinate as well as in momentum space are given by Smith (1984) and Antonov, Hodgson, and Petkov (1988) together with some functional relations between $\rho(\mathbf{r})$ and $n(\mathbf{k})$.

2.4 Local-scale point transformations

The nuclear Hamiltonian contains no external fields of the type of $\omega(\mathbf{r})$ exploited within the Hohenberg-Kohn theorem. Therefore, it is important to consider the approach given by Petkov and Stoitsov (1980a, 1981, 1983a), which makes an essential use of the local-scale point transformations of the three dimensional vector space R^3. This approach will allow us later in Chapter 7 to develop a rigorous and constructive proof of the Hohenberg-Kohn theorem, whereby a set of completely definite approximate energy density functionals results.

Group of local-scale point transformations

The local-scale point transformations of three-dimensional vector space (2.14) are defined by vector functions of the form

$$\mathbf{f}(\mathbf{r}) = \mathbf{r}_0\, f(\mathbf{r}). \tag{2.31}$$

The mapping $\mathbf{r} \to \mathbf{f}(\mathbf{r})$ establishes a one-to-one correspondence between every vector $\mathbf{r} \in R^3$ and the vector $\mathbf{r}' \equiv \mathbf{f}(\mathbf{r}) \in R^3$. The latter possesses the same direction $\mathbf{r}_0 \in \Omega^1$, but a changed length $r' = f(\mathbf{r}) \in R^1 \equiv [0,\infty)$ depending on the scalar local-scale transformation function $f(\mathbf{r}) \equiv f(r,\mathbf{r}_0) \in R^1$.

Let us assume the following rather general mathematical conditions with regard to the scalar function $f(\mathbf{r})$:

(i) $f(r,\mathbf{r}_0) \in R^1$ for $r \in R^1$ and $\mathbf{r}_0 \in \Omega^1$;

(ii) $f(r,\mathbf{r}_0) \in C^2(R^1) \times C^2(\Omega^1)$;

(iii) $f(r_1,\mathbf{r}_0) < f(r_2,\mathbf{r}_0)$ for all $r_1 < r_2$ and $\mathbf{r}_0 \in \Omega^1$;

(iv) $D(\mathbf{r}) \neq 0$ and $D(\mathbf{r}) < \infty$ for all $\mathbf{r} \in R^1$,

where $D(\mathbf{r})$ is the Jacobian of the local-scale point transformation

$$D(\mathbf{r}) \equiv D(\mathbf{f}(\mathbf{r});\mathbf{r}) = \frac{D(f_x,f_y,f_z)}{D(x,y,z)} = \left(\frac{f(\mathbf{r})}{r}\right)^2 \frac{\partial f(\mathbf{r})}{\partial r}, \tag{2.32}$$

and $\partial f(\mathbf{r})/\partial r$ stands for the partial derivative of $f(r,\mathbf{r}_0)$ with respect to $r \in R^1$.

The set $\mathcal{F}$ of all vector functions (2.31) also satisfying the conditions (i)–(iv) forms the so-called local-scale point transformation group (Petkov and Stoitsov 1983a) with respect to the composition law (the group product)

$$f_1 f_2 \to f(\mathbf{r}) = f_2(f_1(\mathbf{r})). \tag{2.34}$$

According to eqn (2.34), the element $f = f_1 f_2 \in \mathcal{F}$ corresponds to the group product of any two elements f_1 and f_2 belonging to $\mathcal{F}$ and it is fairly easy to prove that the composition law (2.34) is associative. The unit element of the group $\mathcal{F}$ is the identity transformation $f_e \equiv \mathbf{r}$ defined for all $\mathbf{r} \in R^3$. For every element $f \in \mathcal{F}$ there exists an inverse element $f^{-1} \in \mathcal{F}$ which is uniquely determined so that

$$f(f^{-1}(\mathbf{r})) = f^{-1}(f(\mathbf{r})) = f_e(\mathbf{r}). \tag{2.35}$$

One can isolate in $\mathcal{F}$ the important subgroup $\mathcal{F}_S \subset \mathcal{F}$ of pure scaling point transformations, the well known group of dilatations (Simon 1979). Each element $f \in \mathcal{F}_S$ is defined by eqn (2.31) with the restriction that the scalar transformation function is of the form

$$f(\mathbf{r}) = \alpha \mathbf{r}, \tag{2.36}$$

where α is an arbitrary constant. The unit element in $\mathcal{F}_S$ is obviously the one with $\alpha = 1$, while the element $f^{-1}(\mathbf{r}) \equiv \mathbf{r}/\alpha \in \mathcal{F}_S$ is the inverse one to the element $f(\mathbf{r}) \equiv \alpha\mathbf{r} \in \mathcal{F}_S$ which is $f^{-1}(f(\mathbf{r})) \equiv \mathbf{r}$.

Unitary representation in the Hilbert space

With every element $f \in \mathcal{F}$ we associate the operator $\mathcal{U}_f(\mathbf{r}_1,\mathbf{r}_2\ldots,\mathbf{r}_N)$ acting in the Hilbert space $\mathcal{H}^N$ according to the rule

$$\mathcal{U}_f \Psi = D^{1/2}(f(\mathbf{r}_1);\mathbf{r}_1)\ldots D^{1/2}(f(\mathbf{r}_N);\mathbf{r}_N)\ \Psi(f(\mathbf{r}_1),\ldots,f(\mathbf{r}_N)), \tag{2.37}$$

where Ψ is an arbitrary function from $\mathcal{H}^N$. The operator

$\mathcal{U}_f \equiv \mathcal{U}_f(\mathbf{r}_1,\mathbf{r}_2\ldots,\mathbf{r}_N)$ itself represents a linear real unitary operator acting in the Hilbert space $\mathcal{H}^N$. In order to obtain its explicit expression, Epstein and Hirschfelder (1961) construct the following unitary operator

$$U_f(\mathbf{r}) = \exp(\, i\hat{W}_f(\sigma)), \tag{2.38}$$

where the Hermitian operator $\hat{W}_f(\sigma) = \hat{W}_f^+(\sigma)$ is of the form

$$\hat{W}_f(\sigma) = -i\left[\frac{\sigma(\mathbf{r})}{r} + \frac{1}{2}\frac{\partial\sigma(\mathbf{r})}{\partial r} + \sigma(\mathbf{r})\frac{\partial}{\partial r}\right]. \tag{2.39}$$

The scalar function $\sigma(\mathbf{r})$ entering eqn (2.39) is related to the local-scale transformation function $f(\mathbf{r})$ by the equation

$$S(f) = S(r) + 1 \tag{2.40}$$

with

$$S(r) = \int^{r} \frac{du}{\sigma(u,\mathbf{r}_0)}, \tag{2.41}$$

or

$$\frac{\sigma(f(r),\mathbf{r}_0)}{\sigma(r,\mathbf{r}_0)} = \frac{\partial f(r,\mathbf{r}_0)}{\partial r}. \tag{2.42}$$

One can verify that for any given $\sigma(\mathbf{r})$ there is a local-scale transformation function $f(\mathbf{r})$ simply defined as

$$f(\mathbf{r}) = U_f(\mathbf{r})\; r\; U_f^+(\mathbf{r}). \tag{2.43}$$

The unitary operator $\mathcal{U}_f$ from eqn (2.37) is then the product

$$\mathcal{U}_f \equiv \mathcal{U}_f(\mathbf{r}_1,\mathbf{r}_2\ldots,\mathbf{r}_N) = \prod_{i=1}^{N} U_f(\mathbf{r}_i) \tag{2.44}$$

of N commuting operators $U_f(\mathbf{r}_i)$ defined by eqn (2.37) for $i = 1,2,\ldots,N$, each of which realizes a local-scale point transformation of the wave function $\Psi \in \mathcal{H}^N$ with respect to the corresponding coordinate $\mathbf{r}_i$. Some topological observations regarding the possible functions $\sigma(\mathbf{r})$ and their properties are collected by Epstein and Hirschfelder (1961).

The operators (2.44) that correspond to all elements $\mathbf{f} \in \mathcal{F}$, form a group $\mathcal{U}_{\mathcal{F}} = \{\mathcal{U}_f;\ \mathbf{f} \in \mathcal{F}\}$ with respect to the ordinary operator multiplication

$$\mathcal{U}_{f_1 f_2} = \mathcal{U}_{f_1} \mathcal{U}_{f_2}, \tag{2.45}$$

$I \equiv \mathcal{U}_{f_e}$ being the unit operator in $\mathcal{U}_{\mathcal{F}}$. Then the morphism $\mathcal{F} \longrightarrow \mathcal{U}_{\mathcal{F}}$ is in fact a homomorphism of the group $\mathcal{F}$ in the space $\mathcal{U}_{\mathcal{F}}$ which defines the group $\mathcal{U}_{\mathcal{F}}$ as unitary representation of $\mathcal{F}$ in $\mathcal{H}^N$.

Orbits

Acting on a given wave function $\bar{\Psi} \in \mathcal{H}^N$ (also called the model or prototypical wave function) with all elements $\mathcal{U}_f \in \mathcal{U}_{\mathcal{F}}$, we construct the so-called orbit of the Hilbert space $\mathcal{H}^N$ with respect to the group $\mathcal{U}_{\mathcal{F}}$

$$\mathcal{O} = \{\ \mathcal{U}_f\bar{\Psi};\ \ \mathcal{U}_f \in \mathcal{U}_{\mathcal{F}}\ \}. \tag{2.46}$$

It follows from group theory (Naimark 1976) that the space $\mathcal{H}^N$ decomposes into pairwise nonintersecting orbits (2.46) associated with different prototypical functions $\bar{\Psi} \in \mathcal{H}^N$. In other words, the Hilbert space $\mathcal{H}^N$ is partitioned into a set of nonintersecting or disjoint orbits of the type of eqn (2.46) with all N-particle wave functions belonging to a given orbit being related via local-scale point transformations.

Petkov and Stoitsov (1983a) have proved that groups $\mathcal{F}$ and $\mathcal{U}_{\mathcal{F}}$ are isomorphic to each other. The isomorphism of $\mathcal{F}$ and $\mathcal{U}_{\mathcal{F}}$ makes it possible, on the one hand, to interpret $\mathcal{U}_{\mathcal{F}}$ as a group of local-scale point transformations of the space $\mathcal{H}^N$ and, on the other, to specify the orbit (2.46) by the elements of the group $\mathcal{F}$; that is,

$$\mathcal{O} = \{ \mathcal{U}_f \bar{\Psi};\ f \in \mathcal{F} \}. \tag{2.47}$$

Obviously, among the all orbits (2.47) of $\mathcal{H}^N$, there is one of them which contains the exact ground-state wave function $\Psi_0 \in \mathcal{H}^N$ of the N-particle system with Hamiltonian $\hat{H}$. So, by an 'eigenorbit' of the space $\mathcal{H}^N$ we shall understand the orbit $\mathcal{O}_H \subset \mathcal{H}^N$ containing the exact ground-state wave function $\Psi_0 \in \mathcal{H}^N$, the solution of the Schrödinger equation (1.37) with the smaller eigenvalue E_0.

Local-scale transformation method

Let the N-particle prototypical or model wave function $\bar{\Psi} \in \mathcal{H}^N$ be given. It generates the orbit $\mathcal{O} \subset \mathcal{H}^N$ according to eqn (2.47). Then, the expectation value of the Hamiltonian $\hat{H}$ taken between states $\Psi_f = \mathcal{U}_f \bar{\Psi} \in \mathcal{O}$ is transformed into the following functional of f(**r**) defined on the orbit $\mathcal{O} \subset \mathcal{H}^N$

$$E\,[f] = \frac{\langle \Psi_f |\ \hat{H}\ | \Psi_f \rangle}{\langle \Psi_f | \Psi_f \rangle}\ . \tag{2.48}$$

It is clear now that the minimization of the functional E[f] with respect to f(**r**) corresponds to a variational method applied within the orbit $\mathcal{O} \subset \mathcal{H}^N$. Therefore, from the point of view of the variational method, the local-scale point transformations define the class of trial wave functions as an orbit $\mathcal{O}$ of the space $\mathcal{H}^N$ and the minimum of E[f] on this orbit, if it exists, determines approximately the ground-state energy of the system.

In conclusion, the so-called local-scale transformation method determining the ground state of the many-particle systems consists in the following operations:

1. Choice of the model wave function $\bar{\Psi} \in \mathcal{H}^N$ which determines the space $\mathcal{O} \subset \mathcal{H}^N$ of the trial wave functions $\Psi_f = \mathcal{U}_f\bar{\Psi}$ according with eqns (2.37) and (2.47).
2. Construction of the functional E[f] by eqn (2.48) on the base of Hamiltonian $\hat{H}$ of the system.
3. Solution of the stationary equation

$$\frac{\delta E\,[f]}{\delta f(\mathbf{r})} = 0, \qquad \mathbf{f} \in \mathcal{F}, \tag{2.49}$$

   where the existence of the global minimum of E [f] on the orbit $\mathcal{O} \subset \mathcal{H}^N$ is assumed.

Solving eqn (2.49), one could obtain the only unknown function f(**r**) which further determines the approximate ground-state wave function

$$\Psi_f(\mathbf{r}_1,\ldots,\mathbf{r}_N) = \mathcal{U}_f(\mathbf{r}_1,\ldots,\mathbf{r}_N)\,\bar{\Psi}(\mathbf{r}_1,\ldots,\mathbf{r}_N), \tag{2.50}$$

and thus all other characteristics of the system.

Universality of the method

As mentioned in Section 1.2, one can assume the model wave function $\bar{\Psi} \in \mathcal{H}^N$ to be normalized to unity, that is, $\langle\bar{\Psi}|\bar{\Psi}\rangle = 1$. Because of the unitarity of the operators $\mathcal{U}_f \in \mathcal{U}_{\mathcal{F}}$ the trial wave functions $\Psi_f \in \mathcal{O}$ are also normalized to unity

$$\langle\Psi_f|\Psi_f\rangle = \langle\bar{\Psi}|\mathcal{U}_f^{+}\mathcal{U}_f|\bar{\Psi}\rangle = \langle\bar{\Psi}|\bar{\Psi}\rangle = 1. \tag{2.51}$$

In this case, the functional (2.48) takes the form

$$E\,[f] = \langle\Psi_f|\,\hat{H}\,|\Psi_f\rangle, \tag{2.52}$$

while the requirement $\mathbf{f} \in \mathcal{F}$ in eqn (2.49) takes into account the conservation of the norm (2.51) during the process of variation.

Let us introduce the effective Hamiltonian

$$\hat{H}[f] = \mathcal{U}_f^{+}(\mathbf{r}_1,\ldots,\mathbf{r}_N)\ \hat{H}\ \ \mathcal{U}_f(\mathbf{r}_1,\ldots,\mathbf{r}_N). \tag{2.53}$$

Now, from eqn (2.49) and eqn (2.52), one can realize that the f-dependence of both, the energy functional

$$E\ [f] = \langle\bar{\Psi}|\hat{H}[f]|\bar{\Psi}\rangle \tag{2.54}$$

and the stationary equation,

$$\langle\bar{\Psi}|\delta\hat{H}[f]/\delta f(\mathbf{r})|\bar{\Psi}\rangle = 0, \tag{2.55}$$

is universal in the sense that it does not depend on the orbit $\mathcal{O} \subset \mathcal{H}^N$. In other words, the f-dependence of eqn (2.54) and eqn (2.55) may be derived once and for all since the effective Hamiltonian (2.53) does not depend on the particular orbit $\mathcal{O} \subset \mathcal{H}^N$ on which the local-scale transformation method is applied.

Reformulation of the variational principle

Petkov and Stoitsov (1983a) have demonstrated that the functional E[f] derived within the eigenorbit $\mathcal{O}_H \subset \mathcal{H}^N$ is a unique functional of f(**r**). Its absolute minimum coincides with the exact ground-state energy of the system E_0 since the eigenorbit $\mathcal{O}_H \subset \mathcal{H}^N$ contains the exact wave function $\Psi_0 \in \mathcal{H}^N$. Therefore, taking into account that $\mathcal{H}^N$ is a union of all pairwise nonintersecting orbits $\mathcal{O} \subset \mathcal{H}^N$, one can reduce the investigation of all wave functions $\Psi \in \mathcal{H}^N$ with respect to E[Ψ], eqn (1.32), to the investigation of all orbits $\mathcal{O} \subset \mathcal{H}^N$ with respect to the functional E[f] by finding its minimum for each of these orbits, that is,

$$E_0 = \min_{\Psi\in\mathcal{H}^N} \{\ E[\Psi]\ \} = \min_{\mathcal{O}\subset\mathcal{H}^N} \{\ \min_{f\in\mathcal{F}} \{\ E\ [f]\ \}\ \}. \tag{2.56}$$

In other words, eqn (2.56) completes the possible reformulation of the ground-state variational principle considered in Section 1.2 in terms of the local-scale point transformations.

Local density description

There is a quite natural transition from the local-scale transformation method to the density functional theory which will be illustrated below. For this purpose, let us consider the orbit (2.47) generated by a given model function $\bar{\Psi} \in \mathcal{H}^N$. The model local density distribution associated with $\bar{\Psi} \in \mathcal{H}^N$,

$$\bar{\rho}(\mathbf{r}) = N \int |\bar{\Psi}(\mathbf{r},\mathbf{r}_2,\ldots,\mathbf{r}_N)|^2 \, d\mathbf{r}_2 \ldots d\mathbf{r}_N \tag{2.57}$$

belongs to the space D_N introduced in Section 2.1. Let us further consider a certain transformation function $\mathbf{f}(\mathbf{r}) \in \mathcal{F}$ and denote by $\rho(\mathbf{r})$ the local density distribution associated with the trial wave function $\Psi_f = \mathcal{U}_f \bar{\Psi} \in \mathcal{O}$. Then the relation

$$\rho(\mathbf{r}) = \left(\frac{f(\mathbf{r})}{r}\right)^2 \frac{\partial f(\mathbf{r})}{\partial r} \, \bar{\rho}(\mathbf{f}(\mathbf{r})) \tag{2.58}$$

emerges using the definitions of the quantities entering eqn (2.58).

We designate by $\bar{D}_N \subset D_N$ the subset of D_N containing all local densities $\rho(\mathbf{r})$ defined for all elements $\mathbf{f} \in \mathcal{F}$ according to eqn (2.58). Petkov and Stoitsov (1983a) prove that the relation (2.58) is a one-to-one correspondence within a given orbit $\mathcal{O} \subset \mathcal{H}^N$ between the local-scale transformation functions $f(\mathbf{r})$ and the associated local densities $\rho(\mathbf{r}) \in \bar{D}_N$. Then, one can solve conversely eqn (2.58) with respect to $f(\mathbf{r})$ in order to obtain the function $f(\mathbf{r})$ in terms of $\rho(\mathbf{r})$ taken as given density from $\bar{D}_N \subset D_N$.

Note that eqn (2.58) depends parametrically on $\mathbf{r}_0 \in \Omega^1$. For any chosen direction $\mathbf{r}_0 \in \Omega^1$, eqn (2.58) can be represented as an ordinary first-order nonlinear differential equation

$$\frac{df(r)}{dr} = F(f,r), \tag{2.59}$$

where

$$F(f,r) = \frac{\rho(\mathbf{r})\, r^2}{\bar{\rho}(\mathbf{f}(\mathbf{r}))\, f^2(\mathbf{r})}. \tag{2.60}$$

Its solution with respect to $f(\mathbf{r})$ always exists since eqn (2.60) is a continuous function of $r \in R^1$ under sufficiently general mathematical conditions imposed on the functions $\rho(\mathbf{r})$, $\bar{\rho}(\mathbf{r})$ and $f(\mathbf{r})$. These conditions are provided by the fact that both $\rho(\mathbf{r})$ and $\bar{\rho}(\mathbf{r})$ belong to D_N and $\mathbf{f} \in \mathcal{F}$.

Let us finally consider an arbitrary local density $\rho(\mathbf{r}) \in \bar{D}_N$. One can verify that there exists a certain many-particle wave function belonging to $\mathcal{O} \subset \mathcal{H}^N$

$$\Psi[\rho] \equiv \Psi(\mathbf{r}_1,\ldots,\mathbf{r}_N;[\rho]) = \mathcal{U}_{f[\rho]}(\mathbf{r}_1,\ldots,\mathbf{r}_N)\, \bar{\Psi}(\mathbf{r}_1,\ldots,\mathbf{r}_N), \tag{2.61}$$

such that its local density is exactly $\rho(\mathbf{r})$. For this purpose, one only needs the unique solution $f[\rho] \equiv f_{[\rho]}(\mathbf{r})$ of eqn (2.59) for all $\mathbf{r}_0 \in \Omega^1$. Then eqn (2.61) immediately follows after substituting this solution $f[\rho]$ into eqn (2.50).

The expectation value of the Hamiltonian $\hat{H}$ between states (2.61) takes the form of the local density functional

$$E\,[\rho] = \frac{\langle\Psi[\rho]\,|\hat{H}|\,\Psi[\rho]\rangle}{\langle\Psi[\rho]\,|\,\Psi[\rho]\rangle}; \tag{2.62}$$

that is, the functional $E[f]$ following from the local-scale transformation method goes over into the functional (2.62) after substituting $f[\rho]$ into eqn (2.48). The only unknown function, the local density $\rho(\mathbf{r})$, should then be determined by the Euler-Lagrange equation

$$\frac{\delta E\,[\rho]}{\delta\rho(\mathbf{r})} - \mu = 0, \tag{2.63}$$

solved with the normalization condition

$$N[\rho] \equiv \int dr\ \rho(\mathbf{r}) = N\langle\Psi[\rho]|\Psi[\rho]\rangle = N. \tag{2.64}$$

In this way, translating the local-scale transformation method in terms of local densities $\rho(\mathbf{r})$, one realizes that a local density formulation of the many-body problem is in fact achieved.

2.5 Exact energy density functional for light nuclei

It is useful to examine a limiting case of the nucleus ^{4}He treated within the Hartree-Fock theory with effective Skyrme-type forces, which leads to an exact reformulation of the ground-state problem only in terms of the local density $\rho(\mathbf{r})$. In this way we can gain some preliminary understanding about the structure of the functionals used in nuclear density functional theory.

The case of the nucleus ^{4}He

For a system consisting of one occupied state only (the nucleus ^{4}He) the Hartree-Fock ground-state local density $\rho_q(\mathbf{r})$ is also determined by only one wave function, namely that of the occupied $1s_{1/2}$ state $\varphi_q(\mathbf{r})$. Therefore, one can write

$$\varphi_q(\mathbf{r}) = [\tfrac{1}{2}\ \rho_q(\mathbf{r})]^{1/2}, \tag{2.65}$$

which represents the one-body density matrix (1.94) in the form

$$\rho_q(\mathbf{r},\mathbf{r}') = \sqrt{\rho_q(\mathbf{r})\ \rho_q(\mathbf{r}')}\ . \tag{2.66}$$

The index q as before stands for neutrons (q = n) or protons (q = p) with the associated particle numbers N_n and N_p, respectively; $N = N_n + N_p$.

Using eqn (2.66), the kinetic energy density (1.146) becomes

$$\tau_W(\mathbf{r}) = \frac{1}{4}\left[\frac{(\nabla\rho_n(\mathbf{r}))^2}{\rho_n(\mathbf{r})} + \frac{(\nabla\rho_p(\mathbf{r}))^2}{\rho_p(\mathbf{r})}\right], \tag{2.67}$$

which in the proportional local density approximation

$$\rho_n(\mathbf{r}) = \frac{N_n}{N}\,\rho(\mathbf{r}), \qquad \rho_p(\mathbf{r}) = \frac{N_p}{N}\,\rho(\mathbf{r}), \tag{2.68}$$

takes the well known form

$$\tau_W(\mathbf{r}) = \frac{1}{4}\,\frac{(\nabla\rho(\mathbf{r}))^2}{\rho(\mathbf{r})} \tag{2.69}$$

derived by Weizsäcker (1935).

Inserting eqn (2.69) into eqn (1.148), one obtains the Skyrme-HF energy (1.144) as an explicit functional of $\rho(\mathbf{r})$

$$E[\rho] = T[\rho] + V[\rho], \tag{2.70}$$

where the first term is the kinetic energy in its Weizsäcker form

$$T[\rho] \equiv T_W[\rho] = \frac{\hbar^2}{8m}\int \frac{(\nabla\rho(\mathbf{r}))^2}{\rho(\mathbf{r})}\,d\mathbf{r}. \tag{2.71}$$

The potential energy term $V[\rho]$ in eqn (2.70) is derived after inserting eqn (2.68) and eqn (2.69) into eqn (1.148). Performing some rearrangements and taking into account that $\mathbf{J}(\mathbf{r}) \equiv 0$ and $N_n = N_p$ for the nucleus ^{4}He (no Coulomb interaction is included), one obtains

$$V[\rho] \equiv V_W[\rho] = \int d\mathbf{r}\left[\frac{3}{8}t_0\rho^2 + \frac{1}{16}t_3\,\rho^{\alpha+2} + \frac{3}{16}t_1\,(\nabla\rho)^2\right]. \tag{2.72}$$

The variational derivatives of eqn (2.71) and eqn (2.72) are

$$\frac{\delta T_W[\rho]}{\delta\rho(\mathbf{r})} = -\frac{\hbar^2}{8m}\int d\mathbf{r}'\left[\frac{2\Delta\rho(\mathbf{r}')}{\rho(\mathbf{r}')} - \left(\frac{\nabla\rho(\mathbf{r}')}{\rho(\mathbf{r}')}\right)^2\right]\frac{\delta\rho(\mathbf{r}')}{\delta\rho(\mathbf{r})} \tag{2.73}$$

and

$$\frac{\delta V_W[\rho]}{\delta\rho(\mathbf{r})} = \int d\mathbf{r}'\left[\frac{3}{4}t_0\rho + \frac{\alpha+2}{16}t_3\rho^{\alpha+1} - \frac{3}{8}t_1\Delta\rho\right]\frac{\delta\rho(\mathbf{r}')}{\delta\rho(\mathbf{r})}, \quad (2.74)$$

respectively. The Euler–Lagrange equation (2.13) then becomes

$$-\frac{\hbar^2}{4m}\Delta\rho(\mathbf{r}) + \left[\frac{3}{4}t_0\rho + \frac{\alpha+2}{16}t_3\rho^{\alpha+1} - \frac{3}{8}t_1\Delta\rho + \frac{\hbar^2}{8m}\left(\frac{\nabla\rho(\mathbf{r})}{\rho(\mathbf{r})}\right)^2\right]\rho(\mathbf{r}) = \mu\rho(\mathbf{r}). \quad (2.75)$$

It defines an eigenvalue problem whose eigensolution corresponds to the ground-state local density $\rho(\mathbf{r})$. The associated eigenvalue, the chemical potential of the system μ, is the separation energy of the last nucleon in the Hartree–Fock approximation; that is, $\mu = \varepsilon_{1s1/2}$ for the nucleus ^{4}He.

It is not difficult for eqn (2.75) to be transformed into the form of the Hartree–Fock equation (1.151). The result is

$$\left\{-\nabla\frac{\hbar^2}{2m^*(\mathbf{r})}\cdot\nabla + U(\mathbf{r})\right\}\sqrt{\rho(\mathbf{r})} = \varepsilon_{1s1/2}\sqrt{\rho(\mathbf{r})}, \quad (2.76)$$

where both quantities, eqn (1.152) and eqn (1.154), are explicit functionals of the local density, namely

$$U(\mathbf{r}) = \frac{3}{4}t_0\rho + \frac{\alpha+2}{16}t_3\rho^{\alpha+1} + \frac{1}{32}(5t_2 - 9t_1 + 4t_2x_2)\,\Delta\rho$$

$$+ \frac{1}{64}(5t_2 + 3t_1 + 4t_2x_2)\,\frac{(\nabla\rho)^2}{\rho} \quad (2.77)$$

and

$$\frac{\hbar^2}{2m^*(\mathbf{r})} = \frac{\hbar^2}{2m} + \frac{1}{16}\,(3t_1 + 5t_2 + 4t_2x_2)\rho(\mathbf{r}). \quad (2.78)$$

At large distance $r \in R^1$, the asymptotic solution of eqn (2.75) or its equivalent eqn (2.76) leads to the correct exponential fall-off of the density

$$\rho(r) \xrightarrow[r\,\infty]{} (1/r^2)\,\exp(-r/a), \quad (2.79)$$

which is completely determined by the Weizsäcker term of the kinetic energy functional. The range a is given by the Fermi energy $\mu = \varepsilon_{1s1/2}$ ($\mu < 0$) as

$$a = \sqrt{-\frac{1}{4}(\hbar^2/2m)\frac{1}{\mu}}\,. \tag{2.80}$$

At small distances $r \to 0$, eqn (2.75) leads to correct constant behaviour for the density, while the complete solution of this equation may be obtained numerically.

Exact kinetic energy density functional

The above rather simple example is nevertheless important in the nuclear density functional theory because it introduces quite naturally the Weizsäcker kinetic energy term (2.71).

As pointed out by March and Pucci (1981) the exact one-body density matrix (1.94) represented by the Löwdin natural orbitals $\bar{\varphi}_i(\mathbf{r})$, Section 1.4, factorizes for sufficiently large $\mathbf{r}'$ and for all $\mathbf{r} \in R^3$ into

$$\rho(\mathbf{r},\mathbf{r}') = g(\mathbf{r})\sqrt{\rho(\mathbf{r}')}, \tag{2.81}$$

where $\rho(\mathbf{r})$ is the exact local density and $g(\mathbf{r}) = \sum \rho_i\, \bar{\varphi}_i(\mathbf{r})$. According to Morrell, Parr, and Levy (1975) the exponential decay of all natural orbitals becomes identical at large distances and the function $g(\mathbf{r})$ tends to $\sqrt{\rho(\mathbf{r})}$ for sufficiently large $\mathbf{r}$.

Therefore, in the asymptotic region where $r \to \infty$ and $r' \to \infty$, the exact $\rho(\mathbf{r},\mathbf{r}')$ is of the form of eqn (2.66) and the exact kinetic energy density functional is given correctly by the Weizsäcker term (2.67).

2.6 Infinite nuclear matter

Another important example illustrating a possible exact formulation of the many-body problem in terms of the local density

$\rho(\mathbf{r})$ only is that of the infinite nuclear matter, an infinite system of nucleons interacting with the full nuclear interaction but with their electromagnetic interaction switched off. Infinite nuclear matter has been considered mainly to test many-body theories but it presents in itself some real physical interest since it may exist in neutron stars.

Formally, the problem is best defined as that of N nucleons in a box of volume Ω with a wave function $\Psi(x_1,\ldots,x_N) \in \mathcal{H}^N$ required to satisfy the Schrödinger equation (1.4) with periodic boundary conditions and normalized in the box (De Shalit and Feshbach 1974). The results are then evaluated in the limit $N \to \infty$ and $\Omega \to \infty$ taken so that the density

$$\lim_{\substack{\Omega\to\infty \\ N\to\infty}} \frac{N}{\Omega} = \rho_0, \tag{2.82}$$

remains unchanged.

Hartree-Fock approximation

In the HF approximation, the N-particle wave function is given by a Slater determinant with the N lowest states occupied in the ground state. Because we are considering infinite nuclear matter, taken to be homogeneous and isotropic, it is clear that there can be no preferred point in R^3 which one might select as the origin of a coordinate system. The reason lies in the transitional invariance of the nuclear Hamiltonian. Thus, the single-particle states in the Slater determinant are given, even in the presence of interactions, by plane waves

$$\varphi_{\mathbf{k}_\alpha}(\mathbf{r}) = \Omega^{-1/2} \exp(-i\, \mathbf{k}_\alpha \cdot \mathbf{r}), \tag{2.83}$$

which are specified by the three components of the wave vector $\mathbf{k} = (k_x, k_y, k_z)$ related to the momentum $\mathbf{p} = \hbar\mathbf{k}$. We shall not include the additional spin-isospin variables required to specify the nucleon states (2.83).

With the aid of eqn (2.83), the one-body density matrix (1.94) takes its nuclear matter form

$$\rho_{NM}(\mathbf{r},\mathbf{r}') = 4\sum_{\mathbf{k}_\alpha \in \Omega_{k_F}} \varphi^*_{\mathbf{k}_\alpha}(\mathbf{r})\varphi_{\mathbf{k}_\alpha}(\mathbf{r}') = \frac{4\Omega}{(2\pi)^3}\int_{\Omega_{k_F}} d\mathbf{k}\ \varphi^*_{\mathbf{k}}(\mathbf{r})\varphi_{\mathbf{k}}(\mathbf{r}')$$

$$= 4(2\pi)^{-3}\int_{\Omega_{k_F}} d\mathbf{k}\ \exp[-\,i\mathbf{k}.(\mathbf{r}-\mathbf{r}')] = 3\rho\ \frac{j_1(k_F|\mathbf{r}-\mathbf{r}'|)}{k_F|\mathbf{r}-\mathbf{r}'|}, \qquad (2.84)$$

where $j_1(x) = (\sin(x)/x^2 - \cos(x)/x)$ is the spherical Bessel function and the summation in eqn (2.84) is replaced according to the rule

$$\sum_{\mathbf{k}_\alpha \in \Omega_{k_F}} \longrightarrow (2\pi)^{-3}\ \Omega \int_{\Omega_{k_F}} d\mathbf{k} \qquad (2.85)$$

which leads to correct results in the limit of infinite volume. The factor 4 in eqn (2.84) takes into account the fact that each level contains a spin-up and a spin-down nucleon for both protons and neutrons. The single-particle states (2.83) are, therefore, filled according to eqn (2.84) up to a level designated as the Fermi level with the associated Fermi momentum $k_F=|\mathbf{k}_F|$. The occupied levels ($|\mathbf{k}|\le k_F$) are referred to as the Fermi sea which forms the so-called Fermi sphere with volume $\Omega_{k_F} = (4\pi k_F^3/3)$ in momentum space.

It follows from eqn (2.84) that the particle density $\rho \equiv \rho_{NM}(\mathbf{r},\mathbf{r})$ is a constant which satisfies the well known Fermi-gas relation between the density ρ and the Fermi momentum k_F (see for instance Huang (1963))

$$\rho = (2/3\pi^2)\ k_F^3. \qquad (2.86)$$

From eqn (1.91) and eqn (2.85), one can also obtain the kinetic energy density of nuclear matter

$$\tau_{NM}(\mathbf{r}) = \nabla \cdot \nabla' \rho_{NM}(\mathbf{r},\mathbf{r}')\Big|_{\mathbf{r}'=\mathbf{r}} = \frac{3}{5}\, k_F^2\, \rho = \frac{3}{5}\left(\frac{3\pi^2}{2}\right)^{2/3} \rho^{5/3}. \qquad (2.87)$$

In the same manner, as in the case of the nucleus ^4He, one obtains the energy per nucleon for symmetric nuclear matter

$$\frac{E_{NM}}{N} \equiv W(\rho) = \frac{3\hbar^2}{10m} k_F^2 + \frac{3}{8} t_0 \rho + \frac{1}{16} t_3 \rho^{\alpha+1} + \frac{3}{80}(3t_1 + 5t_2 + 4t_2 x_2) k_F^2 \rho, \qquad (2.88)$$

which is again an explicit functional of the local density (2.86). Due to the specifically homogeneous character of the system, this functional is simply an algebraic function of ρ, the so-called nuclear matter saturation curve $W(\rho)$. The associated Euler-Lagrange equation

$$\frac{\partial \rho W(\rho)}{\partial \rho} - \mu = \rho \frac{\partial W(\rho)}{\partial \rho} + W(\rho) - \mu = 0, \qquad (2.89)$$

can be expressed explicitly from eqn (2.88) in the form

$$\frac{\hbar^2}{5m} k_F^2 + \frac{3}{8} t_0 \rho + \frac{(\alpha+1)}{16} t_3 \rho^{\alpha+1} + \frac{1}{16}(3t_1 + 5t_2 + 4t_2 x_2) k_F^2 \rho = 0. \qquad (2.90)$$

Its solution determines the nuclear matter saturation density ρ_0, while the minimum $W(\rho_0) = \mu$ defines the binding energy per nucleon in nuclear matter. Therefore, we actually want to solve eqn (2.89) in order to obtain the nuclear matter density ρ_0 which then determines all other characteristics of the system.

Nuclear matter characteristics

In Figure 2.1 we plot the binding energy per nucleon as a function of the Fermi momentum k_F for the Skyrme-type forces SIII and SkE2 given in Table 1.1. At small densities $\rho \to 0$, the kinetic energy dominates and the system is unbound $W(\rho) > 0$; then, as one increases the density, the energy starts to decrease, reaching a

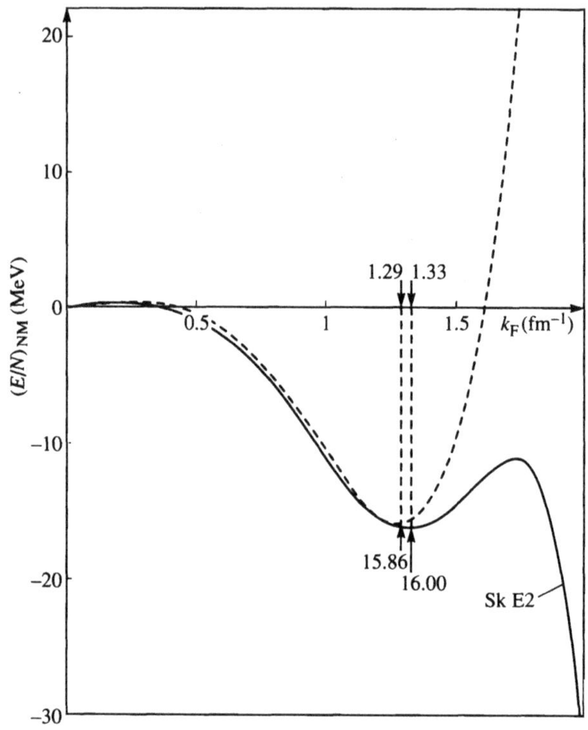

Figure 2.1. Binding energy per nucleon versus Fermi momentum in symmetric infinite nuclear matter (Waroquier, Heyde, and Wenes 1983). The SkE2 force yields a Hartree–Fock solution which gives rise to a local rather than to an absolute minimum in the energy.

minimum and then increasing again due to the repulsive part of the interaction. The only exception to this behaviour occurs at high momenta. The SkE2 force gives rise to a strong decrease in the saturation curve for densities much higher than the nuclear density due to the predominant contribution of the momentum-dependent three-body term.

In Table 2.1 we list some properties of nuclear matter calculated at the saturation point $\rho = \rho_0$ for all interactions given in Table 1.1. As well as the saturation values for the density ρ_0, the energy per nucleon $E_{NM}/N \equiv W(\rho_0)$ and the Fermi

Table 2.1. Characteristics of infinite nuclear matter for the Skyrme-type effective forces given in Table 1.1.

Force	ρ_0, fm^{-3}	k_F^0, fm^{-1}	E_{NM}/N, MeV	K_{NM}, MeV	a_τ, MeV	$(m^*/m)_s$
SI	0.15536	1.32	-16.000	370.0	29.3	0.91
SII	0.14840	1.30	-16.000	342.0	34.2	0.58
SIII	0.14530	1.29	-15.870	356.0	28.16	0.76
SIV	0.15185	1.31	-15.980	325.0	31.22	0.47
SV	0.15536	1.32	-16.060	306.0	32.72	0.38
SVI	0.14530	1.29	-15.770	364.0	26.89	0.95
SkE2	0.15891	1.33	-16.000	200.0	29.7	0.72
Ska	0.15540	1.32	-16.000	263.0	32.91	0.61
Skb	0.15540	1.32	-16.000	263.0	33.80	0.61
T	0.16130	1.337	-15.930	235.6	28.35	1.00
T1	0.16100	1.336	-15.980	236.0	32.02	1.00
SkM	0.16030	1.334	-15.776	216.7	30.75	0.79
SkM*	0.16030	1.334	-15.776	216.7	30.03	0.79

momentum k_F^0, we list also the infinite nuclear matter incompressibility

$$K_{NM} \equiv 9\rho_0\left(\frac{\partial^2 W(\rho)}{\partial\rho^2}\right)_{\rho=\rho_0} = -\frac{3\hbar^2}{5m}k_F^2+\frac{3}{8}(3t_1+5t_2+4t_2x_2)k_F^2\rho + \frac{9\alpha(\alpha+1)}{16}t_3\rho^{\alpha+1}, \tag{2.91}$$

and the isoscalar effective mass

$$\left(\frac{\hbar^2}{2m^*(r)}\right)_s = \frac{\hbar^2}{2m} + \frac{1}{16}(3t_1+5t_2+4t_2x_2)\rho_0. \tag{2.92}$$

For asymmetric nuclear matter with different neutron ρ_{0n} and proton ρ_{0p} densities, one can define also the volume asymmetry energy

$$a_\tau \equiv (2\rho_0)^{-1} \left(\frac{\partial^2 \rho_0 W(\rho_{0n}, \rho_{0p})}{\partial \delta^2} \right)_{\delta=0}$$

$$= \frac{\hbar^2}{6m} k_F^2 - \frac{1}{4} t_0 (\frac{1}{2} + x_0)\rho_0 + [\frac{1}{6} t_2 (1 + \frac{5}{4} x_2) - \frac{1}{8} t_1 x_1)] k_F^2 \rho_0 - \frac{1}{24} t_3 \rho_0^{\alpha+}, \quad (2.93)$$

where $\delta = (\rho_{0n} - \rho_{0p}) / \rho_0$ is the asymmetry parameter of the system and $\rho_0 = \rho_{0n} + \rho_{0p}$. The associated isovector effective mass is given as

$$\left(\frac{\hbar^2}{2m^*(r)} \right)_V = \frac{\delta}{16} (t_2 - t_1 - 2t_1 x_1 + 2t_2 x_2) \rho_0. \quad (2.94)$$

There are, of course, additional nuclear matter characteristics which are not given in Table 2.1 but which can be found in the literature (Bethe 1971, Blaizot 1980, Treiner and Krivine 1986). Other important properties of nuclear matter are reflected in the dynamic response of the system, such as the single-particle strength functions, optical potential, Landau parameters, momentum distribution, and so on. These properties may be described within the Green function approach as reviewed by Ramos, Polls and Dickhoff (1988).

Beyond the Hartree-Fock approximation

In order to get a more fundamental description of the infinite nuclear matter, one has to go beyond the Hartree-Fock theory applying more complicated techniques as well as using realistic nucleon-nucleon forces. The present status of the nuclear matter theory, however, is not completely satisfactory in its nonrelativistic limit. The nuclear matter saturation properties cannot be described satisfactorily when only two-body forces are considered (Day 1983, Jackson 1983). This is true in the conventional hole-line expansion (Bethe 1971, Day 1981a, 1981b) as well as in the variational description (Pandaripande and Wiringa 1979, Day and Wiringa 1985).

In recent years there have been attempts to include certain features of relativistic dynamics in the description of nuclear matter (Anastasio *et al.* 1983, Horowiz and Serot 1984, Brockmann and Machleidt 1984, Ter Haar and Malfliet 1987). In this approach, one has no problem in obtaining correct saturation properties for infinite nuclear matter by using only two-body interactions, which makes adequate the HF approach to the problem with effective two-body forces.

3. Nuclear Thomas–Fermi–Weizsacker functionals

According to density functional theory, if the exact functional $E[\rho]$ were known, the local density $\rho(\mathbf{r})$ alone would carry enough information to determine completely the ground state of the system. Obviously, such a capacity of a single functional $E[\rho]$ implies a considerable nontrivial complexity of its structure.

Unfortunately, the derivation of the exact functional $E[\rho]$ seems to be equivalent to the exact solution of the many-body problem (see Chapter 7) which is presently not feasible. Density functional theory is therefore forced to adopt some approximations to the exact functional $E[\rho]$ which are easy to handle in practice and, at the same time, reflect some important features of its nontrivial structure. In this sense, the present chapter provides an introduction to the simplest approximate energy density functionals currently in use in nuclear density functional theory.

3.1 Thomas–Fermi theory of nuclei

Historically, the Thomas–Fermi (TF) approximation was initiated in the very early days of quantum mechanics by the pioneering work of Thomas (1926) and Fermi (1928). They used the relation between the number of electrons per unit volume (the electron density) and the Fermi momentum in a uniform electron gas, but applied this relation in an inhomogeneous situation as it exists in atoms, molecules or solids.

The use of the TF approximation for nuclei is an old idea (Weizsäcker 1935, Bethe and Bacher 1936, Euler 1937). As seen in the previous chapter, infinite nuclear matter is an important uniform system and, therefore, it is quite natural to use nuclear matter results as a starting point for a TF treatment of nuclei. Skyrme (1956), Berg and Wilets (1956) and Wilets (1956,1958) developed earlier phenomenological TF-like theories, while Hara (1960), Kumar, Le Couteur, and Roy (1963) developed statistical TF methods on the basis of the HF theory using nuclear matter results. Seyler and Blanchard (1963) used for this purpose a velocity-dependent interaction, while a TF theory of finite nuclei with realistic nuclear forces and maximum use of nuclear matter results has been analysed by Bethe (1968, 1971).

Thomas-Fermi approximation

The Thomas-Fermi method is an approximation to the independent-particle model or the HF theory. Therefore, one must consider a single-particle Hamiltonian

$$\hat{H}_0 = -\frac{\hbar^2}{2m}\nabla_r^2 + U(\mathbf{r}), \tag{3.1}$$

where $U(\mathbf{r})$ is the self-consistent HF potential (1.136) or, as we shall often assume for the purpose of demonstration, a given potential well, such as the Woods-Saxon or harmonic oscillator potential. The associated one-body density matrix satisfies the idempotent condition ($\rho^2 = \rho$)

$$\int d\mathbf{r}'' \; \rho(\mathbf{r},\mathbf{r}'') \; \rho(\mathbf{r}'',\mathbf{r}') = \rho(\mathbf{r},\mathbf{r}'), \tag{3.2}$$

which is also the necessary and sufficient condition for the many-body wave function to be represented by a single Slater determinant (Thouless 1972).

The one-body density matrix can be introduced in the TF approximation by means of the well known Slater approximation successfully used by Slater (1951) in treating the exchange term in the HF theory of atoms. The single-particle wave functions are considered as locally plane waves in the neighborhood of the centre-of-mass point $\mathbf{R} = (\mathbf{r}_1 + \mathbf{r}_2)/2$. The resulting expression is

$$\rho_{TF}(\mathbf{r},\mathbf{r}') = 3\rho(\mathbf{R}) \frac{j_1(k_F(\mathbf{R})|\mathbf{r}-\mathbf{r}'|)}{k_F(\mathbf{R})|\mathbf{r}-\mathbf{r}'|}, \tag{3.3}$$

where the Fermi momentum $k_F(\mathbf{R})$ appropriate to $\mathbf{R}$ is related to the local density $\rho(\mathbf{R})$ by

$$k_F(\mathbf{R}) = \left(\frac{3\pi^2}{2}\right)^{1/3} \rho^{1/3}(\mathbf{R}). \tag{3.4}$$

For a uniform system with constant density, eqn (3.3) and eqn (3.4) automatically transform into the nuclear matter results given by eqn (2.84) and eqn (2.86), respectively. Therefore, the TF expression (3.3) implies that locally in the neighbourhood of the point $\mathbf{R}$ the nucleus has the properties of a piece of infinite nuclear matter with local density equal to $\rho(\mathbf{R})$.

In condition of varying densities eqn (3.3) leads to the well known TF expression for the kinetic energy density

$$\tau_{TF}[\rho] = \left[\nabla_r . \nabla_{r'} \rho_{TF}(\mathbf{r},\mathbf{r}')\right]_{r'=r} = \frac{3}{5}\left(\frac{3\pi^2}{2}\right)^{2/3} \rho^{5/3}(\mathbf{r}), \tag{3.5}$$

which also coincides with the nuclear matter result (2.87) when the density $\rho(\mathbf{r})$ is constant. From eqn (3.1) and eqn (3.3) we thus obtain the total energy in TF approximation

$$E_{TF}[\rho] = \int d\mathbf{r} \left[C_k\, \rho^{5/3}(\mathbf{r}) + U(\mathbf{r})\rho(\mathbf{r}) \right], \tag{3.6}$$

where $C_k = \frac{3\hbar^2}{10m}\left(\frac{3\pi^2}{2}\right)^{2/3} \cong 75.0$ MeV fm^2. Its variation with respect to the local density $\rho(\mathbf{r})$ leads to the Euler-Lagrange equation

$$\mu = \frac{5}{3} C_k \rho^{2/3}(\mathbf{r}) + U(\mathbf{r}), \tag{3.7}$$

where the nucleons move in the common potential $U(\mathbf{r})$ and μ is the chemical potential of the system.

Equation (3.7) is the basic equation of the TF method. Its solution with respect to the density $\rho(\mathbf{r})$ gives the result

$$\rho_{TF}(\mathbf{r}) \equiv \frac{2}{3\pi^2} k_F^3(\mathbf{r}) = \frac{2}{3\pi^2} \left[\frac{2m}{\hbar^2} (\mu - U(\mathbf{r})) \right]^{3/2}, \tag{3.8}$$

which is obviously valid only for the classically allowed region $r \leq r_{cl}$, where the 'classical turning point' r_{cl} satisfies the condition

$$\mu - U(r_{cl}) = 0. \tag{3.9}$$

In Figure 3.1, we compare the TF density (3.8) with the exact quantum-mechanical density, both derived in a Woods-Saxon potential well. One can see that the exact local density is surprisingly well reproduced by such a simple formula as eqn (3.8). It is even more interesting if one compares the ground-state energy in the TF approximation, calculated numerically from eqn (3.6) and eqn (3.8) with the same parameters as those of Figure 3.1, with the exact value calculated for the same Woods-Saxon potential. The total energy in the TF approximation differs from the exact one only by approximately 3% for heavy nuclei (Durand, Brack and Schuck 1978). Therefore, the pure TF approximation is useful when considering nuclear characteristics sensitive to the interior of the nuclear densities.

The energy density functional (3.6), however, cannot be applied within variational calculations performed in terms of the local densities $\rho(\mathbf{r}) \in D_N$. The TF local density (3.8) in itself does not belong to the space D_N since $\rho_{TF}(\mathbf{r})$ is not defined beyond the classical turning point r_{cl}. As a consequence, the pure TF approximation is of little use for the self-consistency problem in nuclear physics as will be further discussed.

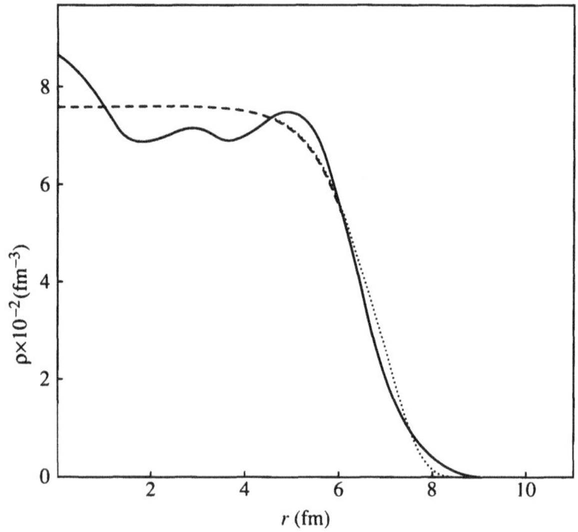

Figure 3.1. Density distributions for N = 184 particles in a spherical Woods-Saxon potential with parameters $V_0 = -4.4$ MeV, $a = 0.76$ fm, $R_0/N^{1/3} = 1.27$ fm (Durand, Brack, and Schuck 1978). The solid line is the exact quantum-mechanical result. Local density distributions calculated in linearized and harmonic approximations (Section 3.1) are shown as dotted and dashed lines, respectively. The Thomas-Fermi density (not explicitly shown) follows almost indistinguishably the 'linearized' density over the entire region, except very close to the classical turning point $r_{cl} \cong 8.2$ fm, where it goes to zero with a zero slope.

Integral Thomas-Fermi equation

Bethe (1968) applied the TF approximation to the density-dependent HF theory by replacing the one-body density matrix in the self-consistent HF energy by its Slater approximation (3.3). The resulting TF energy density functional is of the form (Bethe 1971)

$$E_{TF}[\rho] = \frac{3}{8}\int dr_1 dr_2\, \upsilon_1\, [\rho(r_1)\, \rho_2(r) + |\rho_{TF}(r_1,r_2)|^2\,]$$
$$+ \frac{1}{2}\int dr\, G_s(\rho)\, \rho^2(r) + \frac{\hbar^2}{2m}\int dr\, \tau_{TF}[\rho]. \qquad (3.10)$$

Here v_l means the long-range interaction between nucleons which is assumed to exist only in states of even parity so that the two terms in the square brackets (direct and exchange terms) occur with equal coefficients. The second term in eqn (3.10) includes the short-range interaction $G_s(\rho)$ which depends on the local density $\rho(\mathbf{r})$, while the last term in eqn (3.10) is the kinetic energy in TF approximation.

Bethe (1968, 1971) argues that for constant density, the terms in eqn (3.10) must reproduce the nuclear matter energy $W(\rho)$, and that the only deviation from the nuclear matter is due to the change of the density from $\mathbf{r}_1$ to $\mathbf{r}_2$. Accordingly, eqn (3.10) reads

$$E_{TF}[\rho] = \int d\mathbf{r}\ W(\rho)\rho(\mathbf{r}) + \frac{3}{8}\kappa \int d\mathbf{r}_1 d\mathbf{r}_2\ v_l\ [\rho(\mathbf{r}_1) - \rho\ (\mathbf{r}_2)], \quad (3.11)$$

where $W(\rho)$ is the energy per particle in nuclear matter but depends on the local density $\rho(\mathbf{r})$. The factor $\kappa \cong 0.781$ is inserted in eqn (3.11) in order to take into account the difference between one-particle energy levels in finite nuclei and in nuclear matter. The exchange term in eqn (3.10) may be shown to have exactly the same value as for infinite nuclear matter (Bethe 1968) and it is already included in the first term of eqn (3.11).

The long-range interaction in eqn (3.10) and eqn (3.11) is assumed independent of the local density, which is incorrect due to the saturation of the tensor forces. A correction for density dependence is introduced by Németh and Bethe (1968). Siemens (1970a) further improved the functional (3.11) by replacing v_l by an effective interaction g_D which depends on the density (Siemens 1970b). When account is taken of this density dependence of the long-range interaction, the energy of the nucleus becomes

$$E_{TF}[\rho] = \int d\mathbf{r}\ W(\rho)\rho(\mathbf{r}) + \frac{1}{2}\kappa \int d\mathbf{r}_1 d\mathbf{r}_2\ \rho(\mathbf{r}_1)$$

$$\times\ [\rho(\mathbf{r}_2) g_D(\mathbf{r}_1 - \mathbf{r}_2; \tfrac{1}{2}(\rho(\mathbf{r}_1)+\rho(\mathbf{r}_2)) - \rho(\mathbf{r}_1) g_D(\mathbf{r}_1 - \mathbf{r}_2; \rho(\mathbf{r}_1))]. \quad (3.12)$$

The associated Euler-Lagrange equation yields

$$\frac{d\rho W(\rho)}{d\rho(\mathbf{r})} + \kappa\, U_D(\mathbf{r}) = \mu, \tag{3.13}$$

where $U_D(\mathbf{r})$ is a complicated integral over the density at points $\mathbf{r} \neq \mathbf{r}'$ (Siemens 1970a). The TF equation (3.13) then constitutes an integral equation for $\rho(\mathbf{r})$ with an eigenvalue μ, the chemical potential of the system.

Failure of the nuclear Thomas-Fermi theory

Németh and Bethe (1968) and Lin (1970) have attempted to solve the integral equation (3.13) for a semi-infinite nucleus with a plane surface and a nuclear matter g_D with density dependence. Neither of these attempts was very successful. Siemens (1970a), after trying once more to solve eqn (3.13), concludes that its solution has a discontinuity at some radius $r = R$.

In Figure 3.2 we illustrate the result of Siemens (1970a). The density obtained as a solution of the TF integral equation falls discontinuously from a finite value to zero. Hence, the density in the interior of the nucleus cannot exceed 63% of nuclear matter density which is clearly unacceptable. The reason is that the TF approximation is not well suited to handle many-particle systems with short-range two body forces. This can be clearly seen from the fact that eqn (3.13) even gives no spatial dependence of $\rho(\mathbf{r})$ when the long-range part of the interaction is completely ignored.

Siemens (1970a) tried to remedy the situation by adopting a new approximation to the one-body density matrix which remains valid when $\rho(\mathbf{r})$ changes appreciably with r. He shows that a large fraction of the attractive force is due to the exchange interaction which is treated locally when using eqn (3.3). The attempts to solve the improved TF integral equation for the nucleus ^{208}Pb failed again because the correction due to the exchange terms was not large enough. Thus Bethe (1971) has

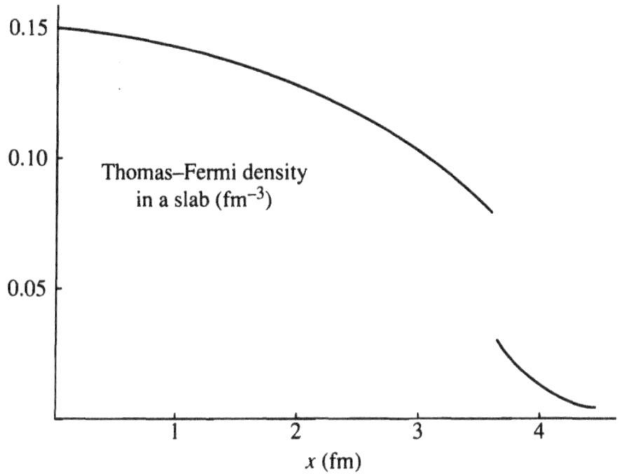

Figure 3.2. Discontinuous solution of a one-dimensional analogue of the integral Thomas-Fermi equation (3.13) according to Siemens (1970a).

concluded that there is no valid TF theory of nuclei, at least with standard realistic nuclear forces.

3.2 Energy density formalism

Due to the great difficulties in the nuclear TF theory, it would be worthwhile to approximate the exact energy density functional $E[\rho]$ according to some other recipe. The energy density formalism (EDF) originally proposed by Brückner, Buchler, and Kelly (1968) operates with an approximate functional $E_{EDF}[\rho]$ postulated on general physical ground with some degrees of phenomenology. Its justification comes merely *a posteriori* by judging its ability to account for the experimental situation. The kind of arbitrariness encountered in such an approach is very similar to the one appearing when choosing an effective interaction to simulate the actual G-matrix within the HF theory. From this point of view both approaches, HF and EDF, contain the same degree of phenomenology which can be visualized in the following way (Lombard 1973)

$$\langle\Psi|\hat{H}|\Psi\rangle \cong \langle\Phi_{HF}|\hat{H}_{eff}|\Phi_{HF}\rangle \cong E_{EDF}[\rho]. \tag{3.14}$$

EDF density functionals

Brückner *et al.* (1968, 1969) postulate the following approximate functional

$$E_{EDF}[\rho] = \int dr\ \mathcal{E}(\rho), \tag{3.15}$$

whose energy density $\mathcal{E}(\rho)$ is of the form

$$\mathcal{E}(\rho) = W(\rho,\alpha)\ \rho + [B_W(\rho) + B_C(\rho)]\ (\nabla\rho)^2 + \mathcal{E}_{Coul}[\rho]. \tag{3.16}$$

Here, as in eqn (3.11), the first term represents the so-called volume energy term, where

$$W(\rho,\alpha) = \mathcal{F}(\rho,\alpha) + \mathcal{V}(\rho,\alpha) \tag{3.17}$$

fits the nuclear matter saturation curve for variable neutron excess

$$\alpha = [\rho_n(r) - \rho_p(r)]/\ \rho(r). \tag{3.18}$$

The kinetic energy density $\mathcal{F}(\rho,\alpha)$ is in the TF approximation

$$\mathcal{F}(\rho,\alpha) = C_k(\alpha)\ \rho^{5/3}(r) \tag{3.19}$$

with a coefficient

$$C_k(\alpha) = \frac{1}{2}\ C_k[(1-\alpha)^{5/3} + (1+\alpha)^{5/3}]\ . \tag{3.20}$$

The potential energy density in eqn (3.17)

$$\mathcal{V}(\rho,\alpha) = b_1(1+a_1\alpha^2)\rho + b_2(1+a_2\alpha^2)\rho^{4/3} + b_3(1+a_3\alpha^2)\rho^{5/3} \tag{3.21}$$

contains six parameters. They allow the analytic expression (3.17) to fit the numerical points of the nuclear matter saturation curve derived by Brückner, Coon, and Dabrowski (1968) with a precision better than 1%.

The second term in eqn (3.16) is the density gradient correction separated into two parts. The first part is the Weizsäcker correction

$$B_W(\rho)(\nabla\rho)^2 = \frac{\hbar^2}{8m}\,\xi\,\frac{(\nabla\rho)^2}{\rho}, \tag{3.22}$$

which differs from the the pure Weizsäcker term (2.69) by the modifying coefficient $\xi = 1/9$. The second part in the second term of eqn (3.16) accounts for the inhomogeneity corrections

$$B_C(\rho)\,(\nabla\rho)^2 = \frac{\hbar^2}{8m}\,\eta\,(\nabla\rho)^2, \tag{3.23}$$

where η is kept as a parameter adjusted to give the correct experimental binding energies.

Equation (3.23) can be approximately derived from the second part of eqn (3.11) after assuming that the density varies slowly so that the difference $\rho(\mathbf{r}_1) - \rho(\mathbf{r}_2)$ may be expanded in a power series. In this way, the parameter η can be related to the long-range part of the nucleon-nucleon interaction (Bethe 1968, Brückner *et al.* 1968). Nevertheless, η is taken as a phenomenological constant in EDF since it is difficult to rely on a two-body potential to make a good estimate of η.

Lombard (1973) makes η to depend on the neutron excess (3.18) by introducing the more general expression

$$\eta = \eta_0\,(1 - k\alpha^2), \tag{3.24}$$

where k should be again considered as a parameter. Equation (3.24) avoids some technical difficulties related to the introduction in eqn (3.23) of an extra term (Brückner *et al.* 1969) accounting for the surface symmetry energy

$$\theta \left(\frac{\hbar^2}{8m} \eta \right) [\nabla(\rho_n - \rho_p)]^2, \tag{3.25}$$

where θ is a new constant. The last term in eqn (3.16) is the Coulomb energy density defined according to eqn (1.149).

In Table 3.1 we list three sets of values for the parameters entering eqn (3.16). The first one (Set 1) fits the nuclear matter results of Brückner, Coon, and Dabrowski (1968) as explained above, while Set 2 is used by Brückner *et al.* (1969). Because of uncertainties in the proposed analytic form of the functional (3.16), Lombard (1973) has disregarded the possibilities of deriving all eight parameters (Set 3) from the condition for an overall good fit of binding energies, proton rms radii and the main nuclear matter characteristics E_{NM}/N and k_F^0. The nuclear matter characteristics associated with the three sets of parameters are also given in Table 3.1.

Differential equations method

According to the general concept of density functional theory, the ground-state local densities $\rho_n(\mathbf{r})$ and $\rho_p(\mathbf{r})$ are obtained by applying the variational principle to

$$E_{EDF}[\rho] = \int d\mathbf{r}\ [\mathcal{E}(\rho) - \mu_n\ \rho_n(\mathbf{r}) - \mu_p\ \rho_p(\mathbf{r})]. \tag{3.26}$$

The variational equations to be solved are then

$$\Delta\rho_n(\mathbf{r}) = \left(\frac{m}{\eta\hbar^2} \right) [D(\rho_n,\rho_p) - (1/\theta)\ F(\rho_n,\rho_p)]$$

and

$$\Delta\rho_p(\mathbf{r}) = \left(\frac{m}{\eta\hbar^2} \right) [D(\rho_n,\rho_p) + (1/\theta)\ F(\rho_n,\rho_p)], \tag{3.27}$$

Table 3.1. Values of parameters of the functional (3.16) used within the energy density formalism. Associated nuclear matter characteristics are also listed.

Parameters	Set 1	Set 2	Set 3
b_1	-717.6	-741.28	-818.25
b_2	1142.2	1179.89	1371.06
b_3	-452.6	-467.54	-556.06
a_1	-0.14	-0.2	-0.316
a_2	0.23	0.316	0.2
a_3	1.2	1.646	1.646
η_0	7.8	11.95(13.0)	15.2
k	0.0	0.0 (2.0)	2.15
E_{NM}/N, [MeV]	-15.23	-16.59	-16.0
k_F^0, [fm^{-3}]	1.433	1.447	1.36
K_{NM}, [MeV]	172.6	184.7	180.0
a_τ, [MeV]	28.8	35.1	33.96

Set 1 is deduced from the nuclear matter calculations by Brückner, Coon, and Dabrowski (1968). Set 2 corresponds to 3% increase of the b_i coefficients of the Set 1. Set 3 is purely phenomenological one derived by Lombard (1973).

where

$$D(\rho_n,\rho_p) = \partial\mathcal{E}[\rho]/\partial\rho_n + \partial\mathcal{E}[\rho]/\partial\rho_p - (\mu_n + \mu_p)$$

and

$$F(\rho_n,\rho_p) = \partial\mathcal{E}[\rho]/\partial\rho_n - \partial\mathcal{E}[\rho]/\partial\rho_p - (\mu_n - \mu_p). \qquad (3.28)$$

In obtaining the differential equations (3.27), Brückner *et al.* (1969) have made the choice (3.25) for the surface symmetry energy and omitted the Weizsäcker correction (3.22); the Lagrange multipliers μ_n and μ_p allow the particle number conditions to be satisfied.

The solution of the coupled equations (3.27) has been found to be extremely unstable and the proper boundary conditions could not be satisfied. This instability is again a consequence of the fact that the TF approximation fails for low densities as explained in the previous section. This imposes the need for cutoffs on the solutions which introduces uncertainties and errors.

Importance of the Weizsäcker correction

The presence of the Weizsäcker correction term in the functional (3.26) ensures the correct behaviour of the nuclear distributions at very low densities as demonstrated by Lombard (1973). He rewrote the total energy by using a simplified functional following from eqn (3.26) in the proportional approximation (2.68)

$$E_{EDF}[\rho] = \int dr \left[F(\rho) + \frac{\hbar^2}{8m}\left(\eta + \xi/\rho\right)(\nabla\rho)^2 - \mu\,\rho(\mathbf{r})\right], \tag{3.29}$$

where $F(\rho)$ contains the volume part of the functional and the Coulomb terms. The associated Euler–Lagrange equation is

$$-2\left(\eta + \xi/\rho\right)\Delta\rho + \xi\,\frac{(\nabla\rho)^2}{\rho} + \frac{8m}{\hbar^2}\,f(\rho) = \frac{8m}{\hbar^2}\,\mu, \tag{3.30}$$

where

$$f(\rho) = \partial F(\rho)/\partial\rho \tag{3.31}$$

is a polynomial form containing only positive powers of $\rho(\mathbf{r})$.

Introducing the asymptotic solution $\rho(r) \cong \exp(-\lambda r)$ of eqn (3.30) at sufficiently large distances r and keeping the leading terms only, we get

$$-\lambda^2\,\xi = \frac{8m}{\hbar^2}\,\mu, \tag{3.32}$$

which shows that the solution for the density at large distances is of the form

$$\rho(r) \cong \exp\left(-r\left[\frac{8m\,\mu}{\hbar^2\xi}\right]^{1/2}\right). \qquad (3.33)$$

Now it is clear that at $\xi = 0$, eqn (3.30) does not admit an exponential decrease of the density at large distances and it does not correspond to the physical problem beyond a certain point.

Variational method

Even when the Weizsäcker term is included, the handling of the differential equation (3.27) is rather tedious due to the large sensitivity of its solution to the eigenvalue μ_q. It is therefore desirable to find a numerically faster way of dealing with the minimization of eqn (3.15).

The variational approach to this problem assumes a restricted trial set $D_{tr} \subset D_N$ generated by a parameterization of the density $\rho(r;\alpha_1,\alpha_2,\ldots,\alpha_k) \in D_{tr}$ and searches for a minimum of $E[\rho]$ within this restricted trial space. This can be done by solving the system of nonlinear algebraic equations

$$\partial E(\alpha_1,\alpha_2,\ldots,\alpha_k)/\partial\alpha_i = 0 \qquad (i = 1,2,3,\ldots,k), \qquad (3.34)$$

where $E(\alpha_1,\alpha_2,\ldots,\alpha_k)$ is the algebraic function obtained after restricting $E[\rho]$ on the trial subspace $D_{tr} \subset D_N$. According to the variational method the following inequality holds

$$\min_{\rho\in D_{tr}} E[\rho] \geq \min_{\rho\in D} E[\rho]. \qquad (3.35)$$

In order to define the variational space $D_{tr} \subset D_N$, Brückner *et al.* (1968) use the following different local density distributions: the modified Fermi-type (n = 1)

$$\rho(r) = \rho_0\, \mathcal{P}(r)\left\{1 + \exp\left[(r^n - R^n)/b^n\right]\right\}^{-1}, \qquad (3.36)$$

or the modified Gaussian (n = 2), both with

$$\mathcal{P}(r) = 1 + pr^2 + qr^3; \tag{3.37}$$

the Gaussian-type

$$\rho(r) = \rho_0 \, \mathcal{P}(r) \exp(- r^2/R^2); \tag{3.38}$$

and the parameterization

$$\rho(r) = \rho_0 \, (1 + pr^2) \left\{1 + \exp\left[(r - R)/b\right]\right\}^{-1}. \tag{3.39}$$

Here R, b, p, q are parameters which have to be varied, while the constant ρ_0 is determined by the normalization condition, such as the condition that the above densities belong to $D_{tr} \subset D_N$.

The variational method has been applied to the TF functionals by Donnelly (1968, 1969) and Pandaripande (1969a,b). Of course, these calculations do not detect the fact that the solution of the associated Euler-Lagrange equations is unstable or discontinuous. Nevertheless, the variational method provides a convenient tool for seeking approximate solutions within nuclear density functional theory.

Numerical results

A review of EDF calculations up to 1973 has been given by Lombard (1973). Typical results for binding energies, charge rms radii and separation energies (obtained as Lagrange multipliers $\varepsilon_F = \mu_q$) are given in Table 3.2. It is seen that the EDF is able to reproduce the experimental binding energies to within approximately 1-10 MeV and experimental charge radii to within approximately 1-4%. At the same time, the results for neutron and proton separation energies are only very rough.

The density profiles obtained within EDF calculations are as a rule rather poor. As an example, we show in Figure 3.3 neutron and proton densities calculated with Set 3 for nuclei ^{40}Ca, ^{90}Zr

Table 3.2. Binding energies E (in MeV), charge rms radii r_c (in fm) and separation energies ε_F (in MeV) obtained from energy density formalism calculations according to Lombard (1973).

	E		r_c		$-\varepsilon_F$(neutrons)		$-\varepsilon_F$(protons)	
	exp	Set 3	exp	Set 3	exp	Set 3	exp	Set 3
^{16}O	127.6	128.8	2.71–2.74	2.66	15.16	10.9	12.13	7.2
^{40}Ca	342.1	342.0	3.50	3.47	15.16	11.9	8.33	4.4
^{48}Ca	416.0	422.0	3.49	3.53	9.95	6.0	15.81	12.6
^{56}Ni	526.9	524.0	3.82	3.80	7.82	9.5	9.86	5.7
^{80}Zr	783.9	780.0	4.265	4.27	11.98	8.4	8.37	6.2
^{90}Zr	988.7	986.0	4.55	4.58	9.57	7.4	9.27	6.2
^{140}Ce	1172.7	1169.0	—	—	9.20	6.0	8.15	7.1
^{208}Pb	1636.5	1627.0	5.50	5.50	7.37	4.9	8.02	6.9
^{235}U	1801.7	1797.0	—	—	6.14	4.8	7.66	6.7

and ^{208}Pb. A large asymmetry in the surface and a rather short tail are evident, features that may be attributed to the TF approximation. Dworzecka and Moszkowski (1975) have found that the EDF can reproduce quite closely the shell-averaged results of the more complicated HF calculations.

Brückner, Chirico, and Meldner (1971) has demonstrated that EDF can be made competitive with conventional mass formulas as far as the bulk properties (binding energies and radii) are concerned and that it is even more reliable for nuclei far from the stability line. EDF has been successfully used also in studying the isotope shift arising from the statistical charge in proton distributions (Brückner, Lin, and Lombard 1969) as well as in investigating some properties of deformed nuclei (Brückner *et al.* 1969, Stock 1975, Stock and Scheid 1975).

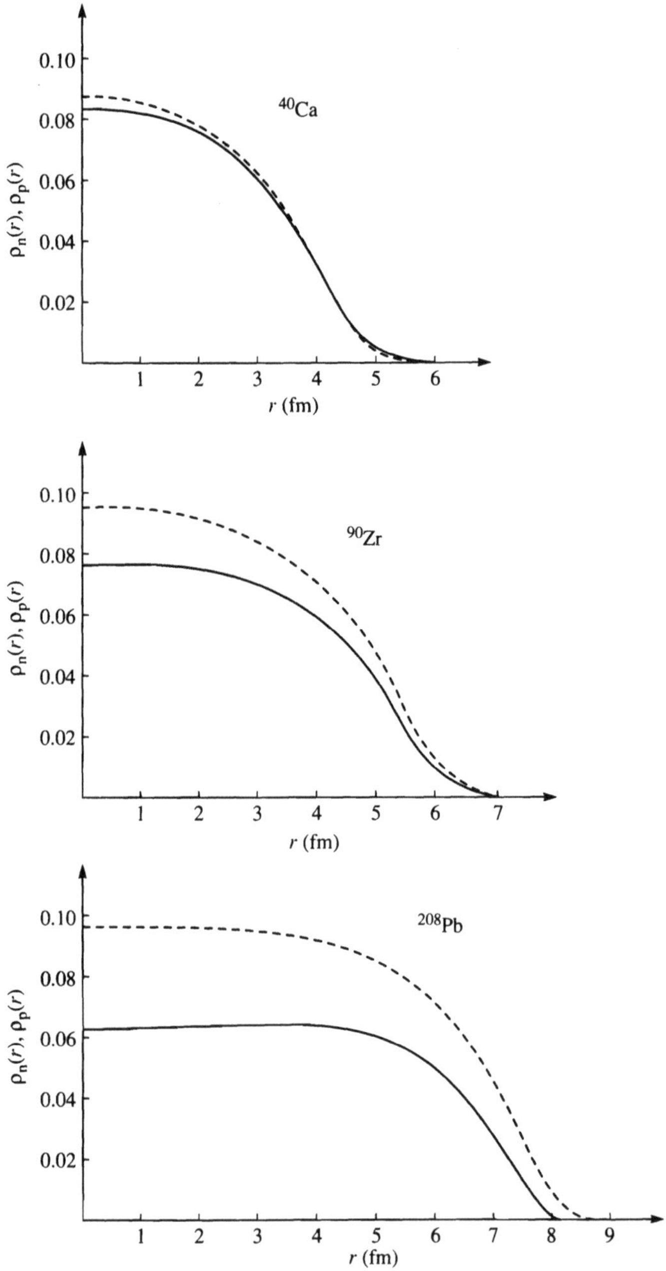

Figure 3.3. Neutron (dotted lines) and proton (full lines) density profiles obtained within the energy density formalism (Set 3) by Lombard (1973).

3.3 Phenomenological Skyrme-Hartree-Fock Functionals

The development of Skyrme-type forces has allowed us to refine the approximate ground-state energy density functionals. In Skyrme-HF theory the total energy of a nucleus is already a functional of the densities $\rho(\mathbf{r})$, $\tau(\mathbf{r})$ and $\mathbf{J}(\mathbf{r})$ (see eqn (1.144) and eqn (1.148)). Therefore, the density functional theory in Skyrme-HF approximation only needs the kinetic as well as spin-orbit densities to be represented as functionals of $\rho(\mathbf{r})$.

Of course, there is no hope of determining both functionals, $\tau[\rho]$ and $\mathbf{J}[\rho]$, exactly. This chapter, however, considers attempts to solve the problem in terms of a purely phenomenological Thomas-Fermi-Weizsäcker representation of the kinetic energy density $\tau[\rho]$, ignoring at the moment contributions coming from the spin-orbit terms.

General representation

In the Skyrme-HF approximation one usually uses the following general relation between $\tau(\mathbf{r})$ and $\rho(\mathbf{r})$

$$\tau[\rho] = \alpha\, \rho^{5/3}(\mathbf{r}) + \frac{1}{4}\,\beta\, \frac{(\nabla\rho)^2}{\rho} + \gamma\, \Delta\rho, \tag{3.40}$$

which generalizes the TF approximation (3.5) and includes gradient-type corrections by means of the Weizsäcker term (3.22).

As we shall see in the next chapter, the expression (3.40) is suggested by a semiclassical expansion of $\tau(\mathbf{r})$ in terms of $\rho(\mathbf{r})$ which also yields explicit predictions for the coefficients $\alpha = \frac{3}{5}(3\pi^2)^{2/3}$, $\beta = 1/9$, and $\gamma = 1/3$. However, there is no reason why the semiclassical values of these parameters should correspond to the best approximation of the exact kinetic energy density functional. Therefore, one might try to adopt eqn (3.40) simply as a phenomenological ansatz in order to facilitate calculations with the density functional theory.

Making use of the Skyrme-HF energy density (1.148) and the ansatz (3.40), one obtains the following approximate energy density functional (Treiner and Krivine 1986)

$$E_{TFW}[\rho] = \int dr \left\{ H(\rho_n,\rho_p) + d_1[(\nabla\rho_n)^2 + (\nabla\rho_p)^2] + d_2 \, \nabla\rho_n \cdot \nabla\rho_p \right.$$

$$+ d_3\left[\rho_p \frac{(\nabla\rho_n)^2}{\rho_n} + \rho_n \frac{(\nabla\rho_p)^2}{\rho_p} \right] + \frac{\hbar^2}{8m} \beta\left[\frac{(\nabla\rho_n)^2}{\rho_n} + \frac{(\nabla\rho_p)^2}{\rho_p} \right]$$

$$+ \frac{1}{16} V_{ls}\left[\rho_n(m_n^*/m)\left((\nabla\rho_n)^2 + \frac{3}{4}(\nabla\rho_p)^2\right) + \rho_p(m_p^*/m)\left((\nabla\rho_p)^2 + \frac{3}{4}(\nabla\rho_n)^2\right)\right.$$

$$\left.\left. + \frac{1}{2} \nabla\rho_n \cdot \nabla\rho_p \left(\rho_n(m_n^*/m) + \rho_p(m_p^*/m)\right) \right] + H_{Coul}(r), \qquad (3.41)$$

with

$$H(\rho_n,\rho_p) = \alpha\left[\frac{\hbar^2}{2m_n^*}\rho_n^{5/3} + \frac{\hbar^2}{2m_p^*}\rho_p^{5/3}\right] + \frac{1}{2} t_0[(1+x_0/2)\rho^2 - (1/2+x_0)(\rho_n^2+\rho_p^2)]$$

$$+ \frac{1}{12} t_3[(1+x_3/2)\rho^2 - (1/2+x_3)(\rho_n^2+\rho_p^2)] \, \rho^{\sigma}, \qquad (3.42)$$

where m_q^*, $(q = p,n)$ is the effective mass (1.154) and $H_{Coul}(\mathbf{r})$ the Coulomb energy density (1.149). The coefficients d_1, d_2, d_3 and V_{ls} in eqn (3.41) are given in terms of the Skyrme parameters as

$$d_1 = \frac{3}{32}[t_1(1-x_1) - t_2(1+x_2)] + \frac{1}{8}[3t_1(1+x_2) + t_2(1-x_1)](\beta/4-\gamma),$$

$$d_2 = \frac{1}{8}[t_1(1+x_1/2) - t_2(1+x_2/2)] - \frac{\gamma}{2}[t_1(1+x_2/2) + t_2(1+x_2/2)],$$

$$d_3 = \frac{1}{16}[t_1(1+x_1/2) + t_2(1+x_2/2)]\beta, \quad V_{ls} = -\frac{9}{32}(W_0)^2 \frac{2m}{\hbar^2}. \qquad (3.43)$$

For completeness, the semiclassical approximation to the

spin-orbit functional $J[\rho]$ (see next chapter) has been included in eqn (3.41). In most cases, this term proportional to V_{ls} is omitted in actual calculations.

In the limit $\rho_n = \rho_p = \rho/2$, the energy density functional (3.40) reads

$$E_{TFW}[\rho] = \int dr \left\{ h(\rho) + \frac{\hbar^2}{8m}\, \beta\, \frac{(\nabla\rho)^2}{\rho} + (d + V_{ls}(m^*/m))\, \Delta\rho \right\}, \quad (3.44)$$

where the Coulomb term is excluded,

$$h(\rho) \equiv H(\rho/2,\rho/2) = \frac{3\hbar^2}{10m^*}\left(\frac{3\pi^2}{2}\right)^{2/3} \rho^{5/3} + \frac{3}{8} t_0 \rho^2 + \frac{1}{16} t_3 \rho^{\sigma+2}, \quad (3.45)$$

and

$$d = \frac{d_1}{2} + \frac{d_2}{4} + \frac{d_3}{2} = \frac{1}{64}[9t_1 - t_2(5 + 4x_2)] + \frac{\beta - 4\gamma}{8}[3t_1 + t_2(5 + 4x_2)]. \quad (3.46)$$

The functional (3.41) should be applied only to spherical nuclei because it is not adequate to describe deformed systems. This can be most simply illustrated for the kinetic energy density (3.40). Krivine, Treiner, and Bohigas (1980) apply the following quadrupole transformation to the single-particle wave functions

$$\varphi_i(x,y,z) \longrightarrow \varphi_i(x/\lambda, y/\lambda, \lambda^2 z), \quad (3.47)$$

where λ is a real parameter. Then the HF kinetic energy for a spherical system transforms as

$$T(\lambda) = \sum_i \int dr \left| \nabla\varphi_i(x/\lambda, y/\lambda, \lambda^2 z) \right|^2 \longrightarrow \frac{1}{3}\left(\frac{2}{\lambda^2} + \lambda^4\right) T(\lambda = 1), \quad (3.48)$$

while the TF term entering eqn (3.40) remains unchanged since eqn (3.47) is a volume conserving transformation. Thus the TF kinetic energy behaves in an essentially different manner compared with the exact one during the process of a quadrupole deformation. At the same time, it is straightforward to show by a simple monopole scaling that the transformed expression (3.40) has the correct behaviour.

Improved Thomas-Fermi method

The so-called improved Thomas-Fermi (ITF) method proposed by Bohigas *et al.* (1976) works within the semiclassical derived values for $\alpha = \frac{3}{5}(3\pi^2)^{2/3}$, $\beta = 1/9$ and $\gamma = 1/3$ in eqn (3.41). For simplicity, we shall illustrate the method within the approximate functional (3.44) given above. The Euler-Lagrange equation corresponding to eqn (3.44) is of the form

$$\frac{\partial h(\rho)}{\partial \rho(\mathbf{r})} + \frac{\hbar^2}{8m} \beta \frac{(\nabla\rho)^2}{\rho} - 2(d + \frac{\hbar^2}{8m} \frac{\beta}{\rho}) \Delta\rho = \mu. \tag{3.49}$$

From the asymptotic form of this equation at large distances

$$\frac{(\nabla\rho)^2}{\rho} - 2\,\Delta\rho = \frac{8m}{\hbar^2} \frac{\mu}{\beta} \rho, \tag{3.50}$$

we obtain a solution that has an exponential fall-off

$$\rho(r) \cong \frac{1}{r^2} \exp\left(-r\left[-\frac{8m}{\hbar^2}\frac{\mu}{\beta}\right]^{1/2}\right). \tag{3.51}$$

The comparison of eqn (3.51) with eqn (3.33) demonstrates that the exponential behaviour of $\rho(\mathbf{r})$ is due to the inhomogeneous Weizsäcker term. Without this term ($\beta = 0$), the asymptotic form of eqn (3.50) would be $\Delta\rho(\mathbf{r})$ = constant, which has no physically acceptable solution over the whole space.

It is appropriate here to compare the asymptotic behaviour (3.51) with the one following from the independent-particle model or the HF theory. From the single-particle wave function representation (1.145) we find that the HF local density $\rho(\mathbf{r})$ corresponds at large distances to the square of the last bound orbital φ_1 with energy $\varepsilon_1 \equiv \varepsilon_F < 0$; that is,

$$\rho(r) \cong |\varphi_1(r)|^2 \cong \frac{1}{r^2} \exp\left(- r\left[\frac{8m}{\hbar^2} \varepsilon_F\right]^{1/2}\right). \tag{3.52}$$

Obviously, eqn (3.51) goes faster to zero with $\beta = 1/9$ than the HF density (3.52) and, therefore, the ITF model leads to a too rapid fall-off of the nuclear density in the surface region.

Numerical results (Bohigas *et al.* 1976) confirm the above conclusion. The obtained ITF proton and neutron densities compare rather well with the HF ones, except that the former have a sharper surface and no local density shell oscillations. The associated ITF results for the neutron and proton rms radii are slightly underestimated. Considering the total energy, the ITF results can be seen to be good to within approximately 0.5 MeV per nucleon giving an overbinding with respect to the HF results obtained with the same Skyrme-type effective forces.

Modified Thomas–Fermi method

Campi and Stringari (1980) have adopted a more pragmatic point of view which consists in keeping the form (3.40) but adjusting all three parameters α,β,γ so that the exact HF kinetic energy density o be reproduced. They found that the values of α,β,γ depend smoothly on the particle number, the asymptotic values being attained for $N \geq 3000$ to within a few per cent.

The same idea is used in the so-called modified Thomas–Fermi (MTF) model proposed by Krivine and Treiner (1979). In this model, the kinetic energy density

$$\tau_{MTF}[\rho] = \alpha(N_q)\, \rho^{5/3}(r) + \frac{1}{4}\, \beta(N_q)\, \frac{(\nabla\rho)^2}{\rho} \tag{3.53}$$

contains two phenomenological functions $\alpha(N_q)$ and $\beta(N_q)$ of the particle number N_q ($q = p,n$). They are determined by a fit to an average description of the bulk properties of nuclei throughout the periodic table. According to the results obtained in Chapter 2, $\alpha(N_q)$ must satisfy the conditions $\alpha(2) = 0$ (the exact result for the nucleus ^{4}He) and $\alpha(N_q \rightarrow \infty) = \frac{3}{5}(3\pi^2)^{2/3}$ (the exact nuclear matter result), while $\beta(2) = 1$ is the appropriate condition used with $\alpha(2) = 0$.

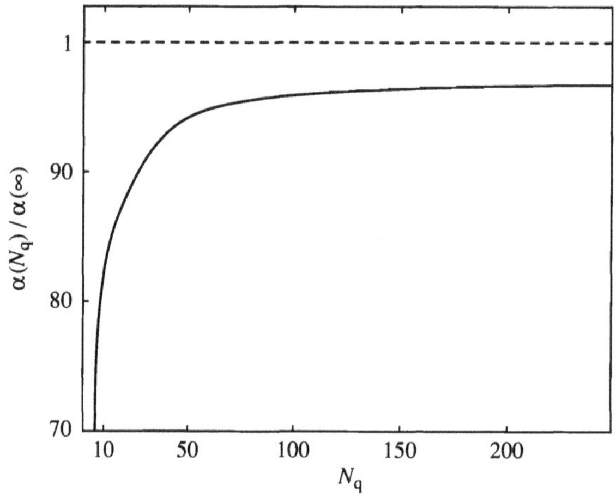

Figure 3.4. Plot of the function $\alpha(N_q)/\alpha(\infty)$ versus N_q (Krivine and Treiner 1979). It can be roughly parameterized by the expression $\alpha(N_q)/\alpha(\infty) = \tanh[7746(N_q - 2)^{1/5}]$.

We plot in Figure 3.4 the resulting function $\alpha(N_q)/\alpha(\infty)$. It is seen that $\alpha(N_q)$ is very rapidly increasing only in the region $N_q \leq 10$, but it differs from the asymptotic value $\alpha(\infty)$ even in the lead region ($N_q \cong 100$) by about 4%. Therefore, the $\rho^{5/3}$ term entering eqn (3.53) is appreciably less than that given by the TF expression ($\alpha = \alpha(\infty)$ and $\beta = 0$). Similar results were obtained by Petkov and Stoitsov (1983c).

The function $\beta(N_q)$ is found to vary rapidly from the value $\beta(2) = 1$ only in the region $N_q \leq 8$ and then tends to a smaller constant value which depends on the interaction. The constant behaviour of $\beta(N_q)$ for $N_q \geq 8$ can be related to the well known fact that the surface thickness is roughly the same in all nuclei.

Having thus determined functions $\alpha(N_q)$ and $\beta(N_q)$, the MTF method fits the kinetic energies of spherical nuclei quite well. It can be used successfully in different physical situations, such as the investigation of monopole resonances in nuclei or the properties of very large systems (up to 5000 nucleons).

Basic dilemma

Eckart and Holzwarth (1977) have investigated the basic dilemma that occurs when using adjustable functionals of the type of eqn (3.41). They confirm the ITF result that a satisfactory agreement with experimental binding energies is found by choosing the strength of the Weizsäcker contribution as given by the semiclassical estimation $\beta = 1/9$. In this case, however, the exponential tail of the density is too steep. Conversely, the original Weizsäcker term without attenuation factor leads to a good description of the density profiles. This can also be seen from eqn (3.51) which ensures precisely the HF fall-off (3.52) in the case $\beta = 1$. However, the choice $\beta = 1$ overestimates the surface energy appreciably. Therefore, the approximate functional (3.40) can either reproduce correctly the energies or the densities, but not both at the same time. This basic dilemma is partially resolved in the MTF model, but the price is that α and β depend on the nucleon mass and on the particular force.

Before concluding it should be mentioned that the ansatz (3.40) can also be used within the density functional theory in a larger context by starting with more realistic finite range forces and using the density matrix expansion suggested by Negele and Vautherin (1972) and by Campi and Bouyssy (1978). The resulting approximate functionals can be substituted for the HF approximation each time one is looking for a gross structure rather than a microscopic description.

3.4 Scattering potentials between two heavy ions

Although the approximate energy density functionals have to be improved to reproduce nuclear properties more consistently they have proved to be very useful in the heavy-ion scattering problem where energy density functionals can be used to calculate the scattering potential between two heavy ions.

Real part of the heavy-ion potential

The optical potential between two colliding heavy nuclei is one of the most useful theoretical tools in the analysis of heavy-ion reactions. The determination of its properties, however, is not unambiguous because of the strong absorption. It is therefore meaningful to develop a simple and transparent theoretical derivation of the optical potential and this has been achieved in the nuclear density functional theory.

Brückner, Buchler, and Kelly (1968) have demonstrated that the heavy-ion potential between two colliding nuclei as a function of the relative distance between them is very simply given by the total energy of the system at the separation distance R minus the total energy of the separated nuclei; that is,

$$U(R) = E[\rho_1+\rho_2] - E[\rho_1] - E[\rho_2]$$

$$= \int d\mathbf{r} \left\{\mathcal{E}(\rho(|\mathbf{r}+\tfrac{1}{2}\mathbf{R}|)+\rho(|\mathbf{r}-\tfrac{1}{2}\mathbf{R}|))-\mathcal{E}(\rho(|\mathbf{r}+\tfrac{1}{2}\mathbf{R}|))-\mathcal{E}(\rho(|\mathbf{r}-\tfrac{1}{2}\mathbf{R}|))\right\}. \quad (3.54)$$

Here $E[\rho]$ is the energy density functional and ρ_1, ρ_2 are the optimal ground-state density distributions as obtained by minimizing $E[\rho]$ for nuclei with particle numbers N_1, N_2, respectively. One has thus assumed that the densities of the interacting nuclei superimpose without distortion during the collision (the so-called sudden approximation). The composite density $\rho(|\mathbf{r} + \frac{1}{2}\mathbf{R}|) + \rho(|\mathbf{r} - \frac{1}{2}\mathbf{R}|)$ is obtained by adding the densities of the two colliding nuclei, whereby the total energy is still expressed by the same functional.

The properties of the heavy-ion scattering potential (3.54) have been systematically investigated within the EDF by Brückner, Buchler, and Kelly (1968), Ngô *et al.* (1975a,b), Ngô and Ngô (1980), Delchev, Marchev, and Petkov (1979). In their calculations Brink and Stancu (1975) and Stancu and Brink (1976) use for the same purposes Skyrme-type forces and the TF approximation to the kinetic energy and investigate the contributions to U(R) of the exchange effects due to the antisymmetrization condition. The effects of the nuclear polarization on U(R) are discussed by Fleckner and Mosel (1977).

Complex heavy-ion potentials

In order to take into account the many open channels during the collision, one should include an imaginary part in the heavy-ion potential representing the loss of flux from the elastic channel due to all the nonelastic channels. A consistent description of both the real and imaginary parts of the complex nucleus-nucleus interaction has been developed by Izumoto, Krewald, and Faessler (1980a,b) starting from a realistic two-nucleon potential and making use of the nuclear matter results. They introduce the difference between the energy of the two nuclei separated by a distance R minus their energy at infinity,

$$U(R,k_r) = E(R,k_r) - E(\infty,k_r). \tag{3.55}$$

The use of eqn (3.55) as an optical potential was justified on the ground of Feshbach theory by Khadkikar *et al.* (1981).

The total energy in eqn (3.55) is obtained as a functional

$$E(R,k_r) = \int d\mathbf{r}\ \mathcal{E}(\mathbf{r};R,k_r), \tag{3.56}$$

where

$$\mathcal{E}(\mathbf{r};\mathbf{R},k_r) = \frac{\hbar^2}{2m}\tau[\rho_T(|\mathbf{r}|),\rho_P(|\mathbf{r}-\mathbf{R}|);k_r] + \pi[\rho_T,\rho_P;k_r] \tag{3.57}$$

contains the kinetic energy density τ and the potential energy density π. They both, τ and π, depend on the average momentum

$$k_r = \left(\frac{\hbar^2}{2m}E_L\ N^{-1}\right)^{1/2},$$

related to the laboratory energy E_L of the projectile with mass number N_P, and the target and projectile local densities ρ_T and ρ_P, respectively.

For a fixed separation distance R, the local densities ρ_T and ρ_P are identified with the radii of two Fermi spheres with centres separated by the average relative momentum k_r. This allows one to consider the collision of two ions as being locally represented by two spheres of nuclear matter passing through each other. This picture is mathematically tractable and takes into account important features of the collision process.

The energy densities τ and π entering eqn (3.57) are introduced in momentum space by Brückner G-matrix calculations in the local density approximation elaborated by Trefz, Faessler, and Dickhoff (1985). With the energy density (3.57) one can then obtain the energy (3.56) and therefore the optical potential (3.55). Its imaginary part turns out to be produced by the 2p–2h excitation energies that occur conserving energy and momentum. An important feature of these calculations is the correct treatment of the Pauli principle and the possible treatment of deformed nuclei as well (Ismail *et al.* 1986, Rashdan *et al.* 1987, 1988).

Within the above procedure, one does not expect to obtain the ground-state energy of the separated nuclei because as we have seen earlier the nuclear matter results impose a pure TF approximation for the kinetic energy. Recently, some modifications have been made to correct the energy density (3.57) by adding to it the Weizsäcker term and symmetry energy corrections with their strengths adjusted to reproduce the ground-state properties of nuclei (Ohtsuka *et al.* 1987, Faessler *et al.* 1987, Rashdan *et al.* 1988).

Sudden and adiabatic approximation

In order to visualize the optical potential discussed in the last section, Ohtsuka *et al.* (1987) and Faessler *et al.* (1987) consider two opposite extreme situations. In the first case, the collision time is shorter than the characteristic time of the internal motion of a nucleon, so that the nuclei remain frozen during the collision (sudden approximation). In the second case, the collision time is long enough for the density to rearrange itself so that the total energy of the composite system is lowest at a given R (adiabatic approximation). Obviously, the actual situation lies between these two cases depending on the incident energy.

In Figure 3.5 we show typical results for the optical potential calculated for two colliding nuclei ^{40}Ca in both sudden and adiabatic approximations. The real part derived at $k_r = 0$ (Figure 3.5a) shows a behaviour which is typical for all calculations based on the nuclear density functional theory. In the sudden approximation, the potential has a repulsive core at small distances due to the faster increase of the kinetic energy with respect to the potential energy. In the adiabatic approximation, the real part is attractive or at least shallow for each value of k_r (Figure 3.5a and Figure 3.5b). For completeness, in Figure 3.5c we plot the imaginary part of the potential calculated for $k_r = 1.0\ fm^{-1}$. At $k_r = 0$, there is no imaginary part in the optical potential.

Clearly, the potentials in both sudden and adiabatic approximations are considerably different at low energies. The strong absorption in this region of the potential, however, makes the cross section insensitive to this difference (Ohtsuka *et al.* 1987). At higher energies, the cross section is more sensitive

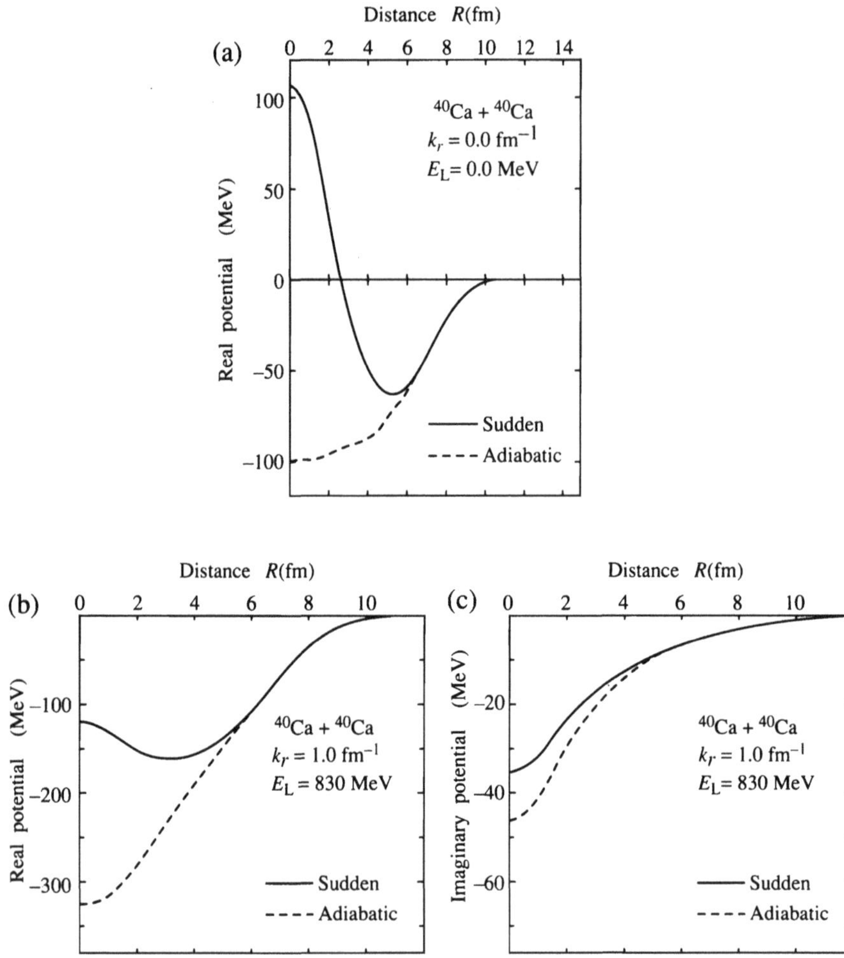

Figure 3.5. Heavy-ion optical potentials at k_r = 0 fm (E_L = 0 MeV) and k_r = 1 fm (E_L = 830 MeV) calculated for colliding nuclei ^{40}Ca+^{40}Ca in both sudden (full line) and adiabatic (dashed line) approximations (Ohtsuka *et al.* 1987).

to the central part of the potential. The difference between the sudden and adiabatic approximation, however, becomes smaller at higher energies and the calculated cross sections are quite similar.

4. Semiclassical nuclear density functional theory

4.1 Introduction

The semiclassical approach to the nuclear many-body problem can be used as an appropriate theoretical background in nuclear density functional theory. In particular, it allows one to clarify the TF approximation and further to derive systematic corrections which take variations of the density into account. The resulting approximate functionals, such as these for $\tau[\rho]$ and $J[\rho]$ are universal ones. They are free of any adjustable parameters and can be calculated once and for all, a feature that accords with the Hohenberg-Kohn theorem (Section 2.1).

The semiclassical approach to the many-body problem proposed by Wigner (1932) and Kirkwood (1933) and recently rediscovered in the context of nuclear density functional theory is based on a semiclassical expansion of the HF one-body density matrix. At present, more than one method exists for making semiclassical $\hbar$-expansions with a leading term derived in the TF approximation. For example, one can work directly with the Wigner representation of the Bloch density matrix as done by Grammaticos and Voros (1979, 1980), or one can work with the Laplace transform of the Kirkwood representation as done by Bhaduri and Ross (1971), Jennings and Bhaduri (1975), Jennings, Bhaduri, and Brack (1975). A different method has also been used by Kirzhnits (1957, 1967) and Hodges (1973). Balazs and Jennings (1984) have shown that all these methods are equivalent and, therefore, the choice of the method is purely one of taste and computation ease.

4.2 Bloch density matrix and its Wigner representation

In order to derive a semiclassical approximation to the HF theory, one may start with the single-particle Hamiltonian (3.1)

$$\hat{H} = -\frac{\hbar^2}{2m}\nabla^2 + U(\mathbf{r}), \tag{4.1}$$

where $U(\mathbf{r})$ is a single-particle potential in which all nucleons move and which should eventually be calculated by the self-consistent HF procedure. The most important case is of course the nuclear ground state in which the N eigenstates $\varphi_i(\mathbf{r})$ of the Hamiltonian (4.1) with the lowest eigenvalues ε_i are occupied.

Bloch density matrix

Let us consider the problem in terms of the single-particle propagator, the Bloch density matrix

$$C^\beta \equiv C(\mathbf{r},\mathbf{r}';\beta) = \exp(-\beta\hat{H}), \tag{4.2}$$

which is the solution of the time-dependent Schrödinger equation, the Bloch equation ($\beta = it/\hbar$)

$$\frac{\partial C^\beta}{\partial\beta} = -\hat{H}\,C^\beta, \tag{4.3}$$

with the initial condition

$$C(\mathbf{r},\mathbf{r}';\beta = 0) = \delta(\mathbf{r}-\mathbf{r}'). \tag{4.4}$$

In terms of the eigenfunctions $\varphi_i(\mathbf{r})$ and eigenvalues ε_i of the Hamiltonian (4.1), the Bloch density matrix can be expressed in the form

$$C(\mathbf{r},\mathbf{r}';\beta) = 2\sum_i \varphi_i^*(\mathbf{r})\varphi_i(\mathbf{r}')\exp(-\beta\varepsilon_i), \tag{4.5}$$

where the factor of two takes care of the two spin orientations.

For simplicity we consider one kind of nucleons (protons or neutrons) only.

The Bloch density matrix (4.5) can also be defined as the Laplace transform

$$C^{\beta} = \int_0^{\infty} d\varepsilon \; \exp(-\beta\varepsilon) \; G(\mathbf{r},\mathbf{r}';\varepsilon) \equiv \mathscr{L}_{\varepsilon\to\beta}[G^{\varepsilon}], \tag{4.6}$$

of the so-called spectral density matrix

$$G^{\varepsilon} \equiv G(\mathbf{r},\mathbf{r}';\varepsilon) = 2\sum_{i} \varphi_i^*(\mathbf{r})\varphi_i(\mathbf{r}') \; \delta(\varepsilon - \varepsilon_i), \tag{4.7}$$

where we assume that the zero of energy is at the bottom of the potential well. Properties of the Laplace transforms can be found in Abramowitz and Stegun (1965).

Then, if we define the Fermi level μ as lying just above the highest occupied single-particle level, we may express the one-body density matrix $\rho(\mathbf{r},\mathbf{r}')$ satisfying eqn (3.2) as an inverse Laplace transform of C^{β}/β such as

$$\rho(\mathbf{r},\mathbf{r}') \;=\; 2\sum_{i}^{N} \varphi_i^*(\mathbf{r})\varphi_i(\mathbf{r}')$$

$$= 2\sum_{i=1}^{\infty} \varphi_i^*(\mathbf{r})\varphi_i(\mathbf{r}') \; \theta(\mu - \varepsilon_i) = \int_0^{\mu} d\varepsilon \; \exp(-\beta\varepsilon) \; G(\mathbf{r},\mathbf{r}';\varepsilon)$$

$$= \frac{1}{2\pi i}\int_{c-i\infty}^{c+i\infty} d\beta \; \exp(\mu\beta) \; C(\mathbf{r},\mathbf{r}';\beta)/\beta \equiv \mathscr{L}^{-1}_{\beta\to\varepsilon}[C^{\beta}/\beta]. \tag{4.8}$$

From eqn (4.8) it becomes clear that if the Bloch density matrix is determined, then the density matrix $\rho(\mathbf{r},\mathbf{r}')$ and therefore the expectation values of all single-particle operators may be defined (see Section 1.4) in terms of $C(\mathbf{r},\mathbf{r}';\beta)$ as well. In particular, introducing the trace of the Bloch matrix, the so-called quantum partition function

$$Z(\beta) = \mathrm{tr}\ C^{\beta} = \int d\mathbf{r}\ C(\mathbf{r},\mathbf{r};\beta), \tag{4.9}$$

one may define, step-by-step, the level density

$$g(\varepsilon) = \mathrm{tr}\ \delta(\varepsilon - \hat{H}) = \mathcal{L}^{-1}_{\beta\to\varepsilon}[Z(\beta)], \tag{4.10}$$

the particle number

$$N \equiv \int_0^{\mu} d\varepsilon\ g(\varepsilon) = \mathcal{L}^{-1}_{\beta\to\mu}[Z(\beta)/\beta], \tag{4.11}$$

or the ground-state energy

$$E \equiv 2\sum_{i}^{N} \varepsilon_i = \int_0^{\mu} d\varepsilon\ \varepsilon\ g(\varepsilon) = \mu N - \mathcal{L}^{-1}_{\beta\to\mu}[Z(\beta)/\beta^2]. \tag{4.12}$$

In the case of a HF potential $U_{HF}(\mathbf{r})$, the latter quantity has to be corrected due to the interaction energy as in eqn (1.140).

Wigner representation

Let a single-particle operator $\hat{A}$ be given through its matrix elements $\langle \mathbf{r}|\hat{A}|\mathbf{r}'\rangle$ in the configuration space. Its Wigner transform (Wigner 1932, Argyres 1965) is given as

$$A_W \equiv A(\mathbf{q},\mathbf{p}) = \int d\mathbf{s}\ \exp\left(-\frac{i\mathbf{p}.\mathbf{s}}{\hbar}\right) \langle \mathbf{q} + \frac{\mathbf{s}}{2}|\hat{A}|\mathbf{q}-\frac{\mathbf{s}}{2}\rangle, \tag{4.13}$$

where

$$\mathbf{q} = \frac{1}{2}(\mathbf{r} + \mathbf{r}') \quad \text{and} \quad \mathbf{s} = (\mathbf{r} - \mathbf{r}'). \tag{4.14}$$

The representation (4.13) is convenient due to the fact that all operations in quantum mechanics can be expressed in terms of Wigner transforms of operators. For example, the Hamiltonian (4.1) in this representation is equal to its classical counterpart

$$H_W \equiv H(p,q) = \frac{p^2}{2m} + U(q). \tag{4.15}$$

Another example is the Wigner transform of the one-body density matrix (4.8), the so-called Wigner distribution function

$$f_W \equiv f(q,p) = \int ds \exp\left(-\frac{ip.s}{\hbar}\right) \rho(q+\frac{s}{2},q-\frac{s}{2}), \tag{4.16}$$

which simply defines the local density distribution

$$\rho(q) = \int \frac{dp}{(2\pi\hbar)^3} f(q,p), \tag{4.17}$$

and the nucleon momentum distribution (see eqn (2.79))

$$n(p) = \int dq\ f(q,p). \tag{4.18}$$

The convex operation of the Wigner transform (4.13) is the Weyl quantization rule (see Groenewald 1946) which associates with a classical observable B(q,p) the operator $\hat{B}$ defined by its matrix elements

$$\langle r|\hat{B}|r'\rangle = \int \frac{dp}{(2\pi\hbar)^3} B(q,p) \exp\left(\frac{ip.s}{\hbar}\right), \tag{4.19}$$

where **q** and **p** are given by eqn (4.14). The noncommutativity of an operator multiplication is reflected in the fundamental expression giving the Wigner transform of a product

$$(\hat{A}\ \hat{B})_W = A(p,q) \exp(\tfrac{1}{2}i\hbar \overleftrightarrow{\Lambda}\ ) B(q,p), \tag{4.20}$$

where

$$\overleftrightarrow{\Lambda} \equiv \overleftarrow{\nabla}_q \cdot \overrightarrow{\nabla}_p - \overleftarrow{\nabla}_p \cdot \overrightarrow{\nabla}_q = \sum_{i=1}^{3} \left(\frac{\overleftarrow{\partial}}{\partial q_i}\frac{\overrightarrow{\partial}}{\partial p_i} - \frac{\overleftarrow{\partial}}{\partial q_i}\frac{\overrightarrow{\partial}}{\partial p_i}\right), \tag{4.21}$$

the arrows indicating in which direction the gradients act. The operator (4.20) corresponds to the classical Poisson bracket (Goldstein 1959).

Equations (4.2) and (4.13) define the Wigner transform of the Bloch density matrix

$$C_W^{\beta} \equiv C(\mathbf{q},\mathbf{p};\beta) = (\exp(-\beta\hat{H}))_W = \left(\sum_{n=0}^{\infty}\frac{1}{n!}(-\beta\hat{H})^n\right)_W, \tag{4.22}$$

where all powers of the single-particle Hamiltonian $\hat{H}$ should be evaluated by eqn (4.20). We have, for example, the expression

$$(\hat{H}^2)_W = H^2(\mathbf{q},\mathbf{p}) - \frac{1}{4}\frac{\hbar^2}{2m}\,\Delta\,U(\mathbf{q}), \tag{4.23}$$

which follows from eqn (4.22) up to second order in β.

TF approximation to the Bloch density matrix

Let us expand the potential $U(\mathbf{r})$ entering eqn (4.1) in a Taylor series around an arbitrary point $\mathbf{q} \in R^3$ considered for the moment as a parameter. Truncating this expansion at the lowest order, we obtain the following approximation to the Bloch equation (4.3)

$$\left\{\frac{\partial}{\partial\beta} - \frac{\hbar^2}{2m}\nabla^2 + U(\mathbf{q})\right\} C_{\mathbf{q}}(\mathbf{r},\mathbf{r}';\beta) = 0. \tag{4.24}$$

Obviously, eqn (4.24) and consequently its solution $C^{(0)}(\mathbf{r},\mathbf{r}';\beta)$ will depend on the parameter $\mathbf{q} \in R^3$. Then, $C^{(0)}(\mathbf{r},\mathbf{r}';\beta)$ will be a good approximation to $C(\mathbf{r},\mathbf{r}';\beta)$ only for values $\mathbf{r},\mathbf{r}'$ close to $\mathbf{q}$ because the potential U is considered locally as a constant at point $\mathbf{q} \in R^3$. Therefore, if one wants to specify the point $\mathbf{q}$, the best that one can usually do is to identify $\mathbf{q}$ with the centre of mass of $\mathbf{r}$ and $\mathbf{r}'$, that is, to write $\mathbf{q} = \frac{1}{2}(\mathbf{r} + \mathbf{r}')$. In this way, one obtains the approximation

$$C_{TF}^{\beta} \equiv C^{(0)}(\mathbf{r},\mathbf{r}';\beta) = \left(\frac{m}{2\pi\hbar^2\beta}\right)^{3/2}\exp\left[-\frac{m}{2\hbar^2\beta}(\mathbf{r}-\mathbf{r}')^2 - \beta U\left(\frac{\mathbf{r}+\mathbf{r}'}{2}\right)\right]. \tag{4.25}$$

which satisfies eqn (4.24) and the correct boundary condition (4.4) when the spatial dependence of the potential U is neglected.

In fact, eqn (4.25) is a TF approximation to the exact Bloch density C^{β} because it generates via eqn (4.8) the TF result (3.3) for the one-body density matrix.

It is further demonstrated that eqn (4.24) is the lowest-order term in the expansion of the exact Bloch density matrix $C(\mathbf{r},\mathbf{r}';\beta)$ in powers of $\hbar$. We have particularly in mind the $\hbar$-dependence of the exact Bloch density matrix $C(\mathbf{r},\mathbf{r}';\beta)$ coming from the noncommutativity of the kinetic and potential energy parts of the Hamiltonian $\hat{H}$. In terms of the Wigner transform (4.22) this $\hbar$-dependence of $C(\mathbf{r},\mathbf{r}';\beta)$ is obviously absorbed in the exponential factor $\exp(\frac{1}{2}i\hbar\overleftrightarrow{\Lambda})$ entering eqn (4.22). Therefore, we have no reason to worry about the hidden $\hbar^{-1}$-dependence which goes along with the powers of β.

Expanding the exponent $\exp(\frac{1}{2}i\hbar\overleftrightarrow{\Lambda})$ entering eqn (4.22) formally in a Taylor series, and keeping only the lowest order terms in $\hbar$, one can easily derive from eqns (4.20), (4.22) and (4.23) the result

$$C^{(0)}(\mathbf{q},\mathbf{p};\beta) = \exp(-\beta H(\mathbf{q},\mathbf{p})), \tag{4.26}$$

where $\mathbf{q}$ and $H(\mathbf{q},\mathbf{p})$ are given from eqn (4.14) and eqn (4.15), respectively. Finally, calculating the inverse Wigner transform (4.19) of eqn (4.26) we find the TF result (4.25). Therefore, the TF approximation gives the lowest order contributions to different quantities in their systematic expansion in powers of $\hbar$.

4.3 Wigner-Kirkwood expansion

We define the semiclassical expansion of the Bloch density matrix $C(\mathbf{r},\mathbf{r}';\beta)$ as an expansion of its Wigner transform in powers of $\hbar$

$$C(\mathbf{p},\mathbf{q};\beta) = \sum_{n=0}^{\infty} C^{(n)}(\mathbf{p},\mathbf{q};\beta)\ \hbar^{n}, \tag{4.27}$$

where the leading term $C^{(0)}(\mathbf{p},\mathbf{q};\beta)$ can be physically interpreted as the classical limit of the operator C^{β}. Obviously, the convergence of the series (4.27) or its quality as an asymptotic series is open to doubt and depends on the particular choice of the leading term $C^{(0)}(\mathbf{p},\mathbf{q};\beta)$.

Expansion based on a TF leading term

In order to derive a semiclassical expansion of the type of eqn (4.27), one can start by taking the Wigner transform of eqn (4.3). Using eqns (4.3), (4.20) and (4.22), the result is

$$\partial C^{\beta}_{W} / \partial\beta = H_{W} \exp\left[\frac{i\hbar}{2}\left(\overleftarrow{\nabla}_{q}.\overrightarrow{\nabla}_{p} - \overleftarrow{\nabla}_{p}.\overrightarrow{\nabla}_{q}\right)\right] C^{\beta}_{W}\,, \tag{4.28}$$

where $H_W = H(\mathbf{q},\mathbf{p})$ is given by eqn (4.15).

According to Balazs and Jennings (1984), one can try a solution of eqn (4.28) of the form of eqn (4.27)

$$C(\mathbf{p},\mathbf{q};\beta) = \exp\left[-\beta(\mathbf{p}^2/2m + U(\mathbf{q}))\right]$$

$$\times\ (1 + \hbar w_1(\mathbf{q},\mathbf{p};\beta) + \hbar^2 w_2(\mathbf{q},\mathbf{p};\beta) + \ldots), \tag{4.29}$$

where the leading term is the TF expression (4.26). Substituting eqn (4.29) into eqn (4.28) and equating equal powers of $\hbar$, one obtains a recursion relation for w_1, w_2, w_3, ... which can be solved using the initial values $w_0 = 1$ and $w_{-1} = 0$. One thus determine explicitly all the terms in the semiclassical $\hbar$-expansion (4.29) which, up to order $\hbar^2$, is of the form (Wigner 1932, 1934, Kirkwood 1933, Kirzhnits 1957)

$$C(\mathbf{p},\mathbf{q};\beta) = \exp\left[-\beta(\mathbf{p}^2/2m + U(\mathbf{q}))\right]\left\{1 + \frac{\hbar^2\beta}{8m}\left[-\Delta U + \frac{\beta}{3}(\nabla U)^2 + \frac{\hbar^2}{2m}(\mathbf{p}\,.\,\nabla)^2 U\right] + O(\hbar^4) + \ldots\right\}. \tag{4.30}$$

The expressions for higher orders in $\hbar$ are fairly length but nevertheless they can be calculated (see for example Grammaticos and Voros 1979).

It is seen from eqn (4.30) that the powers of $\hbar$ are always even (all terms odd in $\hbar$ are imaginary and do not contribute to C_W^β) and equal to the sum of all derivatives of the potential in a specific term. The $\hbar$ expansion (4.30) is therefore simultaneously a gradient expansion. For slowly varying potentials, the gradients vanish and the series (4.30) reduces to the term in lowest order in $\hbar$, the TF expression (4.26).

Furthermore, the physical characteristics of the system such as density $\rho(\mathbf{r})$ and kinetic energy density $\tau(\mathbf{r})$ can be calculated on the basis of eqn (4.30). We quote here the results up to order $\hbar^2$ (Kirzhnits 1967)

$$\rho_{sc}(\mathbf{r}) = \frac{1}{3\pi^2}\left(\frac{2m}{\hbar^2}\right)^{3/2}(\mu - U(\mathbf{r}))^{3/2}\left\{1 - \frac{\hbar^2}{16m}\left[\frac{\Delta U}{(\mu - U(\mathbf{r}))^2} + \frac{1}{4}\frac{(\nabla U)^2}{(\mu - U(\mathbf{r}))^3}\right] + O(\hbar^4) + \ldots\right\}\theta(\mu - U(\mathbf{r})),$$

$$\tau_{sc}(\mathbf{r}) = \frac{1}{5\pi^2}\left(\frac{2m}{\hbar^2}\right)^{5/2}(\mu - U(\mathbf{r}))^{5/2}\left\{1 - \frac{5\hbar^2}{16m}\left[\frac{5^2}{3}\frac{\Delta U}{(\mu - U(\mathbf{r}))^2} - \frac{3}{4}\frac{(\nabla U)^2}{(\mu - U(\mathbf{r}))^3}\right] + O(\hbar^4) + \ldots\right\}\theta(\mu - U(\mathbf{r})), \tag{4.31}$$

where $\theta(x)$ is the unit step function.

From the above equations it is possible to eliminate algebraically the difference $(\mu - U(\mathbf{r}))$ in the expression for the kinetic energy density, thereby consistently retaining all terms of order $\hbar^2$ and neglecting these of higher orders in $\hbar$ (Kirzhnits 1967). The result

$$\tau[\rho] = \frac{3}{5}(3\pi^2)^{2/3}\rho^{5/3}(\mathbf{r}) + \frac{1}{36}\frac{(\nabla\rho)^2}{\rho} + \frac{1}{3}\Delta\rho \tag{4.32}$$

recovers the expression (3.40) for the kinetic energy density used in the improved TF model (see ITF, Section 3.3). In the lowest-order term we recognize the TF expression as it must be, while the $\hbar^2$ corrections lead to the Weizsäcker term (2.69), but with a coefficient smaller by a factor of nine. The last term in eqn (4.32) is the divergence of a vector field and it will not contribute to the kinetic energy which is an integral over $\tau[\rho]$ but it does contribute to the total energy through the term $\rho\tau[\rho]$ as in the case of the Skyrme-HF functionals presented in Section 3.3.

Classical turning point problem

It is obvious from eqn (4.31) that the semiclassical expansion has fault at the classical turning point r_{cl} defined according to eqn (3.9). The density $\rho_{SC}(\mathbf{r})$ as well as the kinetic energy density $\tau_{SC}(\mathbf{r})$ diverges at the classical turning point, both densities being identically zero outside the classically allowed region due to the step function $\Theta(\mu - U(\mathbf{r}))$.

According to the derivation, the functional (4.32) is also strictly valid only for classical allowed values of $\mathbf{r} \in R^3$. This limitation is bypassed in most calculations by assuming in an *ad hoc* manner that an analytical continuation of eqn (4.32) is valid even beyond the classical turning point. This is the reason why we have denoted the densities $\rho_{SC}(\mathbf{r})$ and $\tau_{SC}(\mathbf{r})$ in eqn (4.32), by $\rho(\mathbf{r})$ and $\tau(\mathbf{r})$, respectively. We shall return to this classical turning point problem later in Chapter 10, where it will be demonstrated that the functional (4.32) is actually correct in the whole space R^3.

Partial re-summation of the ħ-expansion

A way to remove the turning point divergences by partially re-summing the Wigner-Kirkwood $\hbar$ expansion (4.26) has been proposed by Bhaduri (1977) and further developed by Durand, Brack, and Schuck (1978, 1980), and Bartel, Durand, and Brack (1984). The main idea is based on different choices of the leading term entering the Wigner-Kirkwood expansion (4.27).

The TF leading term (4.26) consists of re-summing the series (4.28) for all terms containing no derivatives of the potential U. One can then use as leading term the propagator $C^{(1)}(\mathbf{q},\mathbf{p};\beta)$, re-summing all the terms containing the potential and its first derivatives (the so-called 'linearized' approximation), or the propagator $C^{(2)}(\mathbf{q},\mathbf{p};\beta)$, re-summing all the terms containing the potential and its first and second derivatives (the so-called locally harmonic approximation). This partial summation of infinitely many terms may eventually account for important effects, for instance, tunneling into the classically forbidden region. Local density distributions calculated in both linearized and harmonic approximations are illustrated in Figure 3.1.

We do not want to go into details of the different re-summation techniques, but it should be mentioned that it is not possible for an explicit expression for the kinetic energy density $\tau(\mathbf{r})$ as a functional of $\rho(\mathbf{r})$ to be derived as in the case of eqn (4.32). Instead, the densities $\rho(\mathbf{r})$, $\tau(\mathbf{r})$ and the potential $U(\mathbf{r})$ can be iterated directly to obtain self-consistency as proposed by Bartel and Vallieres (1982).

4.4 Extended Thomas-Fermi model

All the results obtained in the previous Section can be generalized also to the case of velocity-dependent Skyrme-type forces (1.114). For this purpose one has to replace the Hamiltonian $\hat{H}$ from eqn (4.1) by the Skyrme-HF single-particle Hamiltonian (1.153) rewritten in the form

$$H_{HF} = -\frac{\hbar^2}{2m} \nabla \cdot f(\mathbf{r})\nabla + U(\mathbf{r}) - i\mathbf{W}(\mathbf{r}) \cdot (\nabla \times \sigma). \tag{4.33}$$

or simplicity, we assume again only one kind of nucleons so that the functions $f(\mathbf{r}) = m/m^*(\mathbf{r})$, $U(\mathbf{r})$, $\mathbf{W}(\mathbf{r})$ stand for either the proton or the neutron parts of eqns (1.154), (1.152), (1.155), respectively.

The quite lengthy procedure for deriving the Wigner-Kirkwood expansion (4.29) with the new Hamiltonian (4.33) has been carried out using a computer and the REDUCE language for symbolic manipulations by Grammaticos and Voros (1980). They calculate explicit expressions for the functionals $\tau_{sc}(\mathbf{r})$ and $\mathbf{J}_{sc}(\mathbf{r})$ up to fourth order with respect to $\hbar$ as well as the associated functionals $\tau[\rho]$ and $\mathbf{J}[\rho]$ following from an elimination procedure similar to the one used in deriving eqn (4.32).

In this way one could calculate, at least in principle, higher- and higher-order terms in the $\hbar$-expansion in order to express the exact HF energy as a functional of the local density $\rho(\mathbf{r})$. This would be a constructive realization of the Hohenberg-Kohn theorem within the HF approximation. As shown in the next chapter, however, this procedure cannot be stopped at an arbitrary point. Guet and Brack (1980) have shown that an acceptable truncation of the semiclassical expressions for $\tau[\rho]$ and $\mathbf{J}[\rho]$ is the one containing the terms $\tau_{ETF}[\rho]$ and $\mathbf{J}_{ETF}[\rho]$, respectively, which are consistent up to the order $\hbar^4$.

ETF energy density functional

The functionals $\tau_{ETF}[\rho]$ and $J_{ETF}[\rho]$ are the main ingredients in the so-called extended Thomas-Fermi (ETF) model which has been extensively elaborated over the last two decades (see the review article by Brack, Guet, and Håkansson (1985) and references therein). It operates with the approximate local density functional

$$E_{ETF}[\rho] = \int d\mathbf{r}\ \mathcal{E}(\rho,\tau_{ETF}[\rho],\mathbf{J}_{ETF}[\rho]) \tag{4.34}$$

obtained by inserting $\tau_{ETF}[\rho]$ and $\mathbf{J}_{ETF}[\rho]$ in the expression for the Skyrme-HF energy density (1.148). In order to clarify the structure of the functional (4.34) let us first consider the structure of the functionals

$$\tau_{ETF}[\rho] = \tau_{TF}[\rho] + \tau_2[\rho] + \tau_4[\rho] \tag{4.35}$$

and

$$J_{ETF}[\rho] = J_2[\rho] + J_4[\rho], \tag{4.36}$$

where the indices 2 and 4 denote the order of $\hbar$ in the corresponding term.

The zeroth order term $\tau_{TF}[\rho]$ in eqn (4.35) stands for the TF kinetic energy density. For one kind of nucleons it differs from eqn (3.5) by a factor of $2^{2/3}$, compare for example eqn (3.5) with the first term in eqn (4.32). At the same time, there is no zeroth order term appearing in the spin-density functional (4.36).

The second order term $\tau_2[\rho]$ in eqn (4.35) is of the form

$$\tau_2[\rho] = \frac{1}{36}\frac{(\nabla\rho)^2}{\rho} + \frac{1}{3}\Delta\rho + \frac{1}{6}\frac{(\nabla\rho.\ \mathbf{f})}{f}$$

$$+ \frac{1}{6}\rho\frac{\Delta f}{f} - \frac{1}{12}\rho\left(\frac{\nabla f}{f}\right)^2 + \frac{1}{2}\left(\frac{2m}{\hbar^2}\right)^2\rho(\mathbf{W}/f)^2, \tag{4.37}$$

where the first two terms, together with $\tau_{TF}[\rho]$, define the kinetic energy density (4.32) already analysed in connection with the ITF model (Section 3.3). The other terms in eqn (4.37) are obviously the effective mass and spin-orbit contributions to $\tau_2[\rho]$. The spin-orbit density in the same approximation is expressed as

$$J_2[\rho] = -\left(\frac{2m}{\hbar^2}\right)\frac{1}{f}\rho\mathbf{W} = -\left(\frac{2m^*}{\hbar^2}\right)\rho\mathbf{W}. \tag{4.38}$$

Equation (4.38) with $m^* = m$ has been derived earlier by Stocker, Süssmann, and Knaak (1972).

Going up to order $\hbar^4$, the following somewhat lengthy expressions for $\tau_4[\rho]$ and $J_4[\rho]$ are obtained:

$$\tau_4[\rho] = (3\pi^2)^{-2/3}\Bigg\{ \rho^{1/3}\Bigg[-\frac{1}{120}\frac{\nabla^4 f}{f} + f^{-2}\left(\frac{\nabla^2(\nabla f)^2}{120} - \frac{(\nabla^2 f)^2}{240}\right)$$

$$- \frac{1}{30} f^{-3}((\nabla f.\nabla)^2 f) + \frac{1}{80} f^{-4}(\nabla f)^4\Bigg] + \rho^{-2/3}\Bigg[-\frac{1}{180}\nabla^4\rho$$

$$+ \frac{1}{f}\left(\frac{1}{360}\nabla^2(\nabla f.\nabla\rho) + \frac{1}{180}(\nabla f.\nabla)\nabla^2\rho - \frac{1}{120}(\nabla\rho.\nabla)\nabla^2 f\right)$$

$$+ f^{-2}\left(\frac{(\nabla f.\nabla)(\nabla\rho.\nabla)f}{180} - \frac{(\nabla f.\nabla\rho)\ \nabla^2 f}{360} - \frac{(\nabla f.\nabla)^2\rho}{180}\right)\Bigg]$$

$$+ \rho^{-5/3}\Bigg[-\frac{1}{72}(\nabla\rho.\nabla)\nabla^2\rho - \frac{7}{1080}(\nabla^2\rho)^2 - \frac{7}{2160}\nabla^2(\nabla\rho)^2$$

$$+ \frac{1}{f}\left(-\frac{11}{1080}(\nabla\rho.\nabla)^2 f - \frac{13}{1080}(\nabla f.\nabla\rho)\nabla^2\rho - \frac{19}{1080}(\nabla f.\nabla)(\nabla\rho.\nabla)\rho\right)$$

$$+ \frac{1}{144} f^{-2}(\nabla f.\nabla\rho)^2\Bigg] + \rho^{-8/3}\Bigg[\frac{7}{324}(\nabla\rho)^2\nabla^2\rho + \frac{23}{810}(\nabla\rho.\nabla)^2\rho$$

$$+ \frac{13}{648}\frac{(\nabla f.\nabla\rho)(\nabla\ \rho)^2}{f}\Bigg] - \rho^{-11/3}\frac{(\nabla\rho)^4}{45}\Bigg\}$$

$$+ \left(\frac{m}{\hbar^2}\right)^2 (3\pi^2)^{-2/3}\Bigg\{\rho^{1/3}\Bigg[4f^{-2}\left(\mathbf{W}.\nabla^2\mathbf{W} + 2(\mathbf{W}.\nabla)(\nabla.\mathbf{W}) + \frac{1}{2}\nabla^2(\mathbf{W}^2) + (\nabla\mathbf{W})^2\right)$$

$$+ \frac{1}{2} f^{-3}\left(-2\mathbf{W}.(\nabla f.\nabla)\mathbf{W} - (\mathbf{W}.\nabla f)(\nabla\mathbf{W}) - \nabla f(\mathbf{W}.\nabla)\mathbf{W} - \mathbf{W}^2\nabla^2 f - (\mathbf{W}.\nabla)^2 f\right)$$

$$+ \frac{3}{4}f^{-4}\left(\mathbf{W}^2(\nabla f)^2 + (\mathbf{W}.\nabla f)^2 - 4(\mathbf{W}^2)^2\right)\Big] + \frac{1}{6}\,\rho^{-2/3}\Big[f^{-2}\Big(\mathbf{W}.(\nabla\rho.\nabla)\mathbf{W}$$

$$+ (\mathbf{W}.\nabla\rho)(\nabla.\mathbf{W})\Big) - f^{-3}\Big((\nabla f.\nabla\rho)\mathbf{W}^2 + (\nabla f.\mathbf{W})(\nabla\rho.\mathbf{W})\Big)\Big]\Big\} \tag{4.39}$$

and

$$J_4[\rho] = \left(\frac{m}{\hbar^2}\right)(3\pi^2)^{-2/3}\Big\{\rho^{1/3}\Big[-\frac{1}{4}f^{-1}(2\nabla^2\mathbf{W} + \nabla\times(\nabla\times\mathbf{W})) + \frac{1}{4}f^{-2}(2\nabla^2 f.\mathbf{W}$$

$$- (\mathbf{W}\times\nabla)\times\nabla f + \nabla f\times(\nabla\times\mathbf{W}) + 2(\nabla f.\nabla)\mathbf{W}) + f^{-3}(-\frac{1}{2}(\nabla f)^2\mathbf{W} + \frac{1}{4}\nabla f\times(\mathbf{W}\times\nabla f)$$

$$+ 2\left(\frac{m}{\hbar^2}\right)^2\mathbf{W}^2\mathbf{W})\Big] - \rho^{-2/3}\Big[\frac{1}{12}f^{-1}((\nabla\rho.\nabla)\mathbf{W} + \nabla\rho(\nabla.\mathbf{W}))$$

$$+ \frac{1}{12}\,f^{-2}(-\nabla f\times(\mathbf{W}\times\nabla\rho) + 2(\nabla f.\nabla\rho)\mathbf{W}\Big]\Big\}. \tag{4.40}$$

We use here the following conventions (Grammaticos and Voros 1979): any free-standing gradient operators act only on the nearest term on the right-hand side; that is, $(\nabla f\ .\ \nabla)(\nabla\rho\ .\ \nabla)\rho$ means $\sum_{j,k}\partial_j f\partial^2_{jk}\rho\partial_k\rho$ and $(\nabla\rho\ .\ \nabla)^2 f$ means $\sum_{j,k}\partial_j\rho\partial_k\rho\partial^2_{jk}f$.

Guet and Brack (1980) have shown that the semiclassical expansion for $\tau[\rho]$ must be truncated after the term $\tau_4[\rho]$. For example, the term $\tau_6[\rho]$ behaves asymptotically like $\rho^{-1/3}(\mathbf{r})$ and thus diverges when the density falls exponentially to zero. Therefore, eqn (4.35) and eqn (4.36) must be considered only as model expressions (ETF model) reflecting a converging part of the asymptotic semiclassical series.

ETF equation for the density

After determining the functional (4.34), one has to solve the associated Euler-Lagrange equation for the densities $\rho_n(\mathbf{r})$ and $\rho_p(\mathbf{r})$. This is a highly nonlinear fourth order differential equation (Guet and Brack 1980)

$$U(\mathbf{r};[\rho]) + \frac{\hbar^2}{2m}\left\{\left(\frac{3\pi^2}{2}\rho\right)^{2/3} + \frac{1}{36}\left[\left(\frac{\nabla\rho}{\rho}\right)^2 - 2\frac{\Delta\rho}{\rho}\right] + \frac{1}{6480}D_4[\rho]\right\} = \mu, \quad (4.41)$$

where $U(\mathbf{r};[\rho])$ is the variational derivative of the potential part entering the functional $E_{ETF}[\rho]$ and

$$D_4[\rho] = \left(\frac{3\pi^2}{2}\rho\right)^{2/3}\left\{-176\left(\frac{\nabla\rho}{\rho}\right)^4 + \frac{544}{3}\frac{\Delta\rho}{\rho}\left(\frac{\nabla\rho}{\rho}\right)^2\right.$$

$$\left. + 112\frac{\nabla\rho.\nabla(\nabla\rho)}{\rho^3} - 66\left(\frac{\Delta\rho}{\rho}\right)^2 - 106\frac{\nabla\rho.\nabla(\Delta\rho)}{\rho^2}\right\}. \quad (4.42)$$

Here the problem is significantly simplified: one assumes that $\rho_n = \rho_p = \rho/2$ and leaves out the effective mass ($f \equiv 1$), the spin-orbit density ($\mathbf{W} \equiv 0$) and the Coulomb contribution. These simplifications, however, do not affect the general conclusions following from the ETF model.

Equation (4.42), or the enormously more complicated one following from the variation of the general functional (4.34) with respect to $\rho_n(\mathbf{r})$ and $\rho_p(\mathbf{r})$, can only be solved numerically. For example, eqn (4.41) has been solved numerically by Bohigas *et al.* (1976) in connection with the ITF model restricted to terms of the second order in $\hbar$ (see Section 3.3). Equation (4.41), including the term (4.42), has been solved numerically by Bartel, Brack, and Durand (1985) for a symmetric semi-infinite nuclear matter, while a similar equation with a Woods-Saxon potential (see Section 5.3) has been solved numerically by Pi *et al.* (1988).

Asymptotic solutions

In the limit of a large nucleus and for a schematic Skyrme-HF energy density, Brack, Guet, and Håkansson (1985) obtain the solution of eqn (4.41) inside the nucleus

$$\rho(\mathbf{r}\to 0) \cong 1 - \exp\,[(r-R)/\alpha], \quad (4.43)$$

where

$$\alpha = \left[\frac{18}{K_{NM}}(\beta\rho_{NM} + \frac{1}{36}\frac{\hbar^2}{2m})\right]^{1/2} \tag{4.44}$$

is expressed in terms of the nuclear matter compressibility K_{NM}, the saturation density ρ_{NM} and the multiplying coefficient $\beta = (2m/\hbar^2)[\frac{1}{4}(3t_1 + 5t_2) + t_2x_2]$ in front of the surface potential energy term. Therefore, the inner asymptotic part of the nuclear surface is essentially determined by a balance between the gradient term of the potential energy ($\cong (\nabla\rho)^2$) and the incompressibility coefficient K_{NM}.

For realistic values of the quantities entering eqn (4.44), one finds that the range α is approximately 1 fm and thus about twice as large as the typical value of the nuclear diffuseness skin-thickness parameter b ($t_{eff} = 4.4b \cong 2$ fm). This tends to make the densities following from the ETF model asymmetric around the half-density distance R ($\rho(R) = \rho(0)/2$), a feature that agrees with the averaged HF densities.

The asymptotic solution of eqn (4.41) for large distances, where both $\rho(\mathbf{r})$ and $U(\mathbf{r};[\rho])$ go to zero, is easily derived. If the fourth order term $D_4[\rho]$ is omitted in eqn (4.41), the asymptotic solution is governed by the Weizsäcker term and it leads to an exponential fall-off of the density as shown in Section 3.3. However, the only way to make $D_4(\mathbf{r};[\rho])$ finite in eqn (4.41) is the asymptotic solution for $\rho(\mathbf{r})$, which behaves like

$$\rho(r \rightarrow \infty) \cong \alpha r^{-6}, \tag{4.45}$$

where

$$\alpha = \left\{-\frac{\hbar^2}{2m}(3\pi^2)^{-2/3}\frac{13}{45\mu}\right\}^{3/2}. \tag{4.46}$$

Therefore, the fourth order terms of the functional $\tau_{ETF}[\rho]$ lead to a much slower fall-off of the density $\rho(\mathbf{r})$. This contradicts the realistic exponential fall-off of the nuclear densities following from the general quantum mechanical estimates.

ETF results for spherical nuclei

In the general case, the exact variation of the ETF functional (4.34) would lead to two coupled, highly nonlinear fourth order integro-differential equations for $\rho_n(\mathbf{r})$ and $\rho_p(\mathbf{r})$ which seem hopeless to solve even numerically. This is the reason why the numerical ETF calculations are usually performed within the variational method presented in Section 3.2, whereby a restricted density variational space is assumed by parameterizing the expected solution for $\rho_n(\mathbf{r})$ and $\rho_p(\mathbf{r})$.

Bartel *et al.* (1982) and Brack, Guet, and Håkansson (1985) choose for spherical shapes the modified Fermi-type functions

$$\rho_q(r) = \frac{\rho_{0q}(1 + \rho_{1q} \exp[-r^2/\beta_q^2 R_q^2])}{(1 + \exp[(r-R_q)/\alpha_q])^{\gamma_q}}, \tag{4.47}$$

where q = n,p denotes neutrons and protons as before. They fix the radius constants R_q by the normalization conditions to N_q nucleon numbers (q = n,p) and thus have in their calculations ten variational parameters $\rho_{0q}, \beta_q, \alpha_q$ and γ_q.

A comparison of the variational ETF results with the self-consistent HF ones for nucleon densities of two spherical nuclei is given in Figure 4.1. The agreement between the two methods is excellent in the nuclear surface region and this is a consequence of the high order gradient corrections included in eqn (4.34). It is also obvious that the ETF densities reproduce no density shell fluctuations following from the HF results.

That the ETF model is well suited to predict the averaged density profiles in nuclei is seen from Figure 4.2 where the ETF charge distributions are compared with the experimental distributions deduced from the elastic electron scattering experiments at Saclay (Bellicard *et al.* 1981).

In Table 4.1 we list binding energies and charge rms radii for some spherical nuclei derived in the ETF model. Comparison is also made with the experimental data and the self-consistent HF

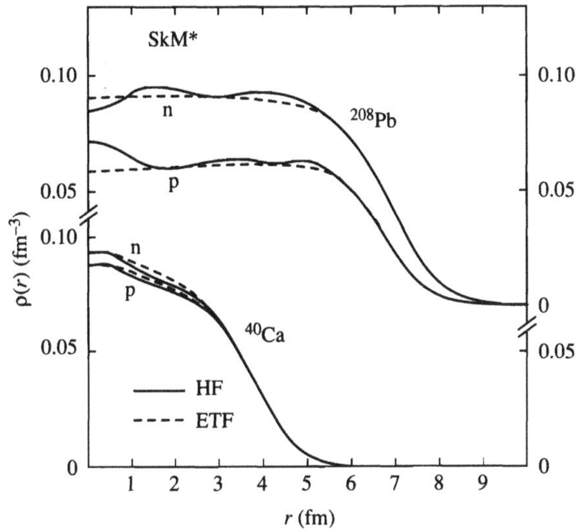

Figure 4.1. Neutron and proton density distributions (in fm^{-3}) calculated within the Hartree-Fock theory (solid lines) and the variational extended Thomas-Fermi method (dashed lines) using SkM^* effective forces (Brack, Guet, and Håkansson 1985).

results. It is seen that the ETF charge radii reproduce the HF ones to better than 2% in light nuclei and well below 1% in heavy nuclei. At first glance, the ETF total binding energies are also comparable with the HF energies E_{HF} which makes the ETF method a useful tool for investigating ground-state properties of nuclei.

As we shall see in the next chapter, however, the ETF energies must be compared with the averaged part of the HF energies only. In this light, adding perturbatively the contributions caused by the nuclear shell effects, the ETF binding energies become too large compared to the HF ones by approximately 2-5 MeV in light nuclei and by approximately 4-10 MeV in heavier ones (Guet, Håkansson, and Brack 1980, Dutta *et al.* 1986a). This overbinding effect is in fact a violation of the variational principle and must be considered as a defect of the approximate ETF functional.

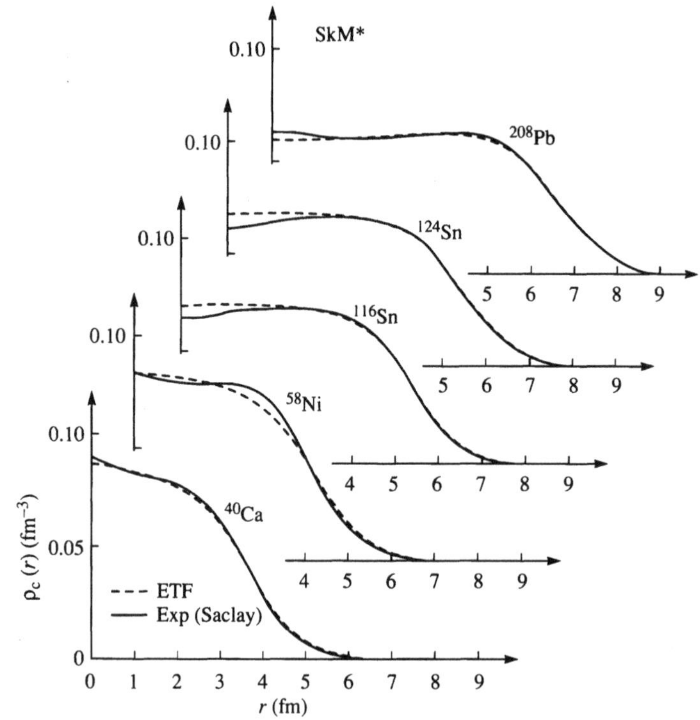

Figure 4.2. Charge density distributions calculated within the variational extended Thomas-Fermi method (dashed lines) compared with experimental results (Brack, Guet, and Håkansson 1985).

4.5 Liquid drop and droplet models of nuclei

The idea of expressing the total binding energy of a nucleus as a functional of the local density $\rho(r)$ has been developed early in the history of nuclear density functional theory in connection with the famous semi-empirical Bethe-Weizsäcker mass formula (Bethe and Bacher 1936, Weizsäcker 1935), the basis of the liquid drop model (LDM) of nuclei. In the LDM, the density distributions of neutrons and protons are considered to be constant inside a sharp boundary and zero outside (Myers and Swiatecki 1966). The nuclear densities, however, are only approximately constant in the

Table 4.1. Binding energies (in MeV) and charge rms radii (in fm) of spherical nuclei obtained with SkM* forces within the extended Thomas-Fermi method (ETF), Hartree-Fock theory (HF) and expectation value method (EVM) according to Brack, Guet, and Håkansson (1985).

Nuclei	Binding energies				Charge rms radii		
	$-E_{EXP}$	$-E_{HF}$	$-E_{ETF}$	$-E_{EVM}$	EXP	HF	ETF
^{16}O	127.6	127.7	128.0	127.4	2.73	2.81	2.78
^{40}Ca	342.1	341.1	345.9	340.1	3.49	3.52	3.47
^{48}Ca	416.0	420.0	421.8	418.4	3.48	3.54	3.54
^{56}Ni	484.0	485.4	483.9	483.1	3.75	3.77	3.81
^{90}Zr	783.9	784.5	786.6	782.7	4.27	4.30	4.29
^{114}Sn	971.6	969.2	976.0	967.9	—	4.60	4.61
^{123}Sn	1102.7	1110.7	1101.5	1108.3	—	4.73	4.75
^{140}Ce	1172.7	1173.9	1174.5	1171.6	4.88*	4.89	4.89
^{208}Pb	1636.5	1636.4	1627.0	1633.7	5.50	5.51	5.53

bulk of the system, and instead of a sharp boundary there is a diffuse surface region in which they decrease smoothly to zero. A generalization and refinement of the LDM of nuclei concerning the finiteness of the thickness of the nuclear surface and the finiteness of the nuclear compressibility is achieved by the droplet model (DM) of Myers and Swiatecki (1969, 1974).

Leptodermous expansion

The liquid drop (or droplet) model can be introduced in the context of the nuclear density functional theory by expanding the energy density functional $E[\rho]$ in powers of the inverse radius R^{-1} ($R = r_0 N^{1/3}$, $r_0 \cong 1.2$ fm), that is, in powers of $N^{-1/3}$ where N stands for the mass number of the nucleus. Such an expansion may be legitimized from the fact that the density distributions are 'leptodermous' in the sense that their surface thickness b is much smaller than the bulk radius R, so that $(b/R) \ll 1$.

In order to obtain some of the DM parameters entering their leptodermous expansion, Myers and Swiatecki (1969, 1974) perform density variational calculations with TF expression for the kinetic energy functional and using forces of Seyler and Blanchard (1961, 1963). The resulting expression for the total energy of a spherical nucleus with Z protons ($C = 3e^2/5r_0$) and (N - Z) neutrons ($I = (N - 2Z)/N$) is of the form

$$E_{DM}(Z,N-Z) = (a_v + J\,\delta^2 - \frac{1}{2}K\,\varepsilon^2 + \frac{1}{2}M\,\delta^4)N$$

$$+ (a_s + \frac{9}{4}\frac{J^2}{Q}\delta^2)N^{2/3} + a_c N^{1/3} + C\left[Z^2N^{-1/3} - \frac{C}{336}\left(\frac{1}{J} + \frac{18}{K}\right)Z^2N^{1/3}\right.$$

$$\left. - \frac{5}{2}\left(\frac{b}{r_0}\right)^2\frac{Z^2}{N} - \frac{5}{4}\left(\frac{3}{2\pi}\right)^{2/3}Z^{4/3}N^{-1/3} - \frac{C}{64}\frac{Z^2}{Q}\right], \qquad (4.48)$$

where

$$\delta = \frac{\left(I + \frac{3}{8}\frac{C}{Q}Z\,N^{-2/3}\right)}{\left(1 + \frac{9}{4}\frac{J}{Q}\,N^{-1/3}\right)} \qquad (4.49)$$

and

$$\varepsilon = K^{-1}(-\,2a_s N^{-1/3} + L\,\delta^2 + CZ^2N^{-4/3}). \qquad (4.50)$$

Initially, most of the parameters entering eqn (4.48), namely

a_v (volume energy coefficient),

a_s (surface energy coefficient),

a_c (curvature correlation coefficient),

J (symmetry energy coefficient),

K (compressibility coefficient),

L (density-symmetry coefficient),

M (symmetry anharmonicity coefficient),

Q (effective surface stiffness coefficient),

were determined by fitting the masses and fission barriers, whereby K, L, M were set at what seemed to be physically reasonable values from the density variational calculations. The undetermined curvature correction coefficient a_c was arbitrarily set to zero (Myers 1976). In the case of deformed nuclei, there are some extra quantities containing various shape dependencies that enter the mass formula (Myers and Swiatecki 1974, Myers 1976).

ETF approach to the DM mass formula

The refinement of nuclear mass formulae goes along with the improvement of approximate energy density functionals in nuclear density functional theory. Brack, Guet, and Håkansson (1985) analyse the DM mass formula in terms of the ETF functional (4.34) and a leptodermous expansion based on the trial local densities (4.47) with $\rho_{1q} = 0$. Treiner and Krivine (1986) apply the same trial densities within the ITF model of the kinetic energy density (4.32).

These calculations are more general than the DM because they automatically contain curvature energy, surface and asymmetry energies, and even higher order contributions which go beyond the usual DM expression (4.48). The problem simplifies, even up to close analytical expressions, in the asymptotic limit of a very heavy nucleus where the density profile across the surface is that of the so-called semi-infinite nuclear matter (Myers and Swiatecki 1969). In particular this allows the DM relations to be tested and also to relate the parameters of the DM to some particular characteristics of the effective nucleon-nucleon interaction in nuclei.

We compare in Table 4.2 the DM parameters as given by Myers and Swiatecki (1974) with some results derived in the ETF calculations with three different sets of Skyrme-type forces. The semi-infinite nuclear matter calculations are performed within the generalized Fermi density profile

$$\rho_q(z) = \frac{\rho_{0q}}{(1 + \exp[(z-Z_q)/\alpha_q])} \qquad (q = n,p), \qquad (4.51)$$

where z measures the distances perpendicular to the surface of the semi-infinite matter.

It is seen from Table 4.2 that the ETF calculations allow results obtained with different forces to be compared. In principle, this can help the parameters of the effective interaction, which do not have obvious physical meaning, to be related to physical quantities of macroscopic character which can be measured experimentally. For example, Köher (1976) has derived the sets Ska and Skb of the Skyrme-type force (Table 1.1) by adjusting the interaction to reproduce the DM values for volume, symmetry and surface energies.

Table 4.2. Comparison of the empirical values (M & S) of droplet model parameters (Myers and Swiatecki 1974) with theoretical estimates of Brack, Guet, and Håkansson (1985) following from extended Thomas-Fermi calculations with Skyrme-type effective forces.

	a_v	a_s	a_c	J	K	L	M	Q
M & S	-15.680	18.56	0.00	28.06	294.8	123.5	2.637	16.04
SIII	-15.857	18.04	7.57	28.16	355.4	9.9	1.66	54.6
Ska	-15.997	18.52	8.86	32.91	263.2	75.3	2.26	29.8
SkM*	-15.776	17.22	10.24	30.03	216.7	45.8	1.88	35.4

Difficulties of the drop(let) model

The considerable effort continuously devoted to the improvement of the nuclear mass formulae of the type of eqn (4.48) is due to the great success of the macroscopic-microscopic approach which provides a good description of ground-state masses, radii, deformation and separation energies, nuclear fission isomers (Nix and Swiatecki 1965, Myers and Swiatecki 1966, Brack *et al.* 1972, Krappe, Nix, and Sierk 1972, Möller and Nix 1981, Myers and Swiatecki 1982, Möller *et al.* 1985). Here, the bulk properties of nuclei are described using a macroscopic model such as the liquid drop(let) one, while quantum effects are then added in a perturbative way using the Strutinski shell correction method (see next chapter).

Further improvements of the DM-type mass formulae is strongly motivated by the need to extrapolate from the available mass data out to unknown nuclei. For example, there is an obvious need to extrapolate towards nuclei lying close to the neutron drip line which play a crucial role in the r-process of nucleo-synthesis in supernova explosions (Seeger 1967, Arnold 1980). Also of interest is extrapolation along the stability line in order to explore the possible existence of superheavy nuclei.

An improvement of the DM due to Möller and Nix (1981) shows that all the volume coefficients can take their nuclear matter values without any renormalization. This model, however, has done nothing to improve the extrapolation properties of the DM in the region of nuclei close to the neutron drip line, which is one of the main objections to the LDM or DM mass formula. In particular, Dutta *et al.* (1986a) have shown that when any of the existing forms of the DM is extrapolated out to the neutron drip line, errors of as much as 15 MeV can result.

This problem might be resolved by inclusion of certain higher order terms in the leptodermous expansion or by analysing the non-leptodermous terms to the liquid drop(let) expansion (Stocker, Gleissl, and Brack 1987). It was shown, however, that this would result in enormous complexity and would need more than 16 (Dutta *et al.* 1986a) or 20 (Brack, Guet and Håkansson 1985) parameters. At the same time, with the Skyrme-type forces great flexibility is achieved with far fewer parameters than these mass formulae. Thus one may feel that it would be worthwhile to consider abandoning the droplet model approach altogether, and to study the computational flexibility of the density functional theory, fitting the parameters of the effective interaction to the mass data (Dutta *et al.* 1986b).

In addition to the above difficulties, a number of important criticisms has been made of the DM concerning those very aspects by which this model improves on the older LDM (Pearson 1982, Tondeur, Pearson, and Farine 1983, Stringari and Lipparini 1982, Brack, Guet, and Håkansson 1985). Some critical analysis of this criticisms also appear (Krivine 1984, Treiner and Krivine 1986, Treiner *et al.* 1986). Obviously further efforts in the macroscopic description of nuclear masses and densities in the context of the nuclear drop(let) model seem to be in order.

5. Averaging nuclear energy density functionals

5.1 Introduction

In this chapter we relate the nuclear density unctional theory in its semiclassical form with the shell-correction method proposed by Strutinski (1967, 1968). This gives an elegant method for calculating not only the ground-state energies but also their dependence on deformation parameters in nuclei.

The total binding energy of nuclei E as a function of the mass number N can be divided into two parts:

$$E = \tilde{E} + \delta E, \tag{5.1}$$

where the part $\tilde{E}$ reproduces the average N-dependence of the binding energy in a smooth way only,and the fluctuating part δE is sensitive to those particular nuclear properties associated with nucleons in the vicinity of the Fermi surface.

According to the Strutinski shell-correction method, the energy $\tilde{E}$ should correspond to the liquid drop(let) model energy E_{DM}, while δE should actually be the same as the energy δE_{sh} derived from phenomenological shell model calculations. We shall show that the above statement is true and even more that the smooth part $\tilde{E}$ can be correctly reproduced by means of the approximate energy density functionals considered so far.

5.2 Shell-correction method

The HF theory reproduces the experimental binding energies to high accuracy (see Section 1.7). It is therefore meaningful to explain the separation (5.1) by splitting the HF energy E_{HF} into two terms $\tilde{E}_{HF}$ and δE_{HF}. Of course, such a procedure is as expensive as the HF one. However, its main purpose is to justify in a microscopic manner the Strutinski shell-correction method.

Self-consistent averaging procedure

Suppose that the self-consistent HF problem is already solved and we know the solution of the HF equations

$$\hat{H}_{HF}[\rho^{(1)}]\ \varphi_i(\mathbf{r}) = \varepsilon_i\ \varphi_i(\mathbf{r}), \tag{5.2}$$

that is, we know the single-particle wave functions $\varphi_i(\mathbf{r})$ and energies ε_i. The HF ground-state energy (1.140) is then defined as

$$E_{HF} \equiv E_{HF}[\rho^{(1)}] = \mathrm{tr}(\hat{t}\ \rho^{(1)}) + \frac{1}{2}\ \mathrm{tr}\ \mathrm{tr}\ (\rho^{(1)}\ \hat{\omega}_A\ \rho^{(1)}), \tag{5.3}$$

where $\hat{\omega}_A$ is the antisymmetrized two-body interaction. We use here a different notation for the HF ground-state energy (1.140), expressing it in terms of the HF one-body density matrix

$$\rho^{(1)} \equiv \rho^{(1)}(\mathbf{r},\mathbf{r}') = \sum_{i=1}^{N} \varphi_i^*(\mathbf{r})\varphi_i(\mathbf{r}') = \sum_{\alpha=1}^{\infty} n_\alpha \varphi_\alpha^*(\mathbf{r})\varphi_\alpha(\mathbf{r}'), \tag{5.4}$$

where n_α are the HF occupation numbers: $n_i = 1$ for $\varepsilon_i \leq \varepsilon_F$ and $n_i = 0$ for $\varepsilon_i \geq \varepsilon_F$, ε_F being the last occupied single-particle state. In the same shorthand notation

$$\hat{H}_{HF} \equiv \hat{H}_{HF}[\rho^{(1)}] = \hat{t} + \mathrm{tr}\ (\hat{\omega}_A\ \rho^{(1)}) \tag{5.5}$$

is the HF single-particle Hamiltonian (1.134) entering eqn (5.2).

The main problem is now to divide up E_{HF} into a smoothly varying part $\tilde{E}_{HF}$ and an oscillating part δE_{HF}; that is, to put eqn (5.3) in the form

$$E_{HF} = \tilde{E}_{HF} + \delta E_{HF}. \tag{5.6}$$

For this purpose, instead of the shell model level density (4.10), $g(\varepsilon) = \sum_{i=1}^{N} \delta(\varepsilon - \varepsilon_i)$, Strutinski (1967) defined a smoothed single-particle level density

$$\tilde{g}(\varepsilon) = \gamma^{-1} \int_{-\infty}^{\infty} d\varepsilon' \; g(\varepsilon') \, f\left(\frac{\varepsilon' - \varepsilon}{\gamma}\right), \tag{5.7}$$

whereby each delta function entering $g(\varepsilon)$ is replaced by an averaging function $f(x)$ and the smoothing parameter γ is chosen to be about $\hbar\omega \cong$ 6–10 MeV, the approximate spacing between two major shells.

The decisive idea concerns the new level density $\tilde{g}(\varepsilon)$ that represents only the smooth part of the level density $g(\varepsilon)$. How to define the averaging function $f(x)$ in an appropriate mathematical way is a well investigated problem in the literature on the shell-correction method (Brack *et al.* 1972, Brack and Pauli 1973). We shall therefore assume that the average part $\tilde{g}(\varepsilon)$ of the level density $g(\varepsilon)$ is given. Then, one can calculate a corresponding Fermi energy $\tilde{\mu}$ by the condition

$$\int_{-\infty}^{\infty} d\varepsilon \; \tilde{g}(\varepsilon) \equiv \sum_{i=1}^{N} \tilde{n}_i = N, \tag{5.8}$$

where N is the number of neutrons (or protons).

The quantities $\tilde{n}_i$ entering eqn (5.8) are important ingredients of the method

$$\tilde{n}_i = \int_{-\infty}^{t_i} dx \; \tilde{g}(x), \qquad t_i = [(\tilde{\mu} - \varepsilon_i)/\gamma], \tag{5.9}$$

which can be considered as generalized occupation numbers because they can take positive as well as negative values in contrast to the usual HF occupation numbers for which $n_i \geq 0$.

The generalized occupation numbers (5.9) define the smooth part of the one-body density matrix

$$\tilde{\rho}^{(1)} \equiv \tilde{\rho}^{(1)}(\mathbf{r},\mathbf{r}') = \sum_{\alpha=1}^{\infty} \tilde{n}_\alpha \varphi_\alpha^*(\mathbf{r})\varphi_\alpha(\mathbf{r}'), \tag{5.10}$$

while its oscillating part is

$$\delta\rho^{(1)} \equiv \delta\rho^{(1)}(\mathbf{r},\mathbf{r}') = \sum_{\alpha=1}^{\infty} \delta n_\alpha \varphi_\alpha^*(\mathbf{r})\varphi_\alpha(\mathbf{r}'), \tag{5.11}$$

where $\delta n_l = n_l - \tilde{n}_l$. Obviously, both density matrices, $\tilde{\rho}^{(1)}$ and $\delta\rho^{(1)}$, form the complete density matrix (5.4) such as

$$\rho^{(1)} = \tilde{\rho}^{(1)} + \delta\rho^{(1)}. \tag{5.12}$$

Inserting this into eqn (5.2) and collecting terms of different order in $\delta\rho^{(1)}$ we get

$$E_{HF} = \tilde{E}_{HF} + \mathrm{tr}(\hat{H}_{HF}[\rho^{(1)}]\ \delta\rho^{(1)}) + O((\delta\rho^{(1)})^2), \tag{5.13}$$

where

$$\tilde{E}_{HF} \equiv E_{HF}[\tilde{\rho}^{(1)}] = \mathrm{tr}(\hat{t}\ \tilde{\rho}^{(1)}) + \frac{1}{2}\ \mathrm{tr}\ \mathrm{tr}\ (\tilde{\rho}^{(1)}\ \hat{\upsilon}_A\ \tilde{\rho}^{(1)}), \tag{5.14}$$

and

$$\hat{H}_{HF}[\tilde{\rho}^{(1)}] = \hat{t} + \mathrm{tr}\ (\hat{\upsilon}_A\ \tilde{\rho}^{(1)}). \tag{5.15}$$

Let us now introduce a new set of single-particle functions $\varphi_1(\mathbf{r})$ and energies ε_1 as a solution of the problem

$$\hat{H}_{HF}[\tilde{\rho}^{(1)}]\ \varphi_1(\mathbf{r}) = \varepsilon_1\ \varphi_1(\mathbf{r}). \tag{5.16}$$

This defines the following iterative process. Using the solution $\varphi_1(\mathbf{r})$ and ε_1 of eqn (5.16) we repeat the procedure which starts with eqn (5.7) and finishes with the new single-particle Hamiltonian (5.15). Then we can solve eqn (5.16) with this new

Hamiltonian again and so on until a convergence is obtained. In other words, the self-consistency of the averaged Hamiltonian (5.15) can be practically achieved by solving eqn (5.16) iteratively and in each step recalculating the averaged density $\tilde{\rho}^{(1)}$ in terms of the new $\varphi_i(\mathbf{r})$ and ε_i until convergence is reached, that is, $\tilde{\rho}^{(1)}$ does not change in two consecutive steps.

At the end of the above procedure, eqn (5.13) reads

$$E_{HF} = \tilde{E}_{HF} + \sum_{\alpha=1}^{\infty} \varepsilon_\alpha \, \delta\tilde{n}_\alpha + O((\delta\rho^{(1)})^2), \tag{5.17}$$

where the second term

$$\delta_1 E_{HF} = \sum_{\alpha=1}^{\infty} \varepsilon_\alpha \, \delta\tilde{n}_\alpha \tag{5.18}$$

follows from the second term of eqn (5.13) after taking into account eqns (5.11) and (5.16). Equation (5.17) is also known as the Strutinski energy theorem (Strutinski 1968, 1974). It states that all shell effects of the first order in $\delta\rho^{(1)}$ are taken into account by summing up single-particle energies of an averaged single-particle Hamiltonian. The above procedure that expresses $\tilde{E}_{HF}$ by eqn (5.14) and δE_{HF} by eqn (5.18) is known as the self-consistent averaging procedure.

Smooth and oscillating parts of the energy

The Strutinski energy theorem has been checked numerically by Brack and Quentin (1975, 1981) within the Skyrme-HF theory. For Skyrme-type forces, the averaged energy (5.14) is calculated in the same way as the Skyrme-HF energy (1.144) but replacing the quantum mechanical densities $\rho(\mathbf{r})$, $\tau(\mathbf{r})$, $\mathbf{J}(\mathbf{r})$ by the averaged densities

$$\tilde{\rho}(\mathbf{r}) = \sum_{\alpha=1}^{\infty} \tilde{n}_\alpha \, |\varphi_\alpha(\mathbf{r})|^2,$$

$$\tilde{\tau}(\mathbf{r}) = \sum_{\alpha=1}^{\infty} \tilde{n}_\alpha \; |\nabla\varphi_\alpha(\mathbf{r})|^2,$$

$$\tilde{J}(\mathbf{r}) = \sum_{\alpha=1}^{\infty} \tilde{n}_\alpha \; (\varphi^*_\alpha(\mathbf{r}) \; \mathbf{l} \, . \, \mathbf{s} \; \varphi_\alpha(\mathbf{r})), \tag{5.19}$$

respectively. In other words, instead of eqn (5.14) one has to calculate

$$\tilde{E}_{HF} = \int d\mathbf{r} \; H(\tilde{\rho}(\mathbf{r}),\tilde{\tau}(\mathbf{r}),\tilde{J}(\mathbf{r})). \tag{5.20}$$

Similarly, eqn (5.16) now reads

$$\hat{H}_{HF}[\tilde{\rho}(\mathbf{r}),\tilde{\tau}(\mathbf{r}),\tilde{J}(\mathbf{r})] \; \varphi_i(\mathbf{r}) = \varepsilon_i \; \varphi_i(\mathbf{r}), \tag{5.21}$$

where the Skyrme-HF single-particle Hamiltonian (1.153) is used in terms of the averaged densities (5.19).

The numerical results obtained by Brack and Quentin (1975, 1981) demonstrate the high degree of accuracy of the self-consistent averaging procedure. This can be seen from Table 5.1 where the HF results for the ground-state energy E_{HF} of some nuclei are listed together with the energies $\tilde{E}_{HF}$ and δE_{HF}. As another example, we show in Figure 5.1 the total energy of the nucleus ^{240}Pu versus the mass quadrupole moment $Q = \langle Q_{20} \rangle$ obtained in both the HF method (full line) and the self-consistent averaging procedure (crosses). The smooth full line in this figure represents the energy $\tilde{E}_{HF}$, eqn (5.20), as a function of Q. The close agreement between E_{HF} and $\tilde{E}_{HF} + \delta E_{HF}$ provides a strong confirmation of the self-consistent averaging procedure.

The second three columns in Table 5.1 illustrate the so-called nonself-consistent results obtained by solving only once eqn (5.16) or its equivalent eqn (5.21). The important result is that the self-consistent condition mainly affects the oscillating part δE_{HF} , leaving the smooth part $\tilde{E}_{HF}$ roughly the same. The differences between self-consistent and nonself- consistent

Table 5.1. Self-consistent and nonself-consistent Strutinski-averaged Hartree-Fock energies for some nuclei in their ground state calculated with SIII forces (Brack and Quentin 1975)

		Self-consistent			Nonself-consistent		
Nuclei	E_{HF}	$\tilde{E}_{HF}$	$\delta_1 E_{HF}$	$\delta_2 E_{HF}$	$\tilde{E}_{HF}$	$\delta_1 E_{HF}$	$\delta_2 E_{HF}$
^{16}O	-126.7	-122.1	-4.7	0.0	-122.7	-5.7	1.6
^{40}Ca	-339.6	-337.5	-2.6	0.5	-338.1	-4.8	3.3
^{56}Ni	-479.9	-473.7	-6.8	0.5	-473.5	-5.5	-0.9
^{90}Zr	-779.2	-774.3	-5.5	0.6	-775.7	-5.8	2.3
^{114}Sn	-960.4	-961.1	0.5	0.2	-961.3	0.2	0.7
^{168}Yb	-1352.2	-1349.6	-3.0	0.4	-1349.9	-4.3	2.0
^{208}Pb	-1625.4	-1606.6	-19.2	0.4	-1607.1	-20.6	2.3

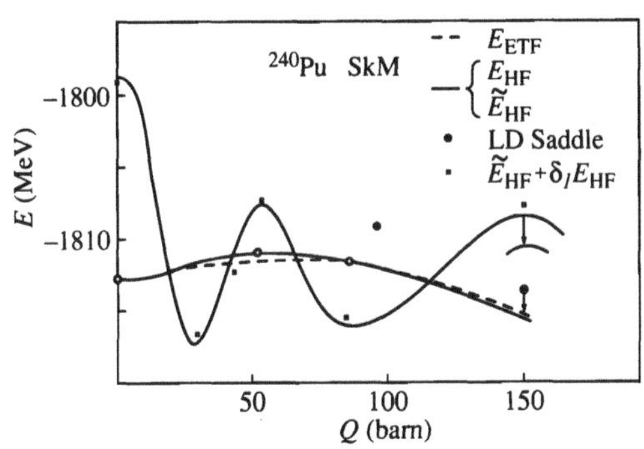

Figure 5.1. Total energy of the nucleus ^{240}Pu as a function of the mass quadrupole moment Q with SkM effective forces. The Hartree-Fock energy E_{HF} and the self-consistent Strutinski-averaged energy $\tilde{E}_{HF}$ are represented by solid lines, while the dashed line represents the variational extended Thomas-Fermi results. The crosses are the results of the self-consistent averaging procedure for $\tilde{E}_{HF} + \delta E_{HF}$. The encircled cross indicates the empirical saddle point (Brack, Guet, and Håkansson 1985).

results for $\tilde{E}_{HF}$ are typically one order of magnitude smaller than for δE_{HF} .

Justification of the shell-correction method

One is able now to justify the shell-correction method in its standard version (Brack *et al.* 1972) which consists in calculating the total binding energy of a nucleus as a sum

$$E = E_{DM} + \delta E_{sh}, \tag{5.22}$$

where the smooth part E_{DM} is the energy resulting from the liquid drop(let) model, and the oscillating part $\delta E_{sh} = \sum_\alpha \varepsilon_\alpha \delta\tilde{n}_\alpha$ is calculated within a phenomenological shell model. For this purpose, Brack and Quentin (1975, 1981) consider the following important numerical observations.

The second order corrections in eqn (5.17) are of the form

$$\delta_2 E_{HF} = -\frac{1}{2} \operatorname{tr} \operatorname{tr} (\delta\rho^{(1)} \hat{\mathfrak{v}}_A \delta\rho^{(1)}). \tag{5.23}$$

They have been checked numerically and indeed turn out to be small corrections (Bunatian, Kolomietz, and Strutinski 1972, Brack *et al.* 1972, Brack and Quentin 1975). Some typical examples have been given in Table 5.1. Therefore, one can write to a good approximation that $\delta E_{HF} \approx \delta_1 E_{HF}$ and

$$E_{HF} \cong \tilde{E}_{HF} + \delta E_{HF}. \tag{5.24}$$

The shell-correction term (5.18) does not change at all if it is calculated with a phenomenological shell model Hamiltonian having completely the same spectrum as the averaged Hamiltonian (5.15). If the phenomenological single-particle spectrum does not reproduce exactly the spectrum of the operator $\hat{H}_{HF}[\tilde{\rho}^{(1)}]$, the above statement nevertheless stays true, because according to the Strutinski energy theorem small deviations again give contributions of second order in eqn (5.17). Therefore, the first order shell-correction term is close to that one following from the phenomenological shell model, that is,

$$\delta_1 E_{HF} \cong \delta E_{sh}. \tag{5.25}$$

The result (5.25) is illustrated in Figure 5.2 where a comparison is made between $\delta_1 E_{HF}$ calculated with two different effective forces and δE_{sh} calculated with a phenomenological Woods-Saxon potential as used in the standard shell-correction calculations (Brack *et al.* 1972). The remarkable agreement among the three curves, within ≅ 1 MeV over the whole range of deformations, strongly support eqn (5.25).

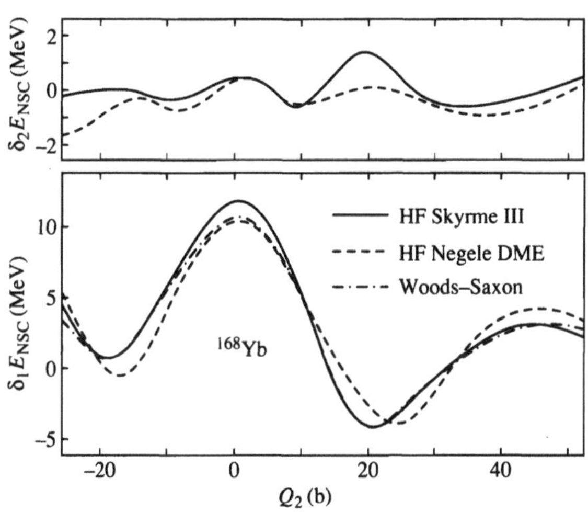

Figure 5.2. The first and second order shell-correction energies along the quadrupole deformation energy curve of ^{168}Y. HF stands for the non-self-consistent (NSC) Hartree-Fock results obtained with SII forces. DME are the results obtained with the Negele effective force within the density matrix expansion of Negele and Vautherin (1972). The first order correction obtained with a Woods-Saxon parameterization of the mean field is also shown (Brack and Quentin 1981).

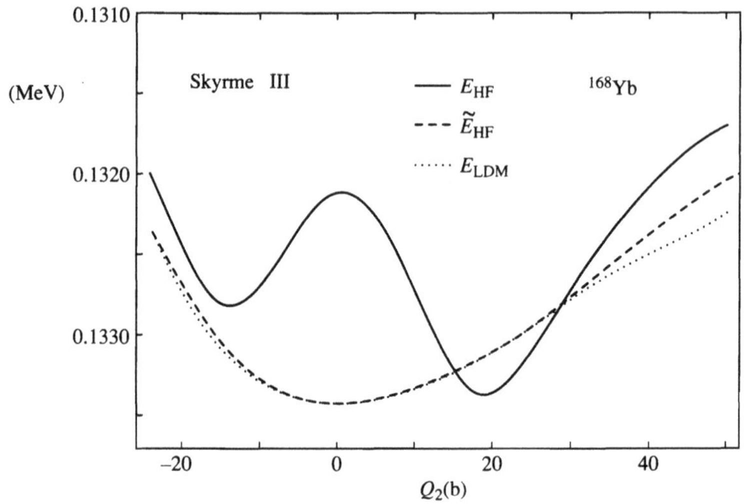

Figure 5.3. The deformation energy of the nucleus ^{168}Yb calculated with SIII effective forces (Brack and Quentin 1975). The solid line represents the exact constrained Hartree-Fock energy E_{HF}, the long-dashed line gives the associated self-consistently averaged part $\tilde{E}_{HF}$. The liquid drop model energy is shown by the short-dashed line (Ring and Schuck 1980).

The averaged energy $\tilde{E}_{HF}$, considered as a function of both the deformation and the nucleon number, has been proved numerically to behave exactly like the liquid drop(let) model energy E_{DM}. For example, we show in Figure 5.3 the deformation energy of the nucleus ^{168}Yb as a function of the quadrupole moment. Both results $\tilde{E}_{HF}$ and E_{DM} are seen to be in remarkable agreement. One can therefore write

$$\tilde{E}_{HF} = \int d\mathbf{r}\; H(\tilde{\rho}(\mathbf{r}),\tilde{\tau}(\mathbf{r}),\tilde{\mathbf{J}}(\mathbf{r})) \cong E_{DM}. \qquad (5.26)$$

Finally, taking into account the fact that the HF energies E_{HF} almost exactly reproduce the experimental binding energies E, together with the results given by eqns (5.24), (5.25), and

(5.26), one gets to the standard shell-correction method, eqn (5.22).

5.3 Semiclassical averaging procedure

In this section we demonstrate that the approximate Thomas-Fermi-Weizsäcker energy density functionals, in particular the functionals following from the ETF method, are in fact an energy density representation of the averaged part $\tilde{E}$ containing no shell effects in nuclei.

Strutinski averaging as a semiclassical limit

We shall first demonstrate the equivalence of the Strutinski-averaged energy $\tilde{E}$ with the semiclassical averaged result E_{sc} derived after summing up terms in the semiclassical expansion of the energy in powers of $\hbar$ (Chapter 4). In the case of an isotropic harmonic oscillator potential $U_{HO}(\mathbf{r}) = \frac{1}{2} m\omega^2 r^2$ the corresponding Bloch equation (4.3) leads to the following exact solution for the Bloch matrix (Krivchenkov 1964)

$$C_{HO}(\mathbf{q},\mathbf{p};\beta) = \left(\frac{2\pi\hbar^2}{m}\right)^{-3/2}\left[\frac{\hbar\omega}{\sinh(\beta\hbar\omega)}\right]^{3/2}$$

$$\times \exp\left\{-\frac{m\omega}{\hbar}\left[q^2\tanh\left(\beta\frac{\hbar\omega}{2}\right) + \frac{s^2}{2}\coth\left(\beta\frac{\hbar\omega}{2}\right)\right]\right\}, \tag{5.27}$$

where $\mathbf{q}$ and $\mathbf{s}$ are defined by eqn (4.14). The associated partition function (4.9) is of the form

$$Z_{HO}(\beta) = \frac{1}{4}\left\{\frac{1}{\sinh(\beta\hbar\omega/2)}\right\}^3. \tag{5.28}$$

It allows the energy (4.12) to be calculated by evaluating the inverse Laplace transform of $\beta^{-2}Z_{HO}(\beta)$. The result is

$$E_{HO} - \mu N \equiv - \mathcal{L}^{-1}_{\beta\to\mu}[Z_{HO}(\beta)/\beta] = - \frac{\mu^4}{12(\hbar\omega)^3} + \frac{\mu^2}{8\hbar\omega} - \frac{17}{960}\hbar\omega$$

$$- \sum_{n=1}^{\infty} \frac{2}{(\hbar\omega)^3}(-1)^n \left\{ \left[\left(\frac{\hbar\omega}{2n\pi}\right)^2 \left((\frac{\hbar\omega}{2})^2 - \mu^2 \right) + 6(n\pi)^{-4}(\frac{\hbar\omega}{2}) \right] \cos\left(\frac{2\pi\nu\mu}{\hbar\omega}\right) \right.$$

$$\left. + 4\mu(n\pi)^{-3}(\frac{\hbar\omega}{2})^2 \sin\left(\frac{2\pi\nu\mu}{\hbar\omega}\right) \right\} = \tilde{E}_{HO} + \delta E_{HO}. \qquad (5.29)$$

It is seen from this equation that the first three terms as a function of μ form a monotonically increasing smooth part $\tilde{E}_{HO}$, whereas all other terms produce an oscillating contribution δE_{HO}.

Let us now substitute the harmonic oscillator potential into the semiclassical expansion (4.30). The result produces exactly the smooth part of the energy (5.29)

$$E^{HO}_{SC} - \mu = - \frac{\mu^4}{12(\hbar\omega)^3} + \frac{\mu^2}{8\hbar\omega} - \frac{17}{960}\hbar\omega = \tilde{E}_{HO}. \qquad (5.30)$$

Therefore, in the case of a harmonic oscillator potential, the semiclassical expansion reproduces the smooth part of the energy only. On the other hand, Brack and Pauli (1973) have shown that in the case of the harmonic oscillator shell model the Strutinski-averaging procedure leads to exactly the same result $\tilde{E}_{HO}$ for the smooth part of the energy.

In the case of a harmonic oscillator potential, the smooth part $\tilde{E}_{HO}$ comes from the singularity of $Z_{HO}(\beta)$ for small β. In the case of an arbitrary potential, this result is usually generalized by assuming that the semiclassical averaging may come again from the behaviour of $Z(\beta)$ for small β, although the analytical form of $Z(\beta)$ is unknown. That the resulting semiclassical-averaged energy is equivalent to the Strutinski-averaged energy was demonstrated numerically by Bhaduri and Ross (1971) for some model examples whose partition functions are known analytically. In the general

Table 5.2. Smooth energies in a spherical Woods-Saxon potential obtained by Jennings (1973). The first column gives the nucleon number N. The second, third and fourth columns give energies corresponding to the contributions of the TF term, the $\hbar^2$-term and the $\hbar^4$-term of E_{sc}, eqn (5.32), respectively. All energies are in MeV. The fifth column gives the total semiclassical energy E_{sc}, while the last one gives the Strutinski-averaged energy $\tilde{E}_{WS}$.

N	$E_{sc}^{(0)}$	$E_{sc}^{(2)}$	$E_{sc}^{(4)}$	E_{sc}	$\tilde{E}_{WS}$
164	-3344.7	88.1	1.0	-3254.9	-3256.0(1.2)
260	-5484.8	119.7	1.1	-5364.0	-5364.4(.08)
416	-9049.1	162.5	1.1	-8881.5	-8882.8(1.2)

case, this result has been proved by Jennings (1973). One thus concludes that the Strutinski-averaged energy $\tilde{E}$ is equivalent to the semiclassical-averaged result E_{sc}. Then it is in order to consider here the different contributions to the energy E_{sc}.

As mentioned above, when calculating E_{sc} one should take into account only contributions that come from the singularity of $Z(\beta)$ at $\beta = 0$. This corresponds to a converging asymptotic series of the semiclassical expansion (4.30), whereas all diverging terms (which usually show up at $\hbar^6$ or higher orders) should be left out as being related to the oscillating part of the energy. Jennings, Bhaduri, and Brack (1975) have demonstrated that the smooth part of the semiclassical expansion (4.30) is a very rapidly convergent series in the particular case of a Woods-Saxon potential. Their results are listed in Table 5.2, where contributions $E_{sc}^{(n)}$ to E_{sc} of the order $\hbar^n$ ($n = 0, 2, 4$) are compared with the Strutinski averaged energy $\tilde{E}_{WS}$. The comparison clearly shows that the semiclassical result E_{sc}, including fourth order in $\hbar$, exhausts most of the smooth energy $\tilde{E}_{WS}$, and one can write

$$\tilde{E}_{WS} \cong E_{sc} = \int dr \left\{ \frac{\hbar^2}{2m} \tau_{sc}(\mathbf{r}) + U(\mathbf{r})\rho_{sc}(\mathbf{r}) \right\} = E_{sc}^{(0)} + E_{sc}^{(2)} + E_{sc}^{(4)}, \quad (5.31)$$

where $\tau_{sc}(\mathbf{r})$ and $\rho_{sc}(\mathbf{r})$ are defined by eqn (4.30) and eqn (4.29), respectively. Now, it is clear that the semiclassical expansion, up to fourth order in $\hbar$, is in fact a way of smoothing the shell effects in nuclei.

Averaging energy density functionals

As seen in Section 4.4, the ETF model represents the ground-state energy as an energy density functional

$$E[\rho] = \int dr \left\{ \frac{\hbar^2}{2m} \tau_{ETF}[\rho] + U(\mathbf{r})\rho_{ETF}[\rho] \right\}, \quad (5.32)$$

which is consistent with eqn (5.31) up to order $\hbar^4$. In contrast to eqn (5.31), which is restricted due to the classical turning point problem, eqn (5.32) is assumed to be valid in the whole space R^3. One is thus interested in checking the validity of the relation (5.31) expressed by means of the energy density functional (5.32) and local densities $\rho(\mathbf{r}) \in D_N$. For this purpose, Guet and Brack (1980) and Pi *et al.* (1988) analyse the following three energies derived from eqn (5.32):

- the energy $E[\tilde{\rho}]$, where $\tilde{\rho}(\mathbf{r})$ is the Strutinski-averaged density (5.19) calculated by means of the single-particle functions and energies of the potential $U(\mathbf{r})$;
- the energy $E[\rho^{tr}]$ where ρ^{tr} is the trial local density (4.47) whose parameters minimize the functional (5.32);
- the energy $E[\rho]$ where $\rho(\mathbf{r})$ is the solution of the Euler–Lagrange equation associated with eqn (5.32).

Some results of these calculations are compared in Table 5.3 for a

Table 5.3. The ability of the extended Thomas-Fermi energy density functional to reproduce the Strutinski-averaging in nuclei. All energies (in MeV) are calculated for a Woods-Saxon potential. E_{WS} and $\tilde{E}_{WS}$ are the exact quantum-mechanical energy and its Strutinski averaged part, respectively. Results $E[\tilde{\rho}]$, $E[\rho^{tr}]$ and $E[\rho]$ correspond to the extended Thomas-Fermi functional when evaluated with quantum-mechanical local density ρ_{WS}, the optimal trial local density distribution (see the text) and the solution of the Euler-Lagrange equation, respectively.

N	E_{WS}	$\tilde{E}_{WS}$	$E[\tilde{\rho}]$	$E[\rho^{tr}]$	$E[\rho]$
78	−1315.4	−1315.4	−1315.0	−1321.4	−1323.2
164	−3239.1	−3256.0	−3254.9	−3267.3	−3269.5
204	−4126.7	−4126.6	−4126.5	−4141.0	−4143.5
260	−5358.1	−5364.4	−5364.0	−5381.6	−5384.3
292	−6081.9	−6078.6	−6078.2	−6097.5	−6100.4
416	−8873.8	−8882.8	−8881.5	−8906.8	−8910.5

Results taken from Bartel, Durand, and Brack (1984) and Pi *et al.* (1988)

Woods-Saxon potential together with the exact quantum mechanical energy E_{WS} and its Strutinski-averaged part $\tilde{E}_{WS}$.

The results in Table 5.3 show that the energy $E[\rho]$ compares with the averaged energy $\tilde{E}_{WS}$ rather well. Therefore, a relation similar to eqn (5.31) should hold with respect to the functional (5.32). This means that the ETF functional contains no shell effects in nuclei. Furthermore, this conclusion can be made with respect to the full ETF functional (4.34) as it stands, because its main features are reflected by eqn (5.32). For example, the term containing $\tau_{ETF}[\tilde{\rho}]$ reproduces the Strutinski-averaged kinetic energy within less than ≅ 1.5 MeV, while the integral of the term

$\tau_{ETF}[\tilde{\rho}]\tilde{\rho}$ as it occurs in eqn (4.33) reproduces the associated averaged value within less than 1 MeV (Guet and Brack 1980).

When minimizing the functional (5.32), variationally or by solving the associated Euler-Lagrange equation, a clear overbinding effect appears (see the results in Table 5.3 for $E[\rho^{tr}]$ and $E[\rho]$, respectively). As mentioned in Section 4.4, the overbinding is a typical feature of the ETF model. Nevertheless, the energies $E[\rho^{tr}]$ and $E[\rho]$ compare the smooth energy $\tilde{E}_{WS}$ within an error less than 0.5%. This is true again for the full energy density functional (4.33), whereby the resulting energies $E[\rho^{tr}]$ compare with the smooth HF energies $\tilde{E}_{HF}$ in approximately the same way (Guet, Håkansson, and Brack 1980). Therefore, returning to eqn (5.26), one could write, with error of about less than 0.5%, the relation

$$\tilde{E}_{HF} \cong E_{ETF}[\rho^{tr}] \cong E_{DM}. \tag{5.33}$$

In this sense, the ETF model approximates only the averaged part of the ground-state energy of nuclei and it can be used as an effective tool for deriving the parameters of the liquid drop(let) model or for applications within the shell-correction method.

Violation of the variational principle

In spite of density functional theory, we may separate the exact functional $E[\rho]$ into a smooth part $\tilde{E}[\rho]$ and an oscillating part, the latter accounting for the shell effects in nuclei. Seen in this light, a variation applied to $\tilde{E}[\rho]$ rather than to the full energy density functional $E[\rho]$ is a violation of the variational principle. As a consequence, the variational equation $\delta\tilde{E}[\rho] = 0$ may show instabilities. The resulting energy may deviate from the value obtained with the exact local density, satisfying equation $\delta_{\rho}E[\rho] = 0$, or its average part.

The semiclassical-averaged energy is nothing but an approximation to $\tilde{E}[\rho]$ that corresponds to the smooth part of the HF ground-state energy of nuclei. Therefore, it becomes clear that the ETF method contains all defects coming from the violation of the variational principle. In particular, the overbinding effect of the ETF variational calculations is one of these defects. As we shall see in the next chapter, the overbinding effect in ETF calculations may be overcome at the price of a HF iteration performed starting from ETF local densities as an input.

5.4 Results for deformed nuclei

We have considered in Section 4.4 some results for spherical nuclei obtained within the ETF model. However, there are about 6000 nuclei lying between the drip lines, very few of which could be supposed a priori to be spherical. In this section we demonstrate that the ETF model is a rather economical tool for describing properties of deformed nuclei as well.

Double-humped fission barrier

The nuclear energy as a function of deformation parameters δ_l, such as quadrupole or hexadecapole deformations, is a multidimensional energy surface. It can be described within the constrained HF theory which asks for particular minima of the energy E_{HF} on this multidimensional energy surface (more detailed discussion on the constrained HF theory is given in Section 6.4). If one plots E_{HF} as a function of deformations δ_l, one may see that more than one minima can eventually develop in constrained HF calculations. An example has been already shown in Figure 5.1 for the nucleus ^{240}Pu where two minima and two maxima (barriers) appear at different quadrupole deformations Q.

The first minimum corresponds to the nuclear ground-state energy. For spherical nuclei, this ground-state minimum is found at the origin, that is at zero deformation $Q_0 = 0$, while for deformed nuclei it occurs at a finite deformation $Q_0 \neq 0$. Actually, this is the minimum that follows from unrestricted HF calculations. In very heavy nuclei, a second minimum may exist at a deformation $Q > Q_0$ which is somewhat higher than the first one and corresponds to the so-called shape isomeric state of nuclei.

If we compare the deformation energy E_{HF} with its smooth part $\tilde{E}_{HF}$ (see Figure 5.1), we realize that the two minima as well as the two barriers in E_{HF} are completely due to nuclear shell effects incorporated in the oscillating part δE_{HF}. The smooth part $\tilde{E}_{HF}$ shows only a broad maximum rather than the double-humped barrier in E_{HF}.

From Strutinski shell-correction calculations we know that it is sufficient to reproduce quantitatively the smooth part of the energy, and especially the empirical height of the LDM barrier, in order to describe successfully the fission processes in nuclei (Brack *et al.* 1972). The oscillating part containing shell effects can then be added in a perturbative way and it is very little influenced by the bulk properties of nuclei.

In this respect the use of the ETF energy density method is a very economical tool for investigating of fission processes in nuclei. It avoids the use of single-particle functions and energies when approximating $\tilde{E}_{HF}$ and still depends on the particular effective interaction in nuclei.

When considering the ground states of deformed nuclei, the smooth part $\tilde{E}_{HF}$ alone is not able to predict the ground-state minimum at all (see Figure 5.1). This is of course also the case for the ETF calculations. All stable nuclei are spherical in their ground state when described in terms of the ETF model. One is able to investigate, however, the smooth part of the deformation energy by adding external constraints in the ETF calculations or simply by minimizing the functional $E_{ETF}[\rho]$ within a restricted variational space of suitably parameterized deformed trial local density distributions.

Parameterization of nuclear densities

In order to define a set of trial local densities for deformed nuclei Brack, Guet, and Håkansson (1985) consider an axially symmetric 'generating' liquid drop type surface with sharp edges which is given by the shape function $\pi(\rho,z;\delta_1) \equiv \rho - f(z) = 0$ in the usual cylindrical coordinates (ρ,z,φ). The deformation trial densities are then defined after replacing the radius vector r by R + u in spherical densities (4.46) , that is,

$$\rho^{tr}(\rho,z,\varphi;\delta_1) = \rho_0[1 + \exp(u/\alpha)]^{-\gamma}, \tag{5.34}$$

where u is the distance normal to the sharp generating surface (compare with eqn (4.46) at $\rho_{1q} = 0$ and one kind of nucleons). In other words, it is assumed that the density varies across the generating surface in the same way as along the radial variable in the spherical case.

The variable u entering eqn (5.34) can be easily defined introducing a curvilinear coordinate system originally proposed by Strutinski and Tyapin (1964). Its coordinates (u,ϑ,φ) are found as shown in Figure 5.4. From any point in the (ρ,z)-plane, one constructs the normal onto the sharp surface. The z-coordinate of this point on the surface is the new coordinate ϑ, while the distance from the surface along the normal is the coordinate u taken to be positive outside and negative inside the sharp surface. Expressions connecting both (ρ,z,φ) and (u,ϑ,φ) are collected in Brack, Guet, and Håkansson (1985).

In the case of different neutron (q = n) and proton (q = p) distributions, eqn (5.34) restricts the number of variational parameters to six: ρ_{0q}, α_q, and γ_q (R_q being determined by the particle conservation conditions). From eqn (4.33) and eqn (5.34), one obtains the energy of the system as an algebraic function

$$E_{\delta_1}(\rho_{0n},\rho_{p0},\alpha_n,\alpha_p,\gamma_n,\gamma_p) = E_{ETF}[\rho_n^{tr},\rho_p^{tr}], \tag{5.35}$$

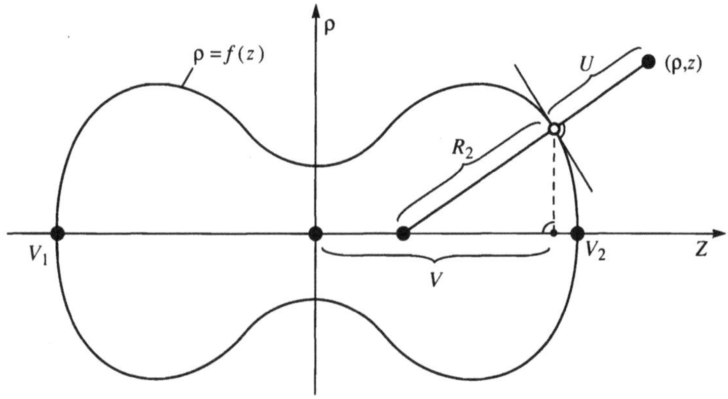

Figure 5.4. The curvilinear coordinate system originally proposed by Strutinski and Tyapin (1964). The curvilinear coordinates (u,v) are connected with the usual cylindrical coordinates (ρ,z) for an axially symmetric deformed nucleus with shape function $\rho = f(z)$ defining the sharp generating surface according to Brack, Guet and Håkansson (1985).

which should be minimized with respect to the variational parameters at certain values of the deformations δ_i.

In this kind of calculation the set of deformation parameters δ_i is defined through the shape function $\pi(\rho,z;\delta_i)$ which is represented by the so-called (c,h) family of fission shapes which is well known from various Strutinski shell-correction calculations (Damgaard *et al.* 1969, Brack *et al.* 1972). The scale factor c depends on the total length of the longer axis of the sharp surface (see Figure 5.4) which is simply equal to 2c. The second parameter h describes the variation of the thickness of the neck without changing the length of the nucleus. Some selected shapes connected with this (c,h) parameterization are shown in Figure 5.5. A different parameterization of eqn (5.35) has been used by Guet, Håkansson, and Brack (1980).

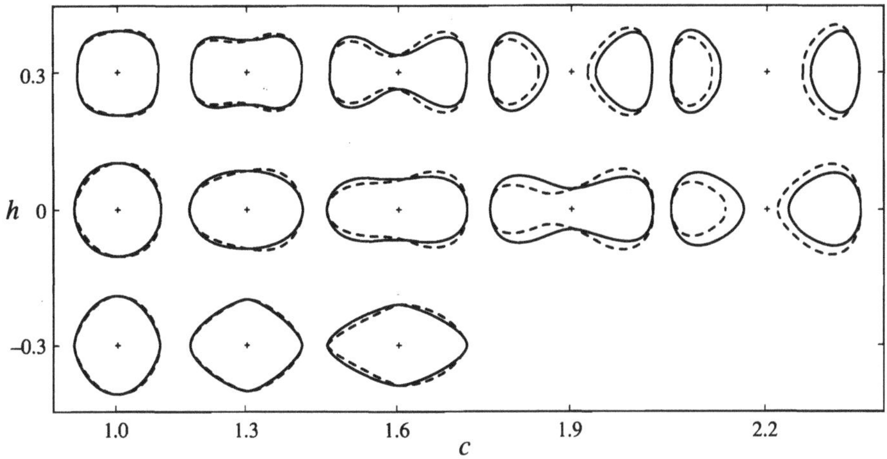

Figure 5.5. Some selected symmetric (solid lines) and asymmetric (dotted lines) shapes in the (c,h) parametrization of Brack *et al.* (1972).

Actual calculations

In order to test the relevance of the calculation procedure, Bartel *et al.* (1982) minimize the energy (5.35) at each value of the quadrupole moment Q with respect to density variational parameters as well as with respect to the parameters c and h used to define the (c,h) family of fission shapes. The result is shown in Figure 5.1. The ETF fission barriers (the dashed line) reproduce the self-consistent averaged energy $\tilde{E}_{HF}$ within approximately 1 MeV for all values of Q, after adjusting the energy at Q = 0 to eliminate the overbinding effect discussed in the previous section. This agreement provides a strong confirmation of the relevance of the ETF energy density functional for calculations of deformed energies in nuclei.

After confirming the relevance of the numerical procedure, many investigations of the nuclear fission processes and properties of deformed nuclei have appeared within the ETF model (Guet, Håkansson, and Brack 1980, Bartel *et al.* 1982, Brack, Guet, and Håkansson 1985, Tondeur *et al.* 1984). For example, the

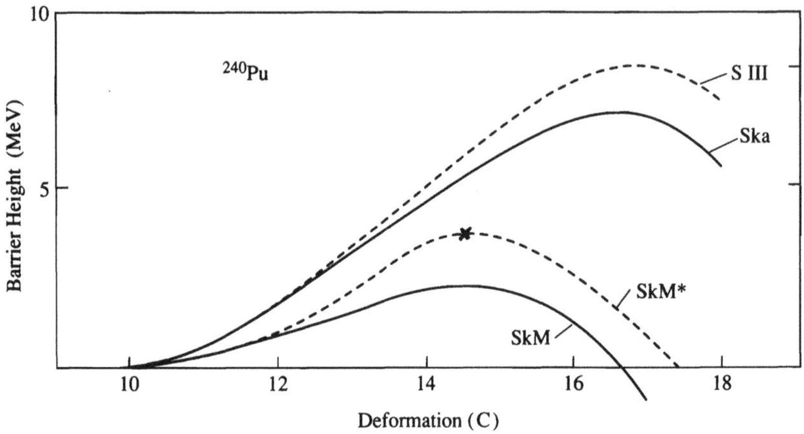

Figure 5.6. Fission barriers for the nucleus ^{240}Pu calculated with the extended Thomas-Fermi method using various Skyrme-type effective forces. The set SkM* is adjusted to reproduce the experimental liquid-drop model barrier height indicated by the cross (Brack, Guet, and Håkansson 1985).

importance of the term of fourth order in $\hbar$ was shown in the ETF energy density functional for obtaining the correct average deformation energies. Another interesting result is that the values of density variational parameters vary very little with deformation. For the force SkM*, for example, one would only make an error of $\cong$ 0.5 MeV in the barrier height if one used constant density parameters extracted at the nuclear ground-state point.

Among many applications to deformed nuclei, the ETF model has allowed the set SkM* of effective Skyrme-type forces to be adjusted to reproduce the empirical LDM barrier height for the nucleus ^{240}Pu while keeping intact the good description of other properties of nuclei. We present in Figure 5.6 the ETF fission barriers of ^{240}Pu calculated with various Skyrme-type forces in order to demonstrate this result. Recently, a full use of potential energy surfaces of triaxial nuclei evaluated by the self-consistent version of the ETF has been made by Delocle *et al.* (1989).

6. The Kohn–Sham theory of nuclei

6.1 Introduction

The aim of this chapter is to show that the approximate energy density functionals considered so far may be exploited within the larger context of density functional theory in order to derive some effective single-particle functions and energies of nuclei. The theoretical background we shall consider is the Kohn-Sham theory originally proposed for the description of the electronic ground state by Kohn and Sham (1965). They introduced a different method for evaluating the exact kinetic energy density functional $T[\rho]$ by separating it into a major part that is known, as for example in the case of a noninteracting system, and a minor correction which can be absorbed by the uncertainties of the approximate functionals. Hence, the minimization of the energy leads to a system of single-particle equations for determining the so-called Kohn-Sham orbitals and energies.

In nuclear physics, the theory of Kohn and Sham is not well known, although there are many calculations which could be viewed in its context, and this is one the goals of the present chapter.

6.2. General Kohn-Sham theory

Let us separate the exact energy density functional $E[\rho]$ into two parts

$$E[\rho] = T[\rho] + V[\rho], \tag{6.1}$$

where $T[\rho]$ is the exact kinetic energy density functional and everything else, including exchange and correlation energies, is included in $V[\rho]$.

As mentioned earlier, we do not know either $T[\rho]$ or $V[\rho]$ exactly. Formally, if we bypass this problem for the moment, we can vary eqn (6.1) with respect to the density $\rho(\mathbf{r})$ in order to obtain the Euler-Lagrange equation

$$\frac{\delta T[\rho]}{\delta \rho(\mathbf{r})} + \frac{\delta V[\rho]}{\delta \rho(\mathbf{r})} = \mu, \tag{6.2}$$

where the chemical potential μ as before is the Lagrange multiplier accounting for the conservation of particle number.

Equation (6.2) is not satisfactory for microscopic calculations because we do not have sufficiently accurate approximations to $T[\rho]$ which eventually reflect the microscopical picture of the system. This problem may be substantially alleviated following the exact but quite formal procedure proposed by Kohn and Sham (1965).

Outline of the theory

Suppose the system of interest contains N particles. Then one can introduce a set of N single-particle functions $\varphi_i(\mathbf{r})$ such that the exact local density $\rho(\mathbf{r})$ is expressed as

$$\rho(\mathbf{r}) = \sum_{i=1}^{N} |\varphi_i(\mathbf{r})|^2. \tag{6.3}$$

Harriman (1981) proves that the exact quantum-mechanical density $\rho(\mathbf{r}) \in D_N$ can be always represented as in eqn (6.3), that is, as a sum of squares of N single-particle wave functions.

Kohn and Sham (1965) express the kinetic energy density functional $T[\rho]$ in terms of the N functions $\varphi_i(\mathbf{r})$ which are the same as those entering eqn (6.3). The result is in either of the equivalent forms

$$T[\rho] = \frac{\hbar^2}{2m} \sum_{i=1}^{N} \int dr \, |\nabla\varphi_i(r)|^2 = - \frac{\hbar^2}{2m} \sum_{i=1}^{N} \int dr \, \varphi_i^*(r)\Delta\varphi_i(r). \qquad (6.4)$$

The expression (6.4) can be considered only as an approximation to $T[\rho]$ because the existence of a set of N single-particle functions simultaneously obeying both eqn (6.3) and eqn (6.4) is not yet proved. Moreover, there is no proof that eqn (6.4) holds for the exact $T[\rho]$. Formally, this complication is ignored within the theory of Kohn and Sham by assuming that the difference between the exact $T[\rho]$ and its approximation (6.4) is small and can be absorbed into the term $V[\rho]$. We suppose this has been already done in eqn (6.1) and do not alter the notation.

The Kohn–Sham idea is then to carry out the variation (6.2) of the functional (6.1) with eqn (6.4) by performing variations with respect to the orbitals $\varphi_i(\mathbf{r})$ entering eqn (6.3). This variation is quite straightforward and yields

$$\left[- \frac{\hbar^2}{2m} \nabla^2 + U(\mathbf{r};[\rho])\right] \varphi_i(\mathbf{r}) = \varepsilon_i \, \varphi_i(\mathbf{r}), \qquad (6.5)$$

where

$$U(\mathbf{r};[\rho]) = \frac{\delta V[\rho]}{\delta\rho(\mathbf{r})} \qquad (6.6)$$

is a single-particle potential which may be nonlocal in the general case. Hence, one obtains a system of Schrödinger-like equations for the single-particle functions $\varphi_i(\mathbf{r})$. These functions may be taken to be orthonormal because the operator in the brackets in eqn (6.5) is Hermitian.

After solving eqn (6.5) with eqn (6.6), known as Kohn–Sham single-particle equations, one obtains the Kohn–Sham orbitals $\varphi_i(\mathbf{r})$ and energies ε_i. The ground-state local density may be then constructed from eqn (6.3) in which the N orbitals $\varphi_i(\mathbf{r})$ associated with the smallest energies ε_i should be included.

Similarly, the total energy of the system may also be determined from eqn (6.1) and eqn (6.4).

Difficulties of the theory

The Kohn-Sham theory is exact as far as we are dealing with the exact expression of the term $V[\rho]$. Nevertheless, this theory is not free of difficulties.

First, the physical interpretation of the Kohn-Sham orbitals is not yet clear. They are introduced by eqn (6.5) which is like a HF equation. It must be solved self-consistently, since $U(\mathbf{r};[\rho])$ depends on $\rho(\mathbf{r})$ according to eqn (6.6). This is the reason why the resulting functions $\varphi_1(\mathbf{r})$ are often considered to be single-particle wave functions in the HF sense. However, this view as no real justification because no assumption has been introduced to represent the many-body wave function as a single Slater determinant.

Secondly, as mentioned above, there is no proof for the simultaneous use of both eqn (6.3) and eqn (6.4) which is also an open problem for the theory.

Thirdly, the physical meaning of the eigenvalues ε_1 of eqn (6.5), associated with the single-particle functions $\varphi_1(\rho)$, is also not clear. Koopmans' theorem, valid in the HF approximation, does not apply to the Kohn-Sham theory.

In spite of the complicated questions raised above, it does appear that the Kohn-Sham theory is capable of giving a reasonable account of single-particle structure at least in various domains of atomic, molecular and condensed matter physics, where the Kohn-Sham theory is widely used (Callaway and March 1984, Kohn and Vashishta 1983).

6.3 The nuclear Kohn-Sham theory

The Kohn-Sham theory does not depend on the particular form of the two-body interaction and should therefore also apply to nuclear systems. One can thus try to define a single-particle potential for atomic nuclei in terms of the Kohn-Sham definition (6.6) and the approximate energy density functionals obtained in nuclear density functional theory.

Earlier calculations

Earlier attempts in this direction have been made within the energy density formalism (EDF) considered in Section 3.2. Brückner, Lin, and Lombard (1969) and later Lombard (1973) have calculated a single-particle potential as a functional derivative of the potential energy part $V[\rho]$ entering the approximate functional (3.15). The resulting potential agrees with the definition (6.6) and is of the form

$$U_n(\mathbf{r};[\rho]) \equiv \frac{\delta V[\rho]}{\delta\rho_n(\mathbf{r})} = U_0(\mathbf{r};[\rho]) + U_\tau(\mathbf{r};[\rho]) + U_s(\mathbf{r};[\rho]) \qquad (6.7)$$

and

$$U_p(\mathbf{r};[\rho]) \equiv \frac{\delta V[\rho]}{\delta\rho_p(\mathbf{r})} = U_0(\mathbf{r};[\rho]) - U_\tau(\mathbf{r};[\rho]) + U_s(\mathbf{r};[\rho]) + \Phi_c(\mathbf{r}) \qquad (6.8)$$

for neutrons and protons, respectively. The explicit expressions for the volume (U_0), symmetry (U_τ) and inhomogeneity (U_s) terms are the following:

$$U_0(\mathbf{r};[\rho]) = \mathcal{V}(\rho,\alpha) + \rho\frac{\partial \mathcal{V}(\rho,\alpha)}{\partial\rho} = 2b_1\rho + \frac{7}{3}b_2\rho^{4/3} + \frac{8}{3}b_3\rho^{5/3}$$

$$+ (c_2\rho^{-2/3} + c_3\rho^{-1/3})(\rho_n - \rho_p)^2, \qquad (6.9)$$

$$U_\tau(\mathbf{r};[\rho]) = 2(c_1 + c_2\rho^{1/3} + c_3\rho^{2/3})(\rho_n - \rho_p), \qquad (6.10)$$

$$U_s(\mathbf{r};[\rho]) = -\,2\eta_0[1 - k\alpha^2]\Delta\rho, \qquad (6.11)$$

where $c_i = a_i b_i$ (i = 1,2,3) using the notations in eqn (3.21). In eqn (6.8), $\Phi_c(\mathbf{r})$ stands for the Coulomb potential

$$\Phi_c(\mathbf{r}) \equiv \frac{\delta E_{Coul}[\rho]}{\delta \rho_p(\mathbf{r})} = e^2 \int d\mathbf{r}' \frac{\rho_p(\mathbf{r}')}{|\mathbf{r}-\mathbf{r}'|} - e^2 (3/\pi)^{1/3} \rho_p^{1/3}(\mathbf{r}). \quad (6.12)$$

A TF-type spin-orbit potential is also added to eqn (6.7) and eqn (6.8),

$$U_q(r) = -\frac{\lambda}{2} \left(\frac{\hbar}{me}\right)^2 \left(\frac{1}{r} \frac{\partial}{\partial r} U_q(r;[\rho])\right), \quad (6.13)$$

where the strength parameter λ is the same for both neutrons (q = n) and protons (q = p). It is determined by fitting the spin-orbit splittings for levels near the Fermi surface.

After solving the Kohn-Sham equations (6.5) with the potentials just derived, one can compare the results of these calculations with the results following from the EDF at the same values of the parameters. In Figure 6.1, we show proton and mass density distributions calculated for some nuclei by Brückner, Lin, and Lombard (1969). It is seen that in contrast to the smooth EDF density distributions, local density shell oscillations appear in the results of the Kohn-Sham calculations as a consequence of the single-particle structure of eqn (6.3) and eqn (6.4).

In Table 6.1, the calculated Kohn-Sham energies ε_i for the nucleus ^{208}Pb are compared with the results obtained from the semi-empirical Saxon-Woods potential of Beiner (1970). The comparison shows that the Kohn-Sham results are sufficient for states close to the Fermi surface in nuclei.

The nuclear energy density method

The energy density method proposed by Beiner and Lombard (1974) is a further refinement of the Kohn-Sham calculations for nuclei. It consists in writing the total energy of any nucleus in a

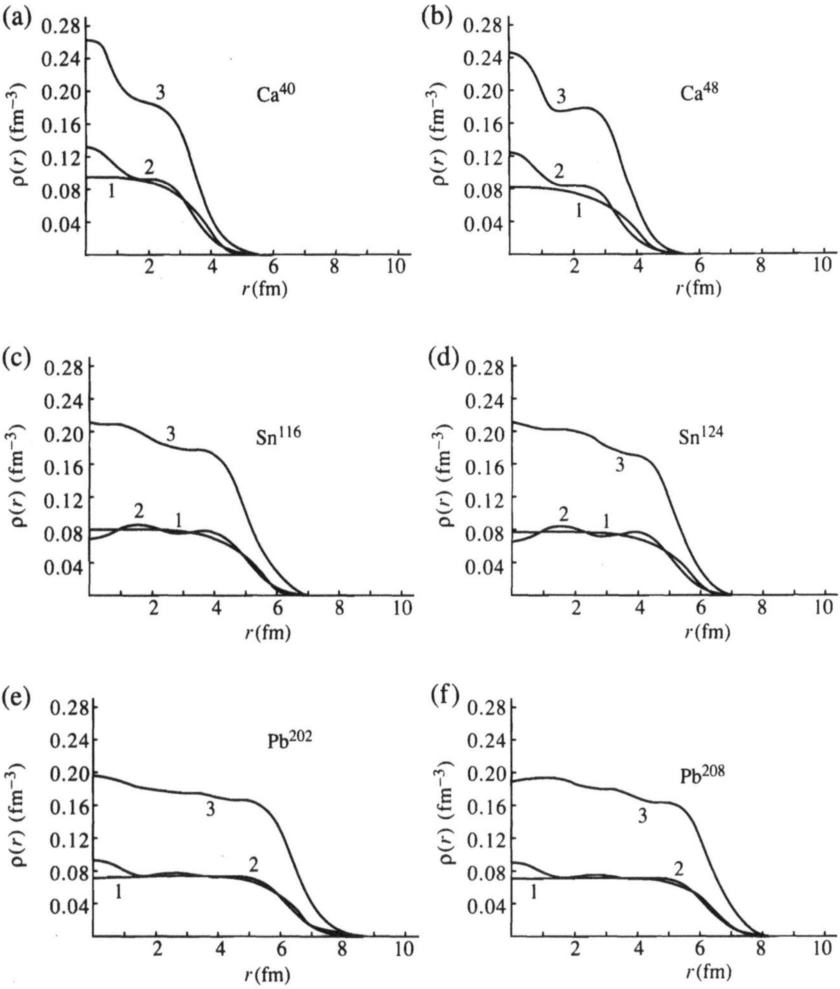

Figure 6.1. Proton and mass density distributions for some nuclei calculated by Brückner, Lin, and Lombard (1969). Curve 1 is the proton local density obtained from the energy density formalism as explained in Section 3.2. Curves 2 and 3 are respectively proton and mass density distributions obtained from Kohn–Sham like calculations.

Table 6.1. Kohn-Sham single-particle energies near Fermi surface in the nucleus ^{208}Pb calculated by Lombard (1973) with the parameters Set 3 from Table 3.1. Comparison is made with results derived in the generalized Saxon-Woods potential of Beiner (1970).

Neutrons			Protons		
States	Set 3	SW	States	Set 3	SW
$2g_{9/2}$	3.68	3.46	$2f_{7/2}$	5.0	4.16
$3p_{1/2}$	0.0	0.0	$3h_{9/2}$	4.79	3.25
$2f_{5/2}$	-0.73	-0.61	$2s_{1/2}$	0.0	0.0
$3p_{3/2}$	-0.99	-0.97	$3d_{3/2}$	-0.33	-0.55
$1i_{13/2}$	-1.11	-1.61	$1h_{11/2}$	0.12	-1.45
$2f_{7/2}$	-3.32	-3.15	$2d_{5/2}$	-1.93	-2.15
$1h_{9/2}$	-3.09	-3.49	$1g_{7/2}$	-2.73	-4.10
$3s_{1/2}$	-8.32	-7.94	$3g_{9/2}$	-5.97	-7.42

functional form, the unknown variables again being a set of single-particle orbitals. The functional in itself is constructed in a close connection with the nuclear matter results and contains as much as twelve adjustable parameters. They are determined from the condition of the closest agreement of the results with experimental data for binding energies, single-particle spectra in the vicinity of the Fermi surface and charge rms radii. The pairing correlations are also taken into account by means of a simple effective interaction and the usual BCS approximation.

A system of equations similar to the HF(+BCS) one is then obtained after minimizing the ground-state energy functional with respect to the single-particle wave functions and occupation numbers. In this sense, the energy density method is equivalent to the parameterization of the interaction when a zero range interaction like Skyrme's force is used. Being Kohn-Sham type theory, however, the energy density method cannot be derived from a simple two-body force.

The energy density method has been successfully applied to describe the masses of spherical nuclei (Beiner, Lombard, and Mass 1976), nuclear shell structure effects (Beiner and Lombard 1974), and to predict ground-state properties of unstable nuclei (Beiner, Lombard, and Mass 1975) and low energy collective states (Lombard and Mass 1986).

Tondeur (1978a,b) further simplifies the energy density method by constructing a rather simple functional

$$E_{Tond} = \int dr \left\{ \frac{\hbar^2}{2m}\tau + a\rho^2 + b\rho^{\gamma} + c\rho^{-1/3}(\rho_n + \rho_p)^2 + \eta(\nabla\rho)^2 + dJ.\nabla\rho \right\}, \quad (6.14)$$

which contains only six adjustable parameters. The associated system of self-consistent equations is of the form

$$\left\{ -\frac{\hbar^2}{2m}\nabla^2 + U_q(r) - iW_q(r) \,.\, (\nabla x\sigma) \right\}\varphi_i(r) = \varepsilon_i\varphi_i(r), \quad (6.15)$$

with

$$U_q(r) = 2a\rho + b\gamma\rho^{\gamma-1} - \frac{1}{3}c\left(\frac{\rho_n - \rho_p}{\rho^{1/3}}\right)^2 \pm 2c(\rho_n - \rho_p)\rho^{-1/3}\rho_q - 2\eta\Delta\rho - d\nabla.J \quad (6.16)$$

where the +/- sign stands for neutrons/protons and

$$\mathbf{W}_q(\mathbf{r}) = d\ \nabla\rho. \quad (6.17)$$

Although the functional (6.14) is quite simple, it proves to give a better prescription of nuclear masses than the other mass formulae having a similar number of parameters. For example, a self-consistent study of the ground-state and the deformation energies of 108 deformed nuclei has shown that the functional (6.14) enables calculations with sufficient accuracy to be made (Tondeur 1978b).

The Kohn-Sham theory and Skyrme forces in nuclei

Neither simple two-body forces in nuclei nor a single Slater determinant wave-function representation have been used in connection with the energy density method. Nevertheless, we have seen that the resulting energy density functionals are rather close to the ones following from the Skyrme-HF theory. The reason is simply that the HF theory with Skyrme-type effective forces is nothing but a nuclear Kohn-Sham theory, extended for spin-orbit and effective mass contributions (Brack 1985b). Hence, one can realize that all kind of correlations could be included in principle in what formally presents itself as the Skyrme-HF ground-state energies and densities. In particular, this explains the spectacular success of the Skyrme-type effective forces in nuclei.

Of course, it is hard to say a priori which correlations and how much of them are taken into account when fitting the parameters of the effective interaction to the experimental data. It is important, however, that the relationship between the Kohn-Sham and Skyrme-HF theories in nuclei gives us confidence in the attempts to built up a good 'multi-purpose' effective force in nuclei.

6.4 Expectation value method

In order to complete the use of approximate energy density functionals in calculations of single-particle properties of nuclei, we consider the so-called expectation value method originally proposed by Ko *et al.* (1974) as an approximation to the constrained HF method in nuclei.

Constrained Hartree-Fock method

It is appropriate here to say a few words about the nuclear constrained Hartree-Fock (CHF) method which is used in the cases when one is interested in other points on the energy surface different from the ground-state HF minimum. In all cases, one is asking for a single Slater determinant $\Phi_Q \equiv \Phi(\mathbf{r}_1,\ldots,\mathbf{r}_N)$ which minimizes the total energy under the constraint that a certain single-particle operator $\hat{Q}$ has a fixed expectation value

$$Q \equiv \langle\hat{Q}\rangle = \langle\Phi_Q|\hat{Q}|\Phi_Q\rangle. \tag{6.18}$$

For different projection techniques, the operator $\hat{Q}$ may be chosen as a particle number operator $\hat{N}$, the total nuclear momentum operator $\hat{P}$ or the angular momentum operator $\hat{J}_x$ (see, for example, Eisenberg and Greiner 1972). When investigating fission processes or the energy surface in the vicinity of the ground state, one could use as constraints the quadrupole deformation $q = \langle\hat{Q}_{20}\rangle$ or the hexadecapole deformation $h = \langle\hat{Q}_{40}\rangle$. Other constraints may be used as a first step towards a derivation of nuclear collective Hamiltonians (see Chapters 8 and 9).

The simplest method for solving the CHF problem consists in adding the condition (6.18) with a Lagrange multiplier λ to the nuclear Hamiltonian $\hat{H}$ and minimizing

$$E_Q = \langle\Phi_Q|\hat{H}|\Phi_Q\rangle - \lambda\,(Q - \langle\hat{Q}\rangle) \tag{6.19}$$

with respect to the single-particle wave functions at a given value of λ. Afterwards, the Lagrange multiplier λ is determined by the condition (6.18). At the end of this self-consistent CHF procedure, called variation with a linear constraint, one obtains

$$\lambda = \frac{d\langle\hat{H}\rangle}{d\lambda}. \tag{6.20}$$

Of course, other (in general non-linear) functions $F(Q, \langle\hat{Q}\rangle)$ may also be used instead of the second term in eqn (6.19). For instance, Flocard *et al.* (1973) apply a quadratic constraint $F(Q, \langle\hat{Q}\rangle) = (Q-\langle\hat{Q}\rangle)^2$ in calculating nuclear deformation energy in the neighbourhood of a saddle point on the fission path.

In the particular case of the constrained Skyrme-HF theory, the appropriate equations to be solved are of the form

$$\left\{-\nabla\frac{\hbar^2}{2m_q^*(\mathbf{r})}.\nabla+U_q(\mathbf{r})-i\mathbf{W}_q(\mathbf{r}).(\nabla\times\sigma)-\lambda\frac{\delta F(Q,\langle\hat{Q}\rangle)}{\delta\varphi_1(\mathbf{r})}\right\}\varphi_1(\mathbf{r}) = \varepsilon_1\varphi_1(\mathbf{r}), \quad (6.21)$$

where a general constrained function $F(Q, \langle\hat{Q}\rangle)$ is used (compare with eqn (1.151)). The solution of eqn (6.21) determines the expectation value

$$E_{CHF} = \langle\Phi_Q|\hat{H}|\Phi_Q\rangle = \int d\mathbf{r}\ H(\rho_{CHF},\tau_{CHF},\mathbf{J}_{CHF}), \quad (6.22)$$

as a minimum of the energy on the hypersurface defined by constrained conditions (in the case of more than one constraint operators) assumed.

Phenomenological realization of the expectation value method

The expectation value method (EVM) is suggested as an appropriate approximation to the much more involved CHF calculations. The first step consists in replacing the quantities $M(\mathbf{r}) = 1/m^*(\mathbf{r})$, $U(\mathbf{r})$ and $\mathbf{W}(\mathbf{r})$ entering eqn (6.21) by some average fields for which a fairly good understanding is developed in many phenomenological shell model calculations.

Ko *et al.* (1974) have adopted to the problem the substitution

$$M(\mathbf{r}) \longleftrightarrow M'(\mathbf{r}) = M_0\left[1 + \exp(\ell(\mathbf{r},R_m)/b_m)\right]^{-1} + 1$$

$$U(\mathbf{r}) \longleftrightarrow U'(\mathbf{r}) = U_0\left[1 + \exp(\ell(\mathbf{r},R_u)/b_u)\right]^{-1} + \delta_{qp}\Phi_c(\mathbf{r}),$$

$$\mathbf{W}(\mathbf{r}) \longleftrightarrow \mathbf{W}'(\mathbf{r}) = X_0\ \nabla\left[1 + \exp(\ell(\mathbf{r},R_s)/b_s)\right]^{-1}, \quad (6.23)$$

applied with different parameters for neutrons and protons, separately. The generalized Woods-Saxon form entering eqn (6.23) is of the type of the deformed phenomenological potential proposed by Damgaard *et al.* (1969) and used in many shell- correction calculations (Brack *et al.* 1972). The shape function $\ell(\mathbf{r},\mathbf{R})$ ensures a deformed diffuse potential well in terms of the (c,h) family of fission shapes considered earlier in Section 5.4. Altogether, there are about 21 phenomenological parameters in eqn (6.23) if one includes there the three parameters characterizing the charge densities entering Coulomb potential $\phi_c(\mathbf{r})$.

As a second step, the phenomenological 'potentials' (6.23) are used to define, similarly to eqn (6.21), the phenomenological single-particle problem

$$\left\{-\nabla M'(\mathbf{r}) \,.\, \nabla + U'(\mathbf{r}) - iW'(\mathbf{r}) \,.\, (\nabla \times \sigma)\right\}\varphi_i(\mathbf{r}) = \varepsilon_i\varphi_i(\mathbf{r}), \qquad (6.24)$$

which is then solved in the configuration space of deformed harmonic oscillator functions. The solution of eqn (6.24) defines a new Slater determinant wave function Φ_{EVM}, the three relevant densities $\rho_{EVM}(\mathbf{r})$, $\tau_{EVM}(\mathbf{r})$ and $J_{EVM}(\mathbf{r})$ as well as the EVM energy

$$E_{EVM} = \langle\Phi_{EVM}|\hat{H}|\Phi_{EVM}\rangle = \int d\mathbf{r}\; H(\rho_{EVM},\tau_{EVM},J_{EVM}). \qquad (6.25)$$

It should be noticed that the energy (6.25) depends on the deformation parameters through the shape function $\ell(\mathbf{r},\mathbf{R})$ which in fact simulates the constraint conditions of the CHF problem.

Actually, the EVM energy (6.25) is expected to give a reasonable approximation to the nuclear deformation energy, although the phenomenological 'potentials' (6.23) may not be realistic in the sense that they may deviate significantly from the self-consistent CHF potentials. The results found in this way were only partially successful (see Brack 1977). Whereas the oscillating part of the deformation energy E_{EVM} was reasonably well reproduced, its averaged part $\tilde{E}_{EVM}$ increased too much at large deformations.

Connection with the density functional theory

The EVM has been further improved by Bohigas *et al.* (1976) and Chu, Jennings and Brack (1977) who combine it with approximate energy density functionals derived in the ITF and ETF models, respectively. In the framework of the ETF model, for instance, the improvement of the EVM consists in performing ETF calculations as described in Section 4.4 and then replacing the variational ETF fields M_{ETF}, U_{ETF} and $\mathbf{W}_{ETF}$ instead of the phenomenological ones M', U' and $\mathbf{W}'$ in eqn (6.24). In this version, the EVM consists in solving the resulting equation which is nothing more than a single HF iteration whose starting fields are taken as the fields generated by the minimization of the ETF energy density functional for the same Skyrme-type force as that entering eqn (6.25).

Brack, Guet, and Håkansson (1985) have shown that within the ETF choice of the starting fields, the convergence of the CHF iterative procedure is so rapid that one iteration (the EVM) suffices to get within 3 to 5 MeV of the final HF energy. Dutta *et al.* (1986a) have found that only three iterations suffice to achieve complete convergence of the lengthy CHF procedure.

Earlier, in the fourth column of Table 4.1, we showed some ground-state results for the energies E_{EVM}. They are smaller than the HF energies E_{HF} by less than approximately 2 MeV in heavy nuclei. It is also seen that the EVM automatically corrects the overbinding effect caused by the violation of the variational principle in the ETF method.

As a second example, we show in Figure 6.2 the results of Bohigas *et al.* (1976) derived for local densities and for kinetic energy densities calculated in the ITF model, the EVM and the HF theory. It is seen that in general the ITF proton and neutron densities compare rather well with the HF ones. The ITF densities, however, have a sharper surface and give no shell-density oscillations as discussed earlier. At the same time, the surface of the EVM densities are almost identical with the HF one including the shell-density oscillations. They are at the correct

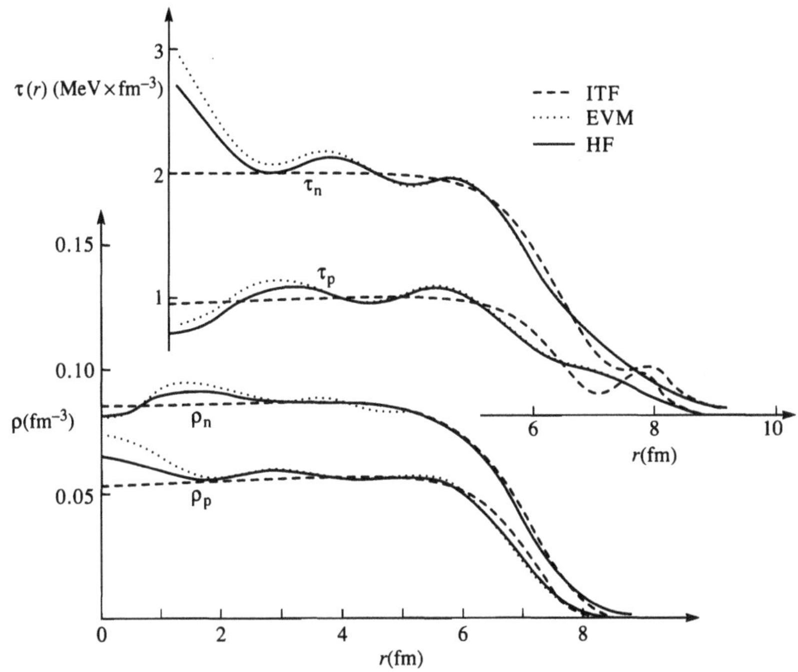

Figure 6.2. Comparison of the kinetic energy density $\tau(r)$ and the local density $\rho(r)$ for neutrons and protons in the nucleus ^{208}Pb calculated within the Hartree-Fock theory (solid line), the improved Thomas-Fermi method (dotted line) and the expectation value method (dashed line) with SVI effective forces (Bohigas *et al.* 1976).

place, although small discrepancies in their amplitudes are present.

The above examples demonstrate that the EVM, at the cost of essentially one microscopical HF step beyond the semiclassical ETF calculation, is a rather accurate approximation to the HF ground-state theory. The EVM also provides a quite accurate approximation to microscopic CHF calculations for the deformation energies of nuclei including shell effects (Tondeur 1980, Tondeur *et al.* 1987, and Bartel *et al.* 1982), although Dutta *et al.* (1986a) noticed that the combination of the ETF model with the shell correction method seems to be a still more adequate technique.

7. Rigorous energy density functional approach

The TF approximation and its phenomenological or semiclassical corrections have been the main ingredients entering the approximate energy density functionals considered so far. It has been shown that all these functionals approximate the smooth part of the ground-state energy only and give no possibility of describing any microscopic properties of nuclei. In order to incorporate nuclear shell effects, one has to add the oscillating part of the energy in a perturbative way by using the shell correction method (Chapter 5) or to solve some self-consistent equations of the Kohn-Sham type (Chapter 6). Both approaches for incorporating of shell effects in nuclei, however, go beyond the main concept of the density functional theory, namely to operate with the local density only.

After the Hohenberg-Kohn theorem is well established (Chapter 2), the density functional theory needs a constructive way of building up approximate functionals. Such a constructive way will be very useful and should give some possibilities for the theory to incorporate nuclear shell effects or even to go beyond the HF approximation. The semiclassical $\hbar$-expansion considered in Chapter 4 was not a possible way to do that. In this chapter we concentrate on such a constructive way of building up approximate energy density functionals which include nuclear shell effects and allow us to describe the averaged as well as microscopic properties of nuclei only in terms of the nuclear local density distribution.

7.1 Exact statements of the approach

We have considered in Section 2.4 a method which makes essential use of the local-scale point transformations of the three-dimensional vector space R^3. It has been demonstrated that a set of invertible transformations of this type forms a local-scale point transformation group $\mathcal{F}$ whose unitary representation $\mathcal{U}_{\mathcal{F}}$ in the Hilbert space $\mathcal{H}^N$ produces the corresponding local-scale point transformation of the N-particle wave functions belonging to $\mathcal{H}^N$. Thus, the Hilbert space $\mathcal{H}^N$ splits into disjoint orbits $\mathcal{O} \subset \mathcal{H}^N$ with all functions in a given orbit being related to local-scale point transformations.

Existence of a set of density functionals

As demonstrated in Section 2.4, for a given prototypical function $\bar{\Psi} \in \mathcal{H}^N$, that is, within the orbit $\mathcal{O} \subset \mathcal{H}^N$ produced by $\bar{\Psi}$ according to eqn (2.47), any arbitrary many-particle wave function belonging to $\mathcal{O} \subset \mathcal{H}^N$ can be represented by means of a given local density distribution $\rho(\mathbf{r}) \in \bar{D}_N$ as

$$\Psi[\rho] \equiv \Psi(\mathbf{r}_1,\ldots,\mathbf{r}_N;[\rho]) = \mathcal{U}_{f[\rho]}(\mathbf{r}_1,\ldots,\mathbf{r}_N)\, \bar{\Psi}(\mathbf{r}_1,\ldots,\mathbf{r}_N)$$

$$= \left\{\prod_{i=1}^{N} D(\mathbf{f}_{[\rho]}(\mathbf{r}_i);\mathbf{r}_i)\right\}^{1/2} \bar{\Psi}(\mathbf{f}_{[\rho]}(\mathbf{r}_1),\ldots,\mathbf{f}_{[\rho]}(\mathbf{r}_N)), \quad (7.1)$$

where $\mathbf{f}_{[\rho]} \equiv \mathbf{r}_0 f_{[\rho]}(\mathbf{r})$ is the implicit solution of eqn (2.59) at a given density $\rho(\mathbf{r}) \in \bar{D}_N$. Indeed, for any given density $\rho(\mathbf{r}) \in \bar{D}_N$, the N-particle wave function (7.1) gives the same local density

$$\rho(\mathbf{r}) = N \int |\Psi(\mathbf{r},\mathbf{r}_2,\ldots,\mathbf{r}_N;[\rho])|^2\, d\mathbf{r}_2 \ldots d\mathbf{r}_N, \quad (7.2)$$

because the local scale transformation function $f_{[\rho]} \equiv f_{[\rho]}(\mathbf{r})$ is the implicit solution of eqn (2.59). Formally, this solution may be represented into the form (Kryachko, Petkov, and Stoitsov 1987a)

$$f_{[\rho]}(\mathbf{r}) = \left\{ 3 \int_0^r dr' \frac{\rho(r', r_0)\, r'^2}{\bar{\rho}(f_{[\rho]}(r', r_0), r_0)} \right\}^{1/3}, \tag{7.3}$$

where $\bar{\rho}(\mathbf{r})$ is the model local density distribution associated with the prototypical function $\bar{\Psi} \subset \mathcal{H}^N$, eqn (2.57).

Further, the many-body wave function (7.1) transforms the mean value of the Hamiltonian $\hat{H}$ of the N-particle system into an energy density functional

$$E\,[\rho] \equiv \langle \Psi[\rho] | \hat{H} | \Psi[\rho] \rangle = T[\rho] + V[\rho], \tag{7.4}$$

where

$$T[\rho] = \langle \Psi[\rho] | \hat{T} | \Psi[\rho] \rangle, \tag{7.5}$$

and

$$V[\rho] = \langle \Psi[\rho] | \hat{V} | \Psi[\rho] \rangle, \tag{7.6}$$

are the kinetic and potential energy density functionals, respectively.

Note that the functional (7.4) is exactly the mean value $E[\Psi]$, eqn (1.46), if one restricts the whole space $\mathcal{H}^N$ onto the different orbits $\mathcal{O} \subset \mathcal{H}^N$. The existence of an energy density functional in any given orbit $\mathcal{O} \subset \mathcal{H}^N$ is thus proved. In fact, one has a set of energy density functionals constructed in the Hilbert space $\mathcal{H}^N$ corresponding to different orbits $\mathcal{O} \subset \mathcal{H}^N$.

Constructive proof of the Hohenberg-Kohn theorem

We shall prove now that the ground-state energy of an N-particle system with Hamiltonian $\hat{H}$ can be represented as an energy density functional of the type of eqn (7.4) which attains its absolute minimum E_0 when evaluated with the exact ground-state local density distribution $\rho_0(\mathbf{r})$. For this purpose, it is sufficient to consider eqn (7.4) within the eigenorbit $\mathcal{O}_H \subset \mathcal{H}^N$ which by definition contains the exact ground-state wave function $\Psi_0 \in \mathcal{H}^N$ (see Section 2.4).

First, we prove that in $\mathcal{O}_H \subset \mathcal{H}^N$ the energy functional (7.4) is a unique functional of $\rho(\mathbf{r}) \in \overline{D}_N$. The proof proceeds by *reductio ad absurdum*. Suppose there exist in $\overline{D}_N$ two different local densities

$$\rho_1(\mathbf{r}) \neq \rho_2(\mathbf{r}) \tag{7.7}$$

that give the same values of the energy density functional (7.4) defined in $\mathcal{O}_H \subset \mathcal{H}^N$

$$E[\rho_1] = E[\rho_2]. \tag{7.8}$$

Without loss of generality, since $\Psi_0 \in \mathcal{O}_H$, we assume that the prototypical function $\overline{\Psi} \in \mathcal{O}_H$ is defined as

$$\overline{\Psi}(\mathbf{r}_1,\ldots,\mathbf{r}_N) = \mathcal{U}_{f^{-1}[\rho_2]}\,\Psi_0(\mathbf{r}_1,\ldots,\mathbf{r}_N) = \mathcal{U}^{+}_{f[\rho_2]}\,\Psi_0, \tag{7.9}$$

where $f^{-1}[\rho_2]$ is the local-scale transformation function inverse to $f[\rho_2]$. The inverse function $f^{-1}[\rho_2]$ does exist since $\mathbf{f}[\rho_2] \in \mathcal{F}$. Then, according to eqn (7.1) and eqn (7.9), the wave functions $\Psi[\rho_1]$ and $\Psi[\rho_2]$ are given by the equations (see also eqn (2.37) and eqn (2.45))

$$\Psi[\rho_1] = \mathcal{U}_{f[\rho_1]}\overline{\Psi} = \mathcal{U}_{f[\rho_1]}\mathcal{U}^{+}_{f[\rho_2]}\Psi_0 = \mathcal{U}_{f[\rho_1]f^{-1}[\rho_2]}\Psi_0 \tag{7.10}$$

and

$$\Psi[\rho_2] = \mathcal{U}_{f[\rho_2]}\overline{\Psi} = \mathcal{U}_{f[\rho_2]}\mathcal{U}^{+}_{f[\rho_2]}\Psi_0 = \Psi_0, \tag{7.11}$$

respectively.

The wave function $\Psi[\rho_2]$ is strictly equal to the exact ground-state wave function Ψ_0 which is a consequence of the particular choice of the prototypical function (7.9). Function $\Psi[\rho_1]$, however, differs from Ψ_0 because the element $f[\rho_1] \cdot f^{-1}[\rho_2] \in \mathcal{F}$ is not equivalent to the identity transformation $f_e \in \mathcal{F}$ since eqn (7.7) holds. Therefore, eqn (7.10) and eqn (7.11) show that the wave functions $\Psi[\rho_1]$ and $\Psi[\rho_2]$ are different, that is, $\Psi[\rho_1] \neq \Psi[\rho_2]$. From the property (1.42) of the nondegenerate ground state Ψ_0 we then obtain

$$\langle\Psi[\rho_1]|\hat{H}|\Psi[\rho_1]\rangle > \langle\Psi[\rho_2]|\hat{H}|\Psi[\rho_2]\rangle, \tag{7.12}$$

since $\Psi[\rho_2] \equiv \Psi_0$ and therefore $\langle\Psi[\rho_2]|\hat{H}|\Psi[\rho_2]\rangle = E_0$. In exactly the same way, interchanging the indices 1 and 2 in the above quantities, one finds

$$\langle\Psi[\rho_1]|\hat{H}|\Psi[\rho_1]\rangle < \langle\Psi[\rho_2]|\hat{H}|\Psi[\rho_2]\rangle, \tag{7.13}$$

which obviously contradicts eqn (7.12). Hence, $\rho_1(\mathbf{r})$ must be identically equal to $\rho_2(\mathbf{r})$ and, therefore, $E[\rho]$ is an unique functional of $\rho(\mathbf{r})$.

Secondly, to establish the minimal property of the functional $E[\rho]$, we simply underline the fact that in general the wave function $\Psi[\rho]$ differs from the exact one Ψ_0 and therefore, taking into account eqn (1.42), we have

$$E[\rho] \geq E_0 \tag{7.14}$$

for all $\rho(\mathbf{r}) \in \overline{D}_N$.

Finally, the attainment of the absolute minimum $E_0 = E[\rho_0]$ follows from the fact that $E[\rho]$ is constructed over the eigenorbit $\mathcal{O}_H \subset \mathcal{H}^N$ containing the exact function $\Psi_0 \in \mathcal{H}^N$. We have thus proved a statement which is close, by its context, to the Hohenberg-Kohn theorem (Section 2.1).

Petkov, Stoitsov, and Kryachko (1986a) and Kryachko, Petkov, and Stoitsov (1987a) have shown that the above analogue of the Hohenberg-Kohn theorem is more general in the sense that it avoids the criticism of the original theorem discussed earlier in Section 2.1. In particular, the exact functional $E[\rho]$, eqn (7.4) within the eigenorbit $\mathcal{O}_H \subset \mathcal{H}^N$, is free of any N- and υ-representability problems. Moreover, in view of the paper of Valone (1980a,b), we recover the one-to-one correspondence between the N-particle wave function $\Psi[\rho]$ and the local density $\rho(\mathbf{r})$ within the orbit $\mathcal{O}_H$. The exact functional $E[\rho]$ also obeys the hyper-virial theorem (Petkov and Stoitsov 1983a) as well as the virial theorem (Kryachko, Petkov, and Stoitsov 1987a).

Constructive two-step variational procedure

As shown in Section 2.4, the full Hilbert space $\mathcal{H}^N$ is a union of the disjoint orbits $\mathcal{O} \subset \mathcal{H}^N$. Therefore, the investigation of the functional $E[\Psi]$ for its absolute minimum E_0 over the whole space $\mathcal{H}^N$ is equivalent, in the context of eqn (7.4) and the statement proved above, to determining the minima of the functional (7.4) defined over all orbits $\mathcal{O} \subset \mathcal{H}^N$ and further comparing them, that is,

$$E_0 = \min_{\mathcal{O}\subset\mathcal{H}^N}\left\{ \min_{\rho\in \overline{D}_N} \left\{E[\rho]\right\}\right\} . \tag{7.15}$$

In other words, the structure of the Hilbert space $\mathcal{H}^N$ is enlarged in the framework of the approach presented and the Schrödinger variational principle (Section 1.2) now proceeds in two steps: (1) minimizing the functional $E[\rho]$ with respect to $\rho(\mathbf{r})$ into orbits $\mathcal{O} \subset \mathcal{H}^N$ and then (2) comparing these minima in order to find the absolute one E_0. This two-step variational procedure closes formally the rigorous formulation of the ground-state problem only in terms of the local density $\rho(\mathbf{r})$ and provides a constructive approach to the nuclear density functional theory.

Of course, one cannot perform the above two-step variational procedure in practice. It is not possible to investigate all orbits $\mathcal{O}$ of the Hilbert space $\mathcal{H}^N$. Such an investigation leading eventually to the exact energy density functional would be equivalent to the exact solution of the many-body problem. However, the energy density functional (7.4) is valid within each definite trial orbit $\mathcal{O} \subset \mathcal{H}^N$, where a constructive procedure for handling the explicit form of the functional is obvious.

Rigorous energy density approach

A general idea of a constructive density functional technique is seen when carrying out only the first step of the two-step variational procedure proposed above. Indeed, when the prototypical function $\bar{\Psi} \in \mathcal{H}^N$ is given as a known model wave function (say the harmonic-oscillator shell model wave function, for instance), the energy density functional (7.4) is well defined and one can obtain the ground-state local density distribution as a solution of the associated Euler-Lagrange equation

$$\delta E[\rho]/\delta\rho(\mathbf{r}) = \mu. \tag{7.16}$$

It could be explicitly derived only taking into account eqn (7.3) whose solution $f[\rho]$ appears in the energy density functional (7.4). Then, the solution of eqn (7.16) defines the approximate ground-state density $\rho(\mathbf{r})$, wave function $\Psi[\rho]$, eqn (7.1), and through it the other ground-state properties of interest can be calculated as well.

We denote the above approximate variational approach to nuclear density functional theory as the rigorous energy density (RED) approach because it leads in principle to the exact results within the eigenorbit $\mathcal{O}_H \subset \mathcal{H}^N$.

One can understand that the RED approach corresponds in fact to a minimization of the functional $E[\Psi]$ within the trial set $\mathcal{O} \subset \mathcal{H}^N$ and therefore we have a particular case of the variational method considered in Section 1.6. According to the variational method (see eqn (1.118)), one hence establishes the inequality

$$\bar{E} \geq E[\rho] \geq E_0, \tag{7.17}$$

where $\bar{E} = \langle\bar{\Psi}|\hat{H}|\bar{\Psi}\rangle$ is the ground-state energy determined by the model function $\bar{\Psi} \in \mathcal{O}$, $E[\rho]$ is the lowest minimum of the functional (7.4) defined on $\mathcal{O} \subset \mathcal{H}^N$ and E_0 is the exact ground-state energy of the system.

According to inequality (7.17) the accuracy of the RED approach in any orbit $\mathcal{O}$ of $\mathcal{H}^N$ different from the eigenorbit $\mathcal{O}_H$ depends strongly on the choice of the model wave function $\bar{\Psi} \in \mathcal{H}^N$. It must be sufficiently simple for actual calculations to be made. At the same time, it should allow the trial wave function $\Psi[\rho] \in \mathcal{O}$ to vary in a sufficiently large or well chosen region for the obtained solution $\Psi[\rho]$ to be near to the exact function $\Psi_0 \in \mathcal{H}^N$. The inequality (7.17) also prevents the RED approach from violating the variational condition and therefore from some overbinding effects which were a characteristic feature of the semiclassical approach.

Possible calculations beyond the HF theory

The most important feature of the RED approach is the extreme simplification of the many-body problem that is mainly achieved by the preliminary choice of the prototypical (or model) wave function $\bar{\Psi} \in \mathcal{H}^N$. For example, when the model wave function is a single Slater determinant built up from known single-particle functions (say harmonic oscillator ones), one can reduce the system of N integro-differential equations of the HF theory to a single equation of the type of eqn (7.16) with respect to $\rho(\mathbf{r})$.

In the same manner, the approach allows one also to construct approximate energy density functionals for calculations beyond the HF approximation. In this case, one has to use model wave functions of Hartree-Fock-Bogolubov, RPA or Hill-Wheeler type. We note that interactions with a strong repulsive core are not ruled out because one can use Jastrow-type model wave functions as well.

7.2 Explicit expressions within the approach

We analyse now the problem of determining the explicit form of the functional $E[\rho]$ in any trial orbit $\mathcal{O} \subset \mathcal{H}^N$ which is of crucial importance for the actual application of the RED approach. For this purpose, it is sufficient to express the first and second order density matrices by means of the local density $\rho(\mathbf{r})$. For simplicity we consider spinless representations given by eqn (1.94) and eqn (1.103), but a generalization to spin- and isospin-dependent cases is straightforward and appears to be trivial.

Density-dependent one- and two-body density matrices

Let us consider a given model wave function $\bar{\Psi} \in \mathcal{H}^N$. It defines both the first and second order model density matrices

$$\bar{\rho}^{(1)} \equiv \bar{\rho}^{(1)}(\mathbf{r};\mathbf{r}') = N\int \bar{\Psi}^*(\mathbf{r}',\mathbf{r}_2,\ldots,\mathbf{r}_N)\bar{\Psi}(\mathbf{r},\mathbf{r}_2,\ldots,\mathbf{r}_N)d\mathbf{r}_2\ldots d\mathbf{r}_N$$

$$= \sum_{\alpha,\beta} n_{\beta\alpha}\, \bar{\varphi}^*_\alpha(\mathbf{r})\, \bar{\varphi}_\beta(\mathbf{r}') \tag{7.18}$$

and

$$\bar{\Gamma} \equiv \bar{\Gamma}(\mathbf{r}_1,\mathbf{r}_2;\mathbf{r}'_1,\mathbf{r}'_2) = \frac{1}{2}N(N-1)\int \bar{\Psi}^*(\mathbf{r}'_1,\mathbf{r}'_2,\ldots,\mathbf{r}_N)\bar{\Psi}(\mathbf{r}_1,\mathbf{r}_2,\ldots,\mathbf{r}_N)d\mathbf{r}_3\ldots d\mathbf{r}_N$$

$$= \sum_{\alpha,\beta,\gamma,\delta} \Gamma_{\alpha\beta\gamma\delta}\, \bar{\varphi}^*_\alpha(\mathbf{r}_1)\, \bar{\varphi}^*_\beta(\mathbf{r}_2)\, \bar{\varphi}_\gamma(\mathbf{r}'_1)\, \bar{\varphi}_\delta(\mathbf{r}'_2), \tag{7.19}$$

respectively. The set of orthonormal single-particle model wave functions $\{\bar{\varphi}_\alpha(\mathbf{r})\}$ as well as the set of coefficients $n_{\alpha\beta}$ and $\Gamma_{\alpha\beta\gamma\delta}$ entering eqn (7.18) and eqn (7.19) are known once the orbit $\mathcal{O} \subset \mathcal{H}^N$ is given. In the particular case of a pure state, the coefficients $n_{\alpha\beta}$ and $\Gamma_{\alpha\beta\gamma\delta}$ fulfil the Pauli conditions (Löwdin 1955, see also

Gupta and Rajagopal 1982)

$$\sum_{\alpha} n_{\alpha\alpha} = N, \qquad 0 \le n_{\alpha\alpha} \le 1 \qquad \text{(for all } \alpha) \tag{7.20}$$

and

$$\sum_{\alpha,\beta} \Gamma_{\alpha\beta\alpha\beta} = N\,(N-1), \qquad 0 \le \Gamma_{\alpha\beta\alpha\beta} \le 1 \quad \text{(for all } \alpha,\beta). \tag{7.21}$$

Performing the local-scale point transformation (7.1), one obtains the density-dependent first and second order density matrices in the associated orbit $\mathcal{O} \subset \mathcal{H}^N$

$$\rho^{(1)}[\rho] \equiv \rho(\mathbf{r};\mathbf{r}';[\rho]) = N\int \Psi^*(\mathbf{r}',\ldots,\mathbf{r}_N;[\rho])\Psi(\mathbf{r},\ldots,\mathbf{r}_N;[\rho])d\mathbf{r}_2\ldots d\mathbf{r}_N$$

$$= D^{1/2}(\mathbf{f}_{[\rho]}(\mathbf{r});\mathbf{r})\;\; D^{1/2}(\mathbf{f}_{[\rho]}(\mathbf{r}');\mathbf{r}')\bar{\rho}^{(1)}(\mathbf{f}_{[\rho]}(\mathbf{r});\mathbf{f}_{[\rho]}(\mathbf{r}'))$$

$$= \sum_{\alpha,\beta} n_{\beta\alpha}\, \varphi^*_{\alpha}(\mathbf{r};[\rho])\; \varphi_{\beta}(\mathbf{r}';[\rho]), \tag{7.22}$$

and

$$\Gamma[\rho] \equiv \Gamma(\mathbf{r}_1,\mathbf{r}_2;\mathbf{r}'_1,\mathbf{r}'_2;[\rho])$$

$$= \frac{1}{2}\, N(N-1)\int \Psi^*(\mathbf{r}'_1,\mathbf{r}'_2,\ldots,\mathbf{r}_N;[\rho])\Psi(\mathbf{r}_1,\mathbf{r}_2,\ldots,\mathbf{r}_N;[\rho])d\mathbf{r}_3\ldots d\mathbf{r}_N$$

$$= \sum_{\alpha,\beta,\gamma,\delta} \Gamma_{\alpha\beta\gamma\delta}\; \varphi_{\alpha}(\mathbf{r}_1;[\rho])\varphi_{\beta}(\mathbf{r}_2;[\rho])\varphi_{\gamma}(\mathbf{r}'_1;[\rho])\varphi_{\delta}(\mathbf{r}'_2;[\rho]), \tag{7.23}$$

where $D(\mathbf{f}_{[\rho]}(\mathbf{r});\mathbf{r})$ is the Jacobian (2.32) and

$$\varphi_{\alpha}[\rho] \equiv \varphi_{\alpha}(\mathbf{r};[\rho]) = D^{1/2}(\mathbf{f}_{[\rho]}(\mathbf{r});\mathbf{r})\; \bar{\varphi}_{\alpha}(\mathbf{f}_{[\rho]}(\mathbf{r})) \tag{7.24}$$

are the local-scale transformed model single-particle wave functions $\bar{\varphi}_{\alpha}(\mathbf{r})$.

In this way both density matrices, $\rho[\rho]$ and $\Gamma[\rho]$, are actual functionals of the local density $\rho(\mathbf{r})$ in any given orbit $\mathcal{O} \subset \mathcal{H}^N$.

Further, the explicit expression for the energy density functional (7.4) within $\mathcal{O} \subset \mathcal{H}^N$ may be derived from eqn (7.5) and eqn (7.6) after expressing them in terms of $\rho[\rho]$ and $\Gamma[\rho]$.

Kinetic energy density functional

The kinetic energy functional (7.5) can be expressed in terms of $\rho(\mathbf{r},\mathbf{r}';[\rho])$, eqn (7.22), as

$$T[\rho] = \frac{\hbar^2}{2m}\int d\mathbf{r}\ \tau[\rho] = \frac{\hbar^2}{2m}\iint d\mathbf{r}\ d\mathbf{r}'\,\delta(\mathbf{r}-\mathbf{r}')\nabla_{\mathbf{r}}.\nabla_{\mathbf{r}'}\,\rho(\mathbf{r},\mathbf{r}';[\rho]), \quad (7.25)$$

where the kinetic energy density $\tau[\rho]$ is of the form

$$\tau[\rho] = \Bigg\{\nabla_{\mathbf{r}}.\nabla_{\mathbf{r}'}\left[D(\mathbf{f}_{[\rho]}(\mathbf{r});\mathbf{r})D(\mathbf{f}_{[\rho]}(\mathbf{r}');\mathbf{r}')\right]^{1/2}$$

$$\times\ \bar{\rho}^{(1)}(\mathbf{f}_{[\rho]}(\mathbf{r});\mathbf{f}_{[\rho]}(\mathbf{r}'))\Bigg\}_{\mathbf{r}'=\mathbf{r}}. \quad (7.26)$$

Carrying out elementary calculations in the right-hand side of eqn (7.26) one obtains

$$\tau[\rho] = \frac{1}{4}\,\frac{[\nabla_{\mathbf{r}}D(\mathbf{f}_{[\rho]}(\mathbf{r});\mathbf{r})]^2}{D(\mathbf{f}_{[\rho]}(\mathbf{r});\mathbf{r})}\,\bar{\rho}(\mathbf{f}_{[\rho]}(\mathbf{r}))$$

$$+\ \frac{1}{2}\nabla_{\mathbf{r}}D(\mathbf{f}_{[\rho]}(\mathbf{r});\mathbf{r}).\left[(\nabla_{\mathbf{r}} + \nabla_{\mathbf{r}'})\ \bar{\rho}^{(1)}(\mathbf{f}_{[\rho]}(\mathbf{r});\mathbf{f}_{[\rho]}(\mathbf{r}'))\right]_{\mathbf{r}'=\mathbf{r}}$$

$$+\ D(\mathbf{f}_{[\rho]}(\mathbf{r});\mathbf{r})\left[(\nabla_{\mathbf{r}}.\nabla_{\mathbf{r}'})\ \bar{\rho}^{(1)}(\mathbf{f}_{[\rho]}(\mathbf{r});\mathbf{f}_{[\rho]}(\mathbf{r}'))\right]_{\mathbf{r}'=\mathbf{r}}, \quad (7.27)$$

where $\bar{\rho}(\mathbf{r})$ is the model local density distribution, the diagonal element of eqn (7.18). The first term of $\tau[\rho]$ can be expressed as

$$\tau^{(1)}[\rho] = \frac{1}{4}\,\frac{[\nabla_{\mathbf{r}}D(\mathbf{f}_{[\rho]}(\mathbf{r});\mathbf{r})]^2}{D(\mathbf{f}_{[\rho]}(\mathbf{r});\mathbf{r})}\,\bar{\rho}(\mathbf{f}_{[\rho]}(\mathbf{r}))$$

$$= \tau_W[\rho] + \tau_1^{(1)}[\rho] + \tau_2^{(1)}[\rho], \quad (7.28)$$

where

$$\tau_W[\rho] = \frac{1}{4}\,\frac{(\nabla_r\rho(\mathbf{r}))^2}{\rho(\mathbf{r})} \tag{7.29}$$

is the Weizsäcker kinetic energy density,

$$\tau_1^{(1)}[\rho] = -\frac{1}{2}\,\nabla\rho(\mathbf{r})\,.\,\frac{\nabla_{\bar{r}}\rho(\mathbf{f}_{[\rho]}(\mathbf{r}))}{\bar{\rho}(\mathbf{f}_{[\rho]}(\mathbf{r}))}, \tag{7.30}$$

and

$$\tau_2^{(1)}[\rho] = -\frac{1}{4}\,\rho(\mathbf{r}))\left|\frac{\nabla_r\bar{\rho}(\mathbf{f}_{[\rho]}(\mathbf{r}))}{\bar{\rho}(\mathbf{f}_{[\rho]}(\mathbf{r}))}\right|^2. \tag{7.31}$$

The second term in eqn (7.27) can be transformed as follows:

$$\tau^{(2)}[\rho] = \frac{1}{2}\,\nabla_r\,D(\mathbf{f}_{[\rho]}(\mathbf{r});\mathbf{r})\,.\left[(\nabla_r + \nabla_{r'})\bar{\rho}^{(1)}(\mathbf{f}_{[\rho]}(\mathbf{r});\mathbf{f}_{[\rho]}(\mathbf{r}'))\right]_{\mathbf{r}'=\mathbf{r}}$$

$$= -\frac{1}{2}\,\tau_1^{(1)}[\rho] - \tau_2^{(1)}[\rho], \tag{7.32}$$

where we use the identity,

$$\left\{\nabla_r\,\bar{\rho}^{(1)}(\mathbf{r},\mathbf{r}') + \nabla_{r'}\bar{\rho}^{(1)}(\mathbf{r};\mathbf{r}')\right\}_{\mathbf{r}'=\mathbf{r}} = \nabla\,\bar{\rho}^{(1)}(\mathbf{r}), \tag{7.33}$$

which can be easily proved in terms of natural orbitals (McWeeny and Sutcliffe 1969). The last term in eqn (7.27) takes the form

$$\tau^{(3)}[\rho] = \frac{\rho(\mathbf{r})}{\bar{\rho}(\mathbf{f}_{[\rho]}(\mathbf{r}))}\left[\nabla_r.\nabla_{r'}\bar{\rho}^{(1)}(\mathbf{f}_{[\rho]}(\mathbf{r});\mathbf{f}_{[\rho]}(\mathbf{r}'))\right]_{\mathbf{r}'=\mathbf{r}}. \tag{7.34}$$

One can express this term as

$$\tau^{(3)}[\rho] = \tau_2^{(1)}[\rho] + \rho(\mathbf{r})\left[\nabla_r.\nabla_{r'}\bar{\gamma}(\mathbf{f}_{[\rho]}(\mathbf{r});\mathbf{f}_{[\rho]}(\mathbf{r}'))\right]_{\mathbf{r}'=\mathbf{r}}, \tag{7.35}$$

after introducing the reduced model first-order density matrix (Gazquer and Ludeña 1981)

$$\bar{\gamma}(\mathbf{r};\mathbf{r}') = \bar{\rho}^{(1)}(\mathbf{r};\mathbf{r}')/(\bar{\rho}(\mathbf{r})\ \bar{\rho}(\mathbf{r}'))^{1/2}, \tag{7.36}$$

which obeys the condition $\bar{\gamma}(\mathbf{r};\mathbf{r}) = 1$ for arbitrary $\mathbf{r} \in R^3$.

Finally, inserting eqns (7.28), (7.32), and (7.35) into eqn (7.27), one obtains

$$\tau[\rho] = \tau_W[\rho] + \tau_\gamma[\rho], \tag{7.37}$$

where

$$\tau_\gamma[\rho] = \rho(\mathbf{r})\left[\nabla_r.\nabla_{r'}\bar{\gamma}(\mathbf{f}_{[\rho]}(\mathbf{r});\mathbf{f}_{[\rho]}(\mathbf{r}'))\right]_{\mathbf{r}'=\mathbf{r}}. \tag{7.38}$$

We thus obtain the kinetic energy density $\tau[\rho]$, and therefore the kinetic energy density functional $T[\rho]$, defined rigorously within $\mathcal{O} \subset \mathcal{H}^N$. It is represented by two terms only. The first one coincides with the original Weizsäcker term $\tau_W[\rho]$. This term emerges independently on the chosen orbit and, therefore, it exists in the exact kinetic energy density functional as well. The second term $\tau_\gamma[\rho]$ is in general a nonlocal one and it depends on the particular orbit via the reduced model density matrix (7.36).

The structure of the kinetic energy density $\tau[\rho]$, eqn (7.37), is remarkable in the sense that there are a number of approaches incorporating the original Weizsäcker term $\tau_W[\rho]$, see for instance March and Young (1958), Ludeña (1982), Acharya (1983). These approaches have been repeatedly discussed with various approximations for the so-called nonlocal term $\tau_{NW} = \tau(\mathbf{r}) - \tau_W[\rho]$. As mentioned by Ludeña (1983) this term is important if one wishes to incorporate in a systematic fashion corrections to $\tau_W[\rho]$, leading eventually to the exact description of the kinetic energy $\tau(\mathbf{r})$ as a functional $\tau[\rho]$ of the density $\rho(\mathbf{r})$. This conclusion is practically seen in eqn (7.37), where the nonlocal term $\tau_\gamma[\rho]$ is an actual functional of $\rho(\mathbf{r})$ and leads to the exact kinetic energy density $\tau[\rho] = \tau_W[\rho] + \tau_\gamma[\rho]$ within the eigenorbit $\mathcal{O}_H \subset \mathcal{H}^N$.

Equation for the local density

Using the same technique as that applied above, one can obtain the potential energy density functional $V[\rho]$ whose explicit form will obviously depend on the particular form of the two-body interaction. Thus, after deriving the explicit form of $E[\rho]$, one could consider the Euler-Lagrange equation

$$\frac{\delta}{\delta\rho(\mathbf{r})} \int d\mathbf{r} \left\{ \frac{\hbar^2}{2m} \tau[\rho] + \upsilon[\rho] - \mu\rho \right\} = 0, \tag{7.39}$$

where $\tau[\rho]$ and $\upsilon[\rho]$ are densities associated with the functionals $T[\rho]$ and $V[\rho]$, respectively.

Taking into account the particular form of the kinetic energy density (7.37), the equation for $\rho(\mathbf{r})$ reads

$$-\frac{\hbar^2}{2m}\nabla^2\rho(\mathbf{r}) + \frac{\hbar^2}{8m}\frac{(\nabla\rho)^2}{\rho(\mathbf{r})} + \rho(\mathbf{r})\left[\nabla_{\mathbf{r}}.\nabla_{\mathbf{r}'}\bar{\gamma}(\mathbf{f}_{[\rho]}(\mathbf{r});\mathbf{f}_{[\rho]}(\mathbf{r}'))\right]_{\mathbf{r}'=\mathbf{r}}$$

$$-\rho^2(\mathbf{r})\frac{\delta}{\delta\rho}\left[\nabla_{\mathbf{r}}.\nabla_{\mathbf{r}'}\bar{\gamma}\right]_{\mathbf{r}'=\mathbf{r}} + \upsilon_p(\mathbf{r};[\rho]) = \mu\,\rho(\mathbf{r}), \tag{7.40}$$

where

$$\upsilon_p(\mathbf{r};[\rho]) = \delta\upsilon[\rho]/\delta\rho(\mathbf{r}) \tag{7.41}$$

stands for the variational derivative of the potential energy density functional. Equation (7.40) is a highly nonlinear second-order integro-differential equation for the density $\rho(\mathbf{r})$. Its boundary conditions are such that for $r = 0$ the local density should be constant $\rho(0) = \rho_0$ and that far outside the system the density should vanish ($\rho(r\to\infty) = 0$).

7.3 RED approach in Hartree-Fock approximation

All results derived in previous section are exact when dealing with the eigenorbit $\mathcal{O}_H \subset \mathcal{H}^N$. Their use in practice needs the particular form of the model function $\bar{\Psi}$ defining the trial orbit $\mathcal{O} \subset \mathcal{H}^N$. In this section we consider the approximate energy density functional (7.4) derived within the orbit $\mathcal{O}_{SL} \subset \mathcal{H}^N_{HF}$ introduced by a model single Slater determinant $\bar{\Phi} \in \mathcal{H}^N_{HF}$.

Derivation of the energy density functional

Let us consider as a model function the single Slater determinant

$$\bar{\Phi}(\mathbf{r}_1,\mathbf{r}_2,\ldots,\mathbf{r}_N) = N^{-1/2} \det|\bar{\varphi}_i(\mathbf{r}_j)| \quad (i,j = 1,2,\ldots,N), \tag{7.42}$$

generated by N in number given (say model) orthonormal single-particle wave functions $\bar{\varphi}_i(\mathbf{r})$. The function $\bar{\Phi} \in \mathcal{H}^N_{HF}$ considered as a prototypical function defines the orbit $\mathcal{O}_{SL} \subset \mathcal{H}^N_{HF}$.

The local-scale point transformations preserve the single determinant character of all functions belonging to $\mathcal{O}_{SL} \subset \mathcal{H}^N_{HF}$. Therefore, substituting eqn (7.42) into eqn (7.1), they can be represented in the form

$$\Phi[\rho] \equiv \Phi(\mathbf{r}_1,\ldots,\mathbf{r}_N;[\rho]) = \mathcal{U}_{f[\rho]}\bar{\Phi} = N^{-1/2}\det|\varphi_i(\mathbf{r}_j;[\rho])|, \tag{7.43}$$

where $\varphi_i(\mathbf{r}_j;[\rho])$ are the density-dependent wave functions (7.24). To obtain these functions $\{\varphi_i[\rho]\}$, one has to solve first the differential equation (2.59) with respect to $f(\mathbf{r})$ at a given $\rho(\mathbf{r}) \in \bar{D}_N$ and using the model density distribution associated with eqn (7.42)

$$\bar{\rho}(\mathbf{r}) = \sum_{i=1}^{N} |\bar{\varphi}_i(\mathbf{r})|^2, \tag{7.44}$$

and then to substitute the solution $f_{[\rho]}(\mathbf{r})$ into eqn (7.24).

The kinetic energy density $\tau[\rho]$ now depends on the orbit $\mathcal{O}_{SL} \subset \mathcal{H}^N_{HF}$ through the reduced one-body matrix $\bar{\gamma} = \gamma_{SL}(\mathbf{r},\mathbf{r}')$ which is of the form

$$\gamma_{SL}(\mathbf{r},\mathbf{r}') = \sum_{l=1}^{N} \frac{\bar{\varphi}_l^*(\mathbf{r}')\bar{\varphi}_l(\mathbf{r})}{(\bar{\rho}(\mathbf{r}')\bar{\rho}(\mathbf{r}))^{1/2}}, \tag{7.45}$$

where $\bar{\rho}(\mathbf{r})$ is the model density (7.44). After substituting eqn (7.45) into eqn (7.37), one obtains the kinetic energy

$$\tau[\rho] = \tau_W[\rho] + \tau_\gamma^{SL}[\rho], \tag{7.46}$$

where the Weizsäcker term $\tau_W[\rho]$ is given by eqn (7.29) and the nonlocal term is of the form

$$\tau_\gamma^{SL}[\rho] = \rho(\mathbf{r}) \sum_{l=1}^{N} \left| \nabla \frac{\bar{\varphi}_l(\mathbf{f}_{[\rho]}(\mathbf{r}))}{(\bar{\rho}(\mathbf{f}_{[\rho]}(\mathbf{r})))^{1/2}} \right|^2 . \tag{7.47}$$

The spin-density (1.147) is also a local density functional in the orbit $\mathcal{O}_{SL} \subset \mathcal{H}^N_{HF}$

$$\mathbf{J}[\rho] = (-i) \sum_{l} \varphi_l^*(\mathbf{r};[\rho]) \left[\nabla\varphi_l(\mathbf{r};[\rho]) \times \sigma\right]. \tag{7.48}$$

Following the technique given in the previous section and the Skyrme-type effective forces (Section 1.5), one obtains the functional (7.4) in the orbit $\mathcal{O}_{SL} \subset \mathcal{H}^N_{HF}$

$$E[\rho] = \langle\Psi[\rho]|\hat{H}|\Psi[\rho]\rangle = \int d\mathbf{r}\ \mathcal{E}(\rho,\tau[\rho],\mathbf{J}[\rho]), \tag{7.49}$$

where the energy density

$$\mathcal{E}[\rho] \equiv \mathcal{E}(\rho,\tau[r],\mathbf{J}[\rho]) = \frac{\hbar^2}{2m}\tau[\rho] + \mathcal{E}_{pot}[\rho] \tag{7.50}$$

as well as the potential energy density

$$\mathcal{E}_{pot}[\rho] = \mathcal{E}_{pot}(\rho,\tau[r],J[\rho]) \tag{7.51}$$

follow from eqn (1.144) after substituting in it the quantities $\tau(\mathbf{r})$ and $\mathbf{J}(\mathbf{r})$ with their local density functionals $\tau[\rho]$ and $\mathbf{J}[\rho]$, respectively.

Limit of light nuclei

The functional (7.49) satisfies the general requirement for a reproduction of the exact (HF) energy in the particular case of four nucleons in the $1s_{1/2}$ state of the nucleus ^{4}He (Section 2.5). Really, one obtains from eqn (7.45) that $\gamma_{SL} \equiv 1$ for the nucleus ^{4}He and the second term in eqn (7.46) vanishes. The kinetic energy density functional $\tau[\rho]$ is represented only by its Weizsäcker term $\tau_W[\rho]$ and, therefore, the results of Section 2.5 are exactly reproduced. This feature shows that in principle the approximate functional (7.49) should reproduce the HF properties in light nuclei.

Limit of superheavy nuclei

In the case of a system with large number of particles, the model single-particle wave functions $\bar{\varphi}_i(\mathbf{r})$ may be represented by plane waves with periodic boundary conditions (Petkov and Stoitsov 1981, Kryachko, Petkov, and Stoitsov 1987a. In the limit of infinite nuclear matter, the local density is constant and the Weizsäcker term $\tau_W[\rho]$ vanishes ($\nabla\rho \equiv 0$) in eqn (7.46), while the term $\tau_\gamma^{SL}[\rho]$ transforms into the nuclear matter kinetic energy (2.87). In this case the functional (7.49) leads to the nuclear matter saturation curve (2.88) and, therefore, the nuclear matter (HF) results of Section 2.6 are exactly reproduced. So one can conclude that the term $\tau_\gamma^{SL}[\rho]$, eqn (7.47), is a kind of modification of the TF kinetic energy density $\tau_{TF}[\rho]$, eqn (3.5), and therefore the functional $E[\rho]$ within $\mathcal{O}_{SL} \subset \mathcal{H}_{HF}^N$ is expected to reproduce the HF properties of superheavy nuclei with a very large number of particles.

Finite nuclei

In the case of finite nuclei, the density $\rho(\mathbf{r})$ is not constant and the Weizsäcker term $\tau_W[\rho]$ appears in eqn (7.46) as well as the term $\tau_\gamma^{SL}[\rho]$. At small distances $r \to 0$, we know that the density $\rho(\mathbf{r})$ is nearly constant and the Weizsäcker term $\tau_W[\rho]$ vanishes. Therefore, the nonlocal term $\tau_\gamma^{SL}[\rho]$ is important only in the inner region and this term is expected to introduce nuclear shell effects due to its particular structure.

On the contrary, for $r \to \infty$, we know that the HF-type densities are proportional to the square of the wavefunction of the last occupied state, that is, one can expect that $\rho(r\to\infty) \sim \bar{\varphi}_1(r)$. In this case $\gamma_{SL} \equiv 1$ from eqn (7.45) and the second term $\tau_\gamma^{SL}[\rho]$ in eqn (7.46) vanishes. Therefore, the Weizsäcker term $\tau_W[\rho]$ is important in the nuclear surface and it is expected to determine the asymptotic behaviour of the density at large distances.

Asymptotic density

In order to analyse the importance of the functional (7.49) in forming the behaviour of $\rho(\mathbf{r})$, one should consider the Euler-Lagrange equation (7.40) within the orbit $\mathcal{O}_{SL} \subset \mathcal{H}_{HF}^N$. Performing the variation of the functional (7.49), one obtains

$$-\frac{\hbar^2}{2m}\nabla^2\rho(\mathbf{r}) + \frac{\hbar^2}{8m}\frac{(\nabla\rho)^2}{\rho(\mathbf{r})} + (U(\mathbf{r};[\rho]) - \mu)\,\rho(\mathbf{r}) = 0, \quad (7.52)$$

where

$$U(\mathbf{r};[\rho]) \equiv U_N[\rho] + U_\gamma[\rho]. \quad (7.53)$$

The variation of $\mathcal{E}_{pot}[\rho]$ gives by definition the nuclear part $U_N[\rho]$ of the potential (7.53), that is, $\delta\mathcal{E}_{pot}[\rho] = U_N[\rho]\delta\rho$. For a force with an effective mass $m^*(\mathbf{r}) \neq m(\mathbf{r})$, $U_N[\rho]$ is not a local potential because the ρ-dependence of $\tau[\rho]$ in the term proportional to $\rho\tau[\rho]$ is also varied and included in $U_N[\rho]$. The potential $U_\gamma[\rho]$ in eqn (7.53) stands for the term coming from the variation $\delta\tau_\gamma[\rho] = U_\gamma[\rho]\delta\rho$.

For small distances $r \to 0$, where $\rho(\mathbf{r})$ goes to constant value $\rho(0) = \rho_0$, the potential (7.53) goes to a constant as well. Thus, one can easily verify that the asymptotic solution of eqn (7.52) for $r\to 0$ is of the form

$$\rho(r\to 0) = \text{const}_1 + \text{const}_2\, r^2. \tag{7.54}$$

For large distances $r\to\infty$, both $\rho(\mathbf{r})$ and $U(\mathbf{r};[\rho])$ go to zero and one obtains the correct asymptotic solution

$$\rho(r\to\infty) = \text{const}\, r^{-2} \exp(-\,kr), \tag{7.55}$$

where

$$k = \left(-\,\frac{8m}{\hbar^2}\,\mu\right)^2 \tag{7.56}$$

and $\mu < 0$ is the Fermi energy of the system.

Equations (7.54) and (7.55) show that the energy density functional (7.49) leads to realistic asymptotic solutions for $\rho(\mathbf{r})$. The actual solution for $\rho(\mathbf{r})$ and $0 < r < \infty$, however, can be obtained only solving eqn (7.52) numerically after specifying the particular choice of the model single-particle functions $\bar{\varphi}_i(\mathbf{r})$.

7.4 Comparison with Hartree–Fock results

Harmonic oscillator single Slater determinant orbit

Petkov and Stoitsov (1983b) have applied the RED approach to spherical even–even nuclei within the so-called harmonic oscillator single Slater determinant orbit $\mathcal{O}^{HO}_{SL} \subset \mathcal{H}^N_{HF}$; that is, they have specified the model states $\bar{\varphi}_i(\mathbf{r})$ entering eqn (7.42) to be the harmonic oscillator single-particle wave functions

$$\bar{\varphi}_i(x) = \bar{\varphi}_{nl}(r)\, \mathcal{Y}_{\ell jm}(\mathbf{r}_0,\sigma)\, \chi_q(\tau). \tag{7.57}$$

Here the subscript i includes the set of quantum numbers: the charge q, the principal quantum number n, the orbital momentum l, the total angular momentum j and the magnetic quantum number m. The spin function $\mathcal{Y}_{\ell jm}(\mathbf{r}_0,\sigma)$ includes spin-coordinates. It is constructed by vector coupling the spin **s** to the orbital momentum l to form the total angular momentum **j**

$$\mathcal{Y}_{\ell jm}(\mathbf{r}_0,\sigma) = \sum_{m_\ell, m_s} (l\ 1/2\ m_\ell\ m_s | m\ j) Y_{\ell m_\ell}(\mathbf{r}_0) \chi_{m_s}(\sigma), \tag{7.58}$$

where $\chi_{m_s}(1/2)$ are spin eigenfunctions and $Y_{\ell m_\ell}(\mathbf{r}_0)$ are spherical harmonics. The radial functions

$$\bar{\varphi}_{n\ell}(r) = \pi^{-1/4}\alpha^{3/2}\left[\frac{2^{l-n+3}(2l+2n-1)}{[(2l+1)!!]^2(n-1)!}\right]^{1/2} (\alpha r)^{1/2} \exp[-\tfrac{1}{2}(\alpha r)^2]$$
$$\times \sum_{k=0}^{n-1} (-2)^k \frac{(n-1)!}{(n-k-1)!\ k!} \frac{(2l+1)!!}{(2l+2k+1)!!} (\alpha r)^{2k}, \tag{7.59}$$

contain the radial oscillator parameter $\alpha = (M\omega/\hbar)^2$, the oscillator width, which is different for neutrons and protons. The model wave functions (7.57) are orthonormalized to give $\langle\bar{\varphi}_i|\bar{\varphi}_j\rangle = \delta_{ij}$ via the orthonormalization conditions

$$\int dr\ r^2\ \varphi_{n\ell}(r)\ \varphi_{n'\ell}(r) = \delta_{nn'} \tag{7.60}$$

and

$$\iint d\theta\ d\varphi\ \sin\theta\ \mathcal{Y}_{\ell jm}(\mathbf{r}_0,\sigma)\mathcal{Y}_{\ell' j' m'}(\mathbf{r}_0,\sigma) = \delta_{\ell\ell'}\delta_{mm'}\delta_{jj'}, \tag{7.61}$$

where the orthogonality of the spin wave functions is used

$$[\chi_{m_s}(1/2), \chi_{m_s'}(1/2)] = \delta_{m_s m_s'}. \tag{7.62}$$

Actually, eqns (7.24), (7.43) and (7.57) determine the orbit $\mathcal{O}^{HO}_{SL} \subset \mathcal{H}^{N}_{HF}$ to which all wave functions $\Phi[\rho]$ belong. For spherical even-even nuclei, the spin functions (7.58) coincide with the angular parts of the exact HF solution. As a consequence, the variables in eqn (7.52) are separable and the local density $\rho(\mathbf{r})$ as well as the function $f_{[\rho]}(\mathbf{r})$ are spherically symmetric, that is, they depend only on the magnitude $r \in R^1$ of the vector $\mathbf{r} \in R^3$.

Numerical results

The Euler-Lagrange equation (7.52) rewritten in terms of functions $f_{[\rho_n]}(r)$ and $f_{[\rho_p]}(r)$ has been solved numerically by Petkov and Stoitsov (1983b) after applying a modification of the core-spline method (Aleksandrov and Karadjov 1980, Aleksandrov, Drenska, and Karadjov 1982a) and using the autoregularized iterative process of Gauss-Newton type (Aleksandrov 1971, 1973) and program system REGN (Aleksandrov, Drenska, and Karadjov 1981, 1982b). The calculated binding energies, single-particle energies, densities and other geometrical characteristics of a reference set of nuclei ^{16}O, ^{40}Ca, ^{90}Zr, ^{208}Pb give a clear idea about the potentialities of the approach in the description of the ground state of nuclei within the orbit $\mathcal{O}^{HO}_{SL} \subset \mathcal{H}^{N}_{HF}$.

In Table 7.1 we give the RED results for the rms radii of neutrons r_n, protons r_p and charge r_c distributions and the rms mass radii r_m as well as the total binding energies per nucleon E/N of nuclei with mass number N. A comparison is made with the exact Skyrme HF results (SHF) obtained by Vautherin and Brink (1972) after integrating the Skyrme-HF equations in coordinate space. It is seen that the SHF geometrical characteristics are reproduced exactly within the orbit $\mathcal{O}^{HO}_{SL} \subset \mathcal{H}^{N}_{HF}$. Considering the

Table 7.1. Root mean square radii (in fm) and binding energies per nucleon (in MeV) calculated with SkI forces

Nuclei	Models	r_m	r_n	r_p	r_{ch}	-E/N
^{16}O	HO	2.57	2.56	2.58	2.70	8.01
	EVM	2.63	2.60	2.66	2.73	7.96
	RED+SF	2.55	2.54	2.56	2.68	8.15
	RED	2.55	2.53	2.56	2.68	8.18
	SHF	2.55	2.53	2.56	2.68	8.22
^{40}Ca	HO	3.34	3.32	3.36	3.46	8.29
	EVM	3.33	3.28	3.39	3.46	8.48
	RED+SF	3.3	3.28	3.32	3.41	8.56
	RED	3.29	3.27	3.31	3.41	8.59
	SHF	3.29	3.27	3.31	3.41	8.64
^{90}Zr	HO	4.24	4.26	4.22	4.30	8.2
	EVM	4.2	4.22	4.16	4.23	8.72
	RED+SF	4.18	4.19	4.16	4.235	8.71
	RED	4.17	4.19	4.14	4.22	8.734
	SHF	4.17	4.19	4.14	4.22	8.81
^{208}Pb	HO	5.54	5.58	5.47	5.53	7.7
	EVM	5.46	5.52	5.36	5.42	7.80
	RED+SF	5.45	5.495	5.39	5.448	7.79
	RED	5.44	5.48	5.38	5.44	7.81
	SHF	5.45	5.49	5.38	5.44	7.89

HO stands for values obtained only within harmonic oscillator shell model calculations; RED are results obtained within the RED approach and the orbit $\mathcal{O}^{HO}_{SL} \subset \mathcal{H}^N$ (both HO and RED results are taken from Petkov and Stoitsov (1983b)); RED+SF are results obtained in variational calculations with SF trial densities, eqn (7.64), by Dimitrova, Petkov, and Stoitsov (1986b); EVM are results with the same forces within the expectation value method of Ko *et al.* (1974); Skyrme-HF results (SHF) are taken from Vautherin and Brink (1972).

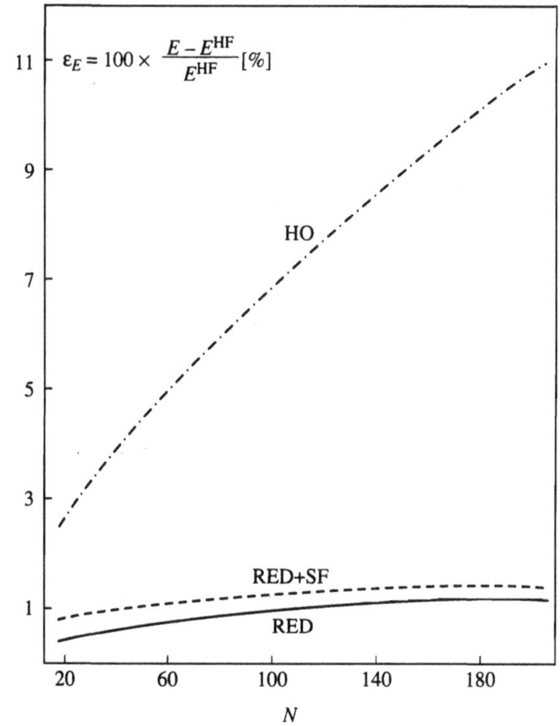

Figure 7.1. Relative error ε_E (in per cent) in the total binding energy with respect to Skyrme-Hartree-Fock results derived by Vautherin and Brink (1972) with SI effective forces. The results following from the rigorous energy density (RED) approach (continuous line) are compared with the results (HO) obtained within the harmonic oscillator shell model (dashed-dotted line). The dashed line represents the RED results obtained in symmetrized Fermi (SF) approximation (Dimitrova, Petkov, and Stoitsov 1986b).

total binding energy, the RED approach is good to within less than 1%. For purpose of comparison, we also show in Table 7.1 the best results (HO) obtained only within the prototypical harmonic-oscillator shell-model wave function.

In Figure 7.1 we show the relative error ε_E (in per cent) within which the HF total energy is reproduced when calculating in

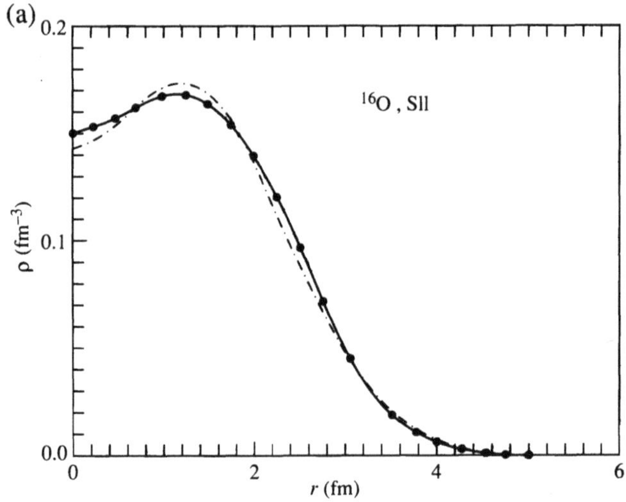
(a)
0.2
0.1
0.0
ρ (fm⁻³)
¹⁶O , SII
0
2
4
6
r (fm)

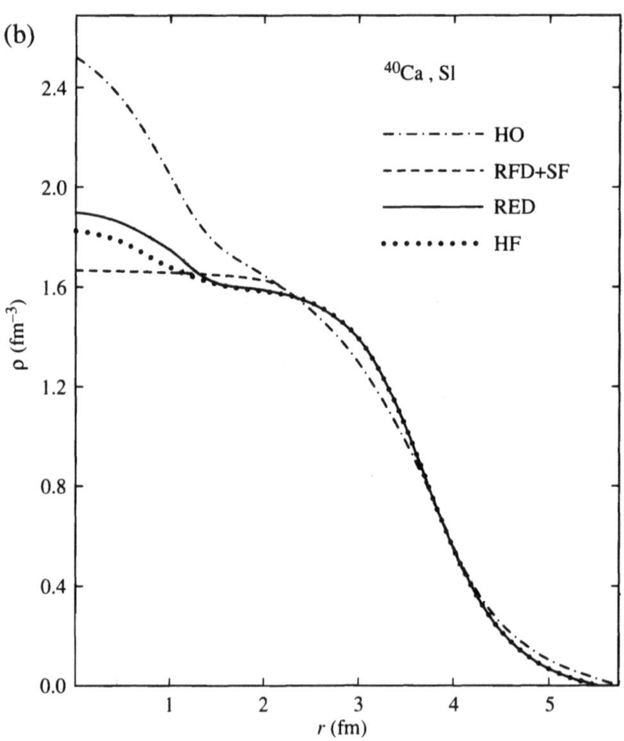
(b)
⁴⁰Ca , SI
HO
RFD+SF
RED
HF
2.4
2.0
1.6
1.2
0.8
0.4
0.0
ρ (fm⁻³)
1
2
3
4
5
r (fm)

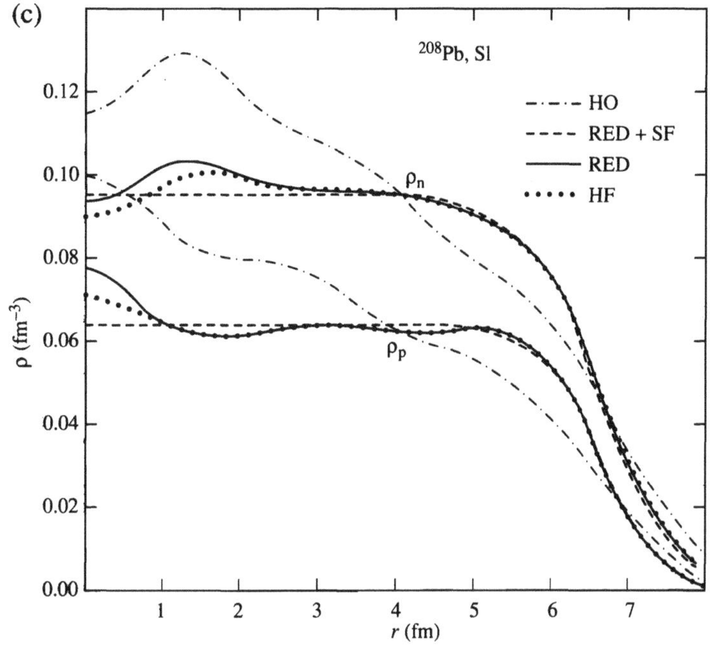

Figure 7.2. Mass density distributions for nuclei ^{16}O (a) and ^{40}Ca (b) and proton and neutron local densities for the nucleus ^{208}Pb (c) calculated with SI forces by Dimitrova, Petkov, and Stoitsov (1986b). Skyrme-Hartree-Fock results (Vautherin and Brink 1972) represented by points are compared with the results from the rigorous energy density (RED) approach (full lines) and harmonic oscillator (HO) shell model (dashed-dotted lines).

the HO and RED approximations. It is seen that in the HO approximation (dashed curve) this error increases (almost linearly) from 3% for ^{16}O to 10–11% for N~208. In contrast, the maximum deviation of the RED energies from E_{HF} (observed for the nucleus ^{208}Pb) is just equal to 1%.

Figures 7.2a–7.2c show typical results obtained for the local density distributions. In the light nuclei region (^{16}O, Figure 7.2a) the RED approach (continuous curve) reproduces exactly the density ρ_{HF} (represented by points). This result is not surprising because even HO results are in good agreement with the HF ones in this region of the mass table (Vautherin and Brink 1972).

For heavier nuclei, however, the densities obtained by harmonic-oscillator shell model wave functions only (dashed-dotted curve) significantly differ from the HF densities (see Figures 7.2a and 7.2b). Nevertheless, the RED approach satisfactorily reproduces the exact HF results for the nuclear surface and the tails of the density distributions. The greatest deviation is observed in the region $r \to 0$, but this region participates with a minimal weight in the determination of the nuclear characteristics (there is an integration with a weight factor of $r^2 dr$).

It should be stressed that the energy density functional $E[\rho]$ within $\mathcal{O}^{HO}_{SL} \subset \mathcal{H}^{N}_{HF}$ also reproduces the HF shell effects. As seen from Figure 7.2, the RED approach reproduces the HF local density shell oscillations quite well including their location and amplitudes. Furthermore, the RED functional $E[\rho]$ allows the single-particle energies to be calculated by using the definition

$$\varepsilon_i[\rho] = \int d\mathbf{r}\ \varphi_i^*(\mathbf{r};[\rho])\ \frac{\delta E[\rho]}{\delta \varphi_i^*}\ \varphi_i(\mathbf{r};[\rho]). \qquad (7.63)$$

We show in Table 7.2 and Table 7.3 the results (in MeV) for the single-particle energies (7.63) of the occupied proton and neutron states, respectively. One gets the impression that the best agreement with the HF single-particle energies is obtained for the states close to the Fermi surface. The maximum deviation from the HF results, however, observed in the deep 1s and 1d states is as low as 1–2 MeV for the nucleus ^{208}Pb. This agreement is rather satisfactory if one takes into account that the experimental data for these states are themselves determined to within ±(7–11) MeV (James *et al.* 1969).

Table 7.2. Comparison of occupied proton single-particle energies (in MeV) calculated with SI forces within the RED approach (Petkov and Stoitsov 1983b) and the Skyrme-HF theory (Vautherin and Brink 1972).

$-\varepsilon_{n\ell j}$	^{208}Pb		^{90}Zr		^{40}Ca		^{16}O	
$n\ell$ j	RED	SHF	RED	SHF	RED	SHF	RED	SHF
1s 1/2	34.9	36.6	35.1	36.3	32.5	32.8	28.7	28.8
1p 3/2	31.8	33.1	30.2	30.8	24.2	24.8	16.7	16.8
1p 1/2	31.4	32.5	28.6	29.0	20.3	20.5	9.9	9.9
1d 5/2	27.7	28.4	23.7	23.9	14.3	14.4		
1d 3/2	26.6	27.0	20.0	20.1	7.2	7.3		
2s 1/2	25.1	24.0	18.7	17.6	8.0	7.9		
1f 7/2	22.6	22.9	15.8	15.9				
1f 5/2	20.1	20.3	9.3	9.6				
2p 3/2	18.3	17.1	8.6	7.8				
2p 1/2	17.0	16.0	6.2	6.0				
1g 9/2	16.7	16.7						
1g 7/2	12.2	12.5						
2d 5/2	10.6	9.7						
1h11/2	9.8	10.0						
2d 3/2	8.1	7.7						
3s 1/2	7.6	6.8						

In general, the result obtained within the RED approach are found to be in a good agreement with the more elaborate HF results. It will be therefore interesting further to simplify the calculations by minimizing the functional $E[\rho]$ within a restricted space of suitably chosen trial densities.

Results of variational calculations

Dimitrova, Petkov, and Stoitsov (1986a,b) parameterize the proton and neutron densities in nuclei by symmetrized Fermi-type (SF) density distributions (Burov, Eldishev, and Lukyanov 1974)

$$\rho_q^{SF}(r) \equiv \rho^{SF}(r;R_q,b_q) = \rho_{0q}\,\mathrm{sh}(R_q/b_q)\left[\mathrm{ch}(R_q/b_q) + \mathrm{ch}(r/b_q)\right]^{-1}$$

$$= \rho_{0q}\left\{\left[1+\exp\left((r-R_q)/b_q\right)\right]^{-1} + \left[1+\exp\left((-r-R_q)/b_q\right)\right]^{-1} + 1\right\}, \quad (7.64)$$

Table 7.3. Comparison of occupied neutron single-particle energies (in MeV) calculated with SI forces within the RED approach (Petkov and Stoitsov 1983b) and the Skyrme-HF theory (Vautherin and Brink 1972).

$-\varepsilon_{n\ell j}$	^{208}Pb		^{90}Zr		^{40}Ca		^{16}O	
$n\ell$ j	RED	SHF	RED	SHF	RED	SHF	RED	SHF
1s 1/2	42.8	44.6	44.1	44.7	40.7	41.0	32.9	33.0
1p 3/2	39.5	40.9	37.7	38.0	32.1	32.2	20.8	20.8
1p 1/2	39.0	40.3	36.3	36.6	28.2	28.4	13.8	13.8
1d 5/2	35.2	36.2	30.1	30.3	22.1	22.2		
1d 3/2	34.1	34.7	26.6	26.9	14.8	15.0		
2s 1/2	33.1	32.5	25.9	25.5	15.8	15.7		
1f 7/2	30.1	30.6	21.8	21.9				
1f 5/2	27.7	28.0	15.4	15.8				
2p 3/2	26.5	25.4	16.2	15.8				
2p 1/2	25.1	24.2	13.7	13.7				
1g 9/2	24.2	24.5	12.8	13.0				
1g 7/2	19.9	20.2						
2d 5/2	19.0	18.1						
1h 11/2	17.6	17.8						
2d 3/2	16.4	15.9						
3s 1/2	16.4	15.6						
1h 9/2	11.0	11.5						
1i 13/2	10.3	10.6						
2f 7/2	11.3	10.6						
2f 5/2	7.4	7.5						
3p 3/2	8.0	7.5						
3p 1/2	6.5	6.4						

where the normalization conditions for neutrons (q = n) and protons (q = p) are satisfied through the relation

$$\rho_{0q} = \rho_0(N_q, R_q, b_q) = \frac{3N_q}{4\pi R_q^3}\left[1 + \left(\frac{\pi b_q}{R_q}\right)^2\right]^{-1}. \tag{7.65}$$

The density parameters, half-radii R_n, R_p and diffuseness parameters b_n, b_p, are then obtained as a solution of the following system of nonlinear algebraic equations

$$\partial E(R_n, R_p, b_n, b_p)/\partial R_q = 0,$$

$$\partial E(R_n, R_p, b_n, b_p)/\partial b_q = 0 \qquad (q = n,p), \tag{7.66}$$

where the function $E(R_n,R_p,b_n,b_p)$ is derived after substituting the densities (7.64) into the energy density functional (7.49), that is, $E(R_n,R_p,b_n,b_p) \equiv E[\rho_n^{SF},\rho_p^{SF}]$.

In order to obtain the function $E(R_n,R_p,b_n,b_p)$, one has to first solve numerically the basic eqn (7.3) rewritten in the orbit $\mathcal{O}_{SL}^{HO}$ as

$$F([\rho_q^{SF}]) \equiv \int_0^r du\, u^2\, \rho_q^{SF}(u) - \int_0^{f[\rho_q^{SF}]} du\, u^2\, \rho_q^{HO}(u) = 0\ . \qquad (7.67)$$

The substitution of the resulting solution $f[\rho_q^{SF}] \equiv f(r;R_q,b_q)$ ($q = n,p$), into eqn (7.43) and the result in eqn (7.49) defines the function $E(R_n,R_p,b_n,b_p)$. Equation (7.66) is then solved numerically with respect to R_n,R_p,b_n,b_p. The whole procedure takes only a few minutes on an IBM-PC computer.

The total binding energy calculated by the above procedure has been already shown in Table 7.1 (results denoted as RED+SF). It is seen that the SF approximation for the local densities is good to within less than 1.3% compared with the self-consistent HF results (see also the dashed-dotted curve in Figure 7.1). In this respect one can note that the results shown in Table 7.1 and Figure 7.1 illustrate the validity of the inequality (7.17) which in the orbits $\mathcal{O}_{SL}^{HO}$ reads

$$E_{HO} \geq E_{RED+SF} \geq E_{RED} \geq E_{SHF}. \qquad (7.68)$$

From Table 7.1 is also seen that the SF radii slightly overestimate (less than about 0.01 fm) the HF values. For comparison, we also show some results following from the expectation value method (EVM) calculations (see Section 6.4) performed by Ko *et al.* (1974) with the same SkI forces.

We illustrate in Figure 7.2b, Figure 7.2c (dashed curves) the resulting SF density distributions which are seen to reproduce the average trend of the HF or RED local density distributions very well. The main difference is the absence of the density shell oscillations in $\rho_{SF}(r)$, but this is the price one has to pay when restricting the variational calculations within smooth SF densities (7.64).

The absence of density shell oscillations does not at all imply the absence of shell effects in the SF results. The shell effects appear in the approximate functional mostly through the particular form of the nonlocal kinetic energy term (7.47) which includes the density-dependent wave functions $\varphi_1(\mathbf{r};[\rho^{SF}])$ defined by eqns (7.24), (7.57) and (7.64). As an illustration, we show in Table 7.4 the agreement of the single-particle energies (7.63) calculated in the SF approximation for the nucleus ^{208}Pb with both RED and HF results.

The SF single-particle energies are still good compared with those of the HF method even for superheavy nuclei. Both SF and HF single-particle energies for a hypothetical superheavy nucleus ^{298}114 are compared in Table 7.4. In this case, the SF results for the binding energy show an underbinding of about 2% with respect to the HF energy, which is in accordance with the variational inequality (7.68). The associated SF and HF mass density distributions for the nucleus ^{298}114 are compared in Figure 7.3. Apart from the density shell oscillations we see that the SF results reproduce the HF local density quite well.

Quasibound and continuum states

Dimitrova, Petkov, and Stoitsov (1988b) have analysed the unoccupied HF single-particle states in terms of the RED approach in its variational SF approximation. In Table 7.5 we compare the single-particle energies around and above the Fermi level derived for the nucleus ^{298}114 with SIII forces in both SF and HF calculations. The HF results are obtained by Brack and Quentin (1974a) in coordinate space (spherical code) as well as in

Table 7.4. Occupied neutron and proton single-particle energies (in MeV) for nuclei $^{298}114$ and ^{208}Pb calculated with SkI forces in Skyrme-HF theory (SHF) by Vautherin and Brink (1972), RED approach by Petkov and Stoitsov (1983b) and in SF approximation by Dimitrova, Petkov, and Stoitsov (1988b).

$-\varepsilon_{n\ell j}$	Neutrons					Protons				
	$^{298}114$		^{208}Pb			$^{298}114$		^{208}Pb		
$n\ell$ j	SF	SHF	SF	RED	SHF	SF	SHF	SF	RED	SHF
1s 1/2	42.9	44.6	42.8	42.8	44.6	30.4	34.6	33.6	33.6	36.6
1p 3/2	39.6	41.8	39.0	39.0	40.9	28.1	31.2	30.1	30.1	33.1
1p 1/2	39.5	41.4	38.7	38.7	40.3	30.0	30.9	30.5	30.5	32.5
1d 5/2	36.0	38.2	34.8	34.8	36.2	35.3	27.2	27.3	27.3	28.4
1d 3/2	35.6	37.2	33.9	33.9	34.7	24.9	26.4	26.2	26.2	27.0
2s 1/2	35.3	34.9	33.4	33.4	32.5	24.2	25.1	25.2	25.2	24.0
1f 7/2	32.2	33.7	30.2	30.2	30.6	21.9	22.6	22.9	22.9	22.9
1f 5/2	31.1	31.9	27.9	27.9	28.0	20.6	21.0	20.4	20.4	20.3
2p 3/2	30.6	29.1	27.2	27.2	25.4	19.3	18.8	18.8	18.8	17.1
2p 1/2	29.9	28.3	26.0	26.0	24.2	18.5	18.1	17.5	17.5	16.0
1g 9/2	28.0	28.7	24.8	24.8	24.5	17.8	17.5	17.5	17.5	16.7
1g 7/2	25.6	25.8	20.6	20.6	20.2	45.0	14.7	13.1	13.1	12.5
2d 5/2	25.0	22.9	19.9	19.9	18.1	13.2	12.1	11.2	11.2	9.7
1h 11/2	23.3	23.1	18.7	18.7	17.8	12.8	11.9	11.0	11.0	10.0
2d 3/2	23.2	21.3	17.4	17.4	15.9	11.4	10.7	8.7	8.7	7.7
3s 1/2	23.0	20.5	17.3	17.3	15.6	10.8	10.1	8.1	8.1	6.8
1h 9/2	19.1	18.8	12.1	12.1	11.5	8.2	7.7			
1i 13/2	17.8	17.1	11.6	11.6	10.6	6.8	5.8			
2f 7/2	18.3	16.4	11.7	11.7	10.6	6.2	5.2			
2f 5/2	15.5	13.8	8.0	8.0	7.5					
3p 3/2	15.1	13.3	8.1	8.1	7.5					
3p 1/2	13.8	12.4	6.7	6.7	6.4					
1i 11/2	11.5	11.1								
1j 15/2	11.5	10.6								
2g 9/2	10.7	9.5								
2g 7/2	6.8	6.2								
3d 5/2	6.7	6.0								
3d 3/2	4.8	4.7								
4s 1/2	4.9	4.6								

configuration space (deformed code). The SF results are seen to reproduce the self-consistent HF spectrum quite well, including its bound, quasibound, and 'continuum' states. This makes the RED+SF approach rather attractive in view of the small amount of numerical work needed.

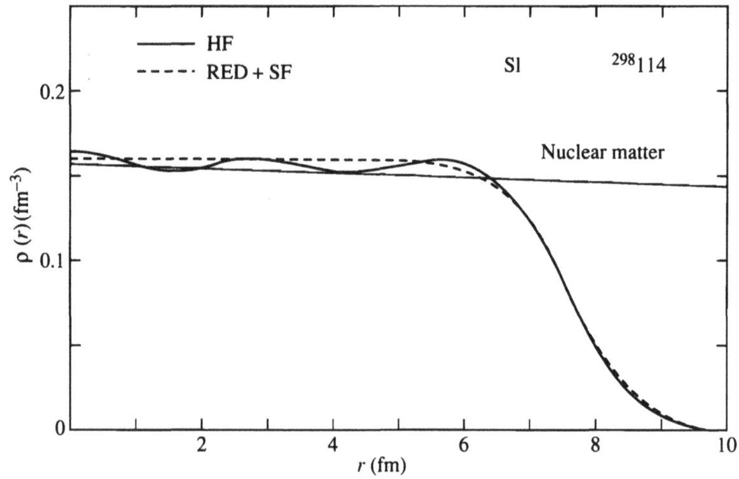

Figure 7.3. Mass density distributions for the hypothetical super-heavy nucleus $^{298}114$ calculated with SI effective forces. The Skyrme-Hartree-Fock results (HF) of Vautherin and Brink (1972) are compared with the results RED+SF following from the rigorous energy density approach in symmetrized Fermi approximation (Dimitrova, Petkov, and Stoitsov 1988b).

Summing up, one can conclude that the RED approach within the orbit $\mathcal{O}^{HO}_{SL}$ as well as its SF approximation provides a quite reasonable approximation to the self-consistent HF calculations not only for average nuclear properties but also for characteristics sensitive to shell effects in nuclei. Thus, a number of essentially many-particle problems, in which it is extremely difficult to use the HF results directly, are amenable to systematic investigations in terms of the approach presented.

7.5 Applications of the approach

Having thus checked the SF approximation of the RED approach in the orbit $\mathcal{O}^{HO}_{SL}$, we consider its application to some particular problems concerning properties of spherical nuclei and effective Skyrme-type forces.

Table 7.5. Comparison between SF and HF single-particle energies (in MeV) for bound, quasibound, and continuum states. SF results are taken from Dimitrova, Petkov, and Stoitsov (1988b) and HF results from Brack and Quentin (1974b).

$^{298}114$, SIII-force		Skyrme HF	
$n\ell$ j	RED+SF	Def.code	Spher.code
$1h_{11/2}$	-13.73	-12.47	-12.34
$2d\ _{3/2}$	-12.94	-11.91	-11.89
$3s\ _{1/2}$	-12.00	-11.14	-11.00
$1h\ _{9/2}$	-9.77	-8.73	-8.60
$1i_{13/2}$	-6.96	-5.62	-5.45
$2f\ _{7/2}$	-6.62	-5.60	-5.44
$2f\ _{5/2}$	-4.16	-3.73	-3.51
$3p\ _{3/2}$	-3.08	-2.80	-2.65
$3p\ _{1/2}$	-2.13	-2.18	-2.01
$1i_{11/2}$	-1.34	-0.44	-0.26
$1j_{15/2}$	0.63	1.60	1.80
$2g\ _{9/2}$	2.06	2.24	2.50
$2g\ _{7/2}$	4.99	4.57	4.93
$3d\ _{5/2}$	7.00	6.27	6.30
$3d\ _{3/2}$	8.27	7.04	7.08
$4s\ _{1/2}$	9.50	8.13	8.11
$1j_{13/2}$	7.74	8.30	8.51
$1k_{17/2}$	8.92	9.11	9.34
$2h_{11/2}$	11.11	10.67	10.67
$2h\ _{9/2}$	14.36	13.25	13.25
$3f\ _{7/2}$	16.36	15.95	14.95
$3f\ _{5/2}$	17.68	46.95	—
$1k_{15/2}$	17.07	17.20	17.48
$4p\ _{3/2}$	18.19	18.16	—
$1l_{19/2}$	17.67	18.25	17.21
$4p\ _{1/2}$	18.69	18.69	—
$2i_{13/2}$	19.73	19.05	18.70

Heavy-ion interaction potential

The possibilities of the approach in calculating the real part of ion-ion interaction potentials (see Section 3.4) have been investigated by Stoitsov (1982), following the method originally proposed by Flieshbach (1971) and further developed by Brink and Stancu (1975) and Göritz and Mosel (1976). Stoitsov (1982) considers the wave function $\Psi \equiv \Psi(\mathbf{r}_1, \mathbf{r}_2, \ldots, \mathbf{r}_{N_1}, \mathbf{r}_{N_1+1}, \ldots, \mathbf{r}_{N_1+N_1})$

of a system of two colliding nuclei with masses N_1 and N_2, and localized at a certain separation distance $R=|\mathbf{R}|$.

The total wave function Ψ can be represented as a product of two functions $\Psi_1(\mathbf{r}_1,\ldots,\mathbf{r}_{N_1})$ and $\Psi_2(\mathbf{r}_{N_1+1},\ldots,\mathbf{r}_{N_1+N_1})$ localized at points $-\mathbf{R}/2$ and $\mathbf{R}/2$, respectively:

$$\Psi(\mathbf{k},\mathbf{R}) = \mathcal{N}\mathcal{A}\left\{\left[\Psi_1(-\mathbf{R}/2)\ \exp(i\mathbf{k}.\mathbf{R}_1)\right]\left[\Psi_2(\mathbf{R}/2)\ \exp(-i\mathbf{k}.\mathbf{R}_2)\right]\right\}, \tag{7.69}$$

where $\mathcal{A}$ is the total antisymmetrization operator, $\mathcal{N}$ is the normalization constant and $\mathbf{R}_1$, $\mathbf{R}_2$ are the centres of mass coordinates of nuclei N_1, N_2, respectively. The relative motion of the two colliding ions is taken into account by a multiplication by a plane wave factor with a relative momentum $\hbar\mathbf{k}$ connected with the scattering energy E_S.

In the sudden approximation (Section 3.4) the wave functions Ψ_1 and Ψ_2 entering eqn (7.69) are represented by their asymptotic forms at $R \to \infty$, where $\Psi_1(\Psi_2)$ is assumed to be the single Slater determinant (7.43) which minimizes the energy of isolated nucleus $N_1(N_2)$. In other words, the total energy of the system at $R\ \ \infty$ is represented simply as the sum

$$E(\mathbf{k},\infty) = E[\rho_1] + E[\rho_2], \tag{7.70}$$

where $E[\rho]$ is the energy density functional (7.49) separately minimized to give the densities $\rho_1(\mathbf{r})$ and $\rho_2(\mathbf{r})$ of nuclei N_1 and N_2, respectively. At finite distances $\mathbf{R}$, the total energy of the interacting system

$$E(\mathbf{k},\mathbf{R}) = \langle\Psi(\mathbf{k},\mathbf{R})|\Psi(\mathbf{k},\mathbf{R})\rangle \tag{7.71}$$

is determined in terms of density-dependent single-particle wave functions $(\varphi_1(\mathbf{r};[\rho_1]);i = 1,\ldots,N_1)$ and $(\varphi_1(\mathbf{r};[\rho_2]);i = 1,\ldots,N_2)$.

According to the Born-Oppenheimer approximation, the heavy-ion interaction potential is defined as the difference

$$V(\mathbf{k},\mathbf{R}) \equiv E(\mathbf{k},\mathbf{R}) - E(\mathbf{k},\infty) = \langle\Psi(\mathbf{k},\mathbf{R})|\Psi(\mathbf{k},\mathbf{R})\rangle - E[\rho_1] - E[\rho_2], \quad (7.72)$$

where the bombarding energy E_S is identified with the energy $E_k = (\hbar\mathbf{k})^2/2M$, $M = N_1N_2/(N_1 + N_2)$ being the reduced mass (Brink and Stancu 1975). A slightly different definition of $V(\mathbf{k},\mathbf{R})$ which takes into account the energy conservation is used by Flieshbach (1971) and Göritz and Mosel (1976). Both definitions have been compared and analysed by Zint (1977).

As an illustration, we compare in Figure 7.4 the potential (7.72) calculated for some pairs of colliding nuclei with the potential

$$\overline{V}(\mathbf{k},\mathbf{R}) = \langle\overline{\Psi}(\mathbf{k},\mathbf{R})|\overline{\Psi}(\mathbf{k},\mathbf{R})\rangle - E[\overline{\rho}_1] - E[\overline{\rho}_2], \quad (7.73)$$

calculated on the basis of pure harmonic oscillator shell model wave functions. In these calculations, an approximate account of the antisymmetrized condition is used (Stoitsov 1982). The results demonstrate the importance of the self-consistent condition which is completely ignored when calculating the heavy-ion potential only within the harmonic oscillator shell model.

Extrapolation to superheavy nuclei

Oulne, Petkov, and Stoitsov (1989) have extended the RED approach to the time-reversal symmetric HF+BCS subspace. In this kind of calculation pairing effects are included by treating them in the BCS approximation (Bardeen, Cooper, and Schrieffer 1957) with a δ-function pairing force (Tondeur 1979 and Tondeur *et al.* 1984)

$$\mathcal{V}_{ij} = -V_0\, \delta(\mathbf{r}_i - \mathbf{r}_j). \quad (7.74)$$

Introducing the matrix elements $-G_{ij}$ of this force taken on the

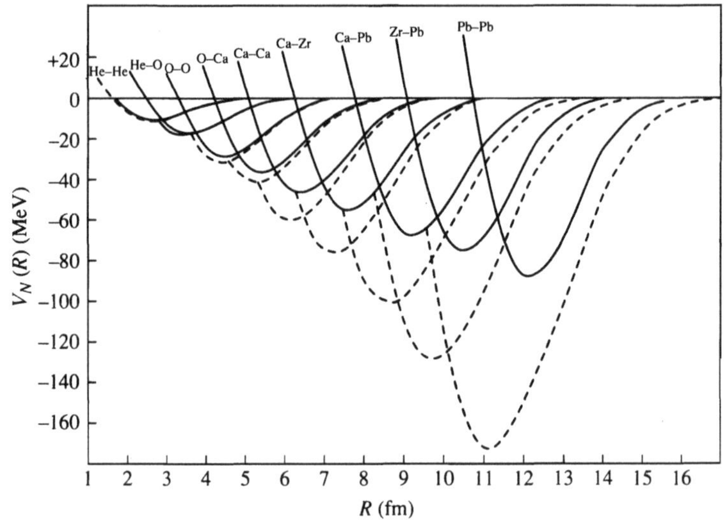

Figure 7.4. The nuclear part of the heavy-ion potential between various pairs of colliding nuclei calculated with SI forces (Stoitsov 1982). The full lines represent the case of eqn (7.72) where the self-consistent condition is taken into account. Results obtained by eqn (7.73) where the self-consistent condition is completely ignored are represented by dashed lines.

radial part of the single-particle wave functions (7.24), that is,

$$G_{ij}[\rho] = \frac{V_0}{4\pi}\int dr\, r^2\, \varphi_i^2(r;[\rho])\, \varphi_j^2(r;[\rho]), \tag{7.75}$$

one obtains the usual BCS equations

$$\Delta_i[\rho] = \frac{1}{4}\sum_j (2j+1)\, \frac{G_{ij}[\rho]\ \Delta_j[\rho]}{\sqrt{(\varepsilon_j[\rho]-\mu_q)^2 + \Delta_j^2[\rho]}}\ , \tag{7.76}$$

which must be solved for μ_q and all $\Delta_i[\rho]$ with the normalization condition

$$\int d\mathbf{r}\ \rho_q(\mathbf{r}) \equiv \sum_i \upsilon_i^2[\rho] = N_q. \tag{7.77}$$

The sums are taken over all single-particle states of $N_p = Z$ protons or $N_n = (N-Z)$ neutrons, according to whether we are treating neutrons ($q = n$) or protons ($q = p$).

The single-particle occupation numbers in eqn (7.77),

$$v_i^2[\rho] = \frac{1}{2}\left[1 - \frac{\varepsilon_i[\rho] - \mu_q}{\sqrt{(\varepsilon_i[\rho]-\mu_q)^2 + \Delta_i^2[\rho]}}\right], \tag{7.78}$$

depend on the chemical potential μ_q and the single-particle energies defined as

$$\varepsilon_i[\rho] = \langle\varphi_i[\rho]|\hat{H}'|\varphi_i[\rho]\rangle. \tag{7.79}$$

The single-particle Hamiltonian

$$\hat{H}' = \sum_{i,k} \langle\varphi_i[\rho]|\hat{T}|\varphi_k[\rho]\rangle\, a_i a_k^+$$
$$+ \sum_{i,k}\sum_j v_j^2[\rho]\, \langle\varphi_i[\rho]\varphi_j[\rho]|\hat{v}|\varphi_k[\rho]\varphi_j[\rho]\rangle_A\, a_i^+ a_k \tag{7.80}$$

includes the kinetic energy operator $\hat{T}$ and the two-particle interaction operator $\hat{v}$ whose antisymmetrized matrix elements appear in eqn (7.80).

In the HF+BCS approximation, the total energy of the system is expressed as

$$E[\rho] = \sum_i v_i^2[\rho]\langle\varphi_i[\rho]|\hat{T}|\varphi_i[\rho]\rangle$$
$$+ \frac{1}{2}\sum_{i,j} v_i^2[\rho]\, v_j^2[\rho]\langle\varphi_i[\rho]\varphi_j[\rho]|\hat{v}|\varphi_i[\rho]\varphi_j[\rho]\rangle_A$$
$$+ E_{pair}[\rho], \tag{7.81}$$

where

$$E_{pair}[\rho] = -\frac{1}{4}\sum_{i,j}(2i+1)(2j+1)G_{ij}[\rho]\ u_i[\rho]u_j[\rho]\upsilon_i[\rho]\upsilon_j[\rho], \quad (7.82)$$

and $u_i^2[\rho]+\upsilon_i^2[\rho] = 1$. It is clear that both sets $(\varphi_i[\rho])$ and $(\upsilon_i^2[\rho])$, and therefore, the total energy (7.81) are functionals of the local density $\rho(\mathbf{r})$, as are all nuclear characteristics of the system under study.

The main steps of the numerical procedure used in the SF approximation are the following. One starts with an initial guess for the parameters R_n, R_p, b_n, b_p; that is, for the SF local densities (7.65). Then, step one, one solves numerically the BCS equations (7.76), adjusting the chemical potentials μ_q to satisfy the corresponding constraint (7.77) to N_q particles (q = n,p). In step two, one solves a nonlinear system of the type of eqn (7.66) with the single-particle occupancies $\upsilon_i^2[\rho_{SF}]$ calculated at step one. The solution obtained with respect to R_n, R_p, b_n, b_p defines new initial values for the iterative procedure at step one, and so on until convergence is achieved.

Discrepancies of a few MeV when compared with HF+BCS binding energies are not a serious problem in the context of the mass formula since the force is, in practice, always fitted to the experimental mass data. Therefore, one can renormalize the effective interaction in order to obtain a better agreement between RED+SF results and the experimental energies in nuclei. The re-fitted set $T1^*$ of Skyrme-type effective forces together with the associated nuclear matter characteristics are compared in Table 7.6 with the original set T1 obtained in HF+BCS calculations (Tondeur *et al.* 1984).

According to the variational method, the HF+BCS total energies obtained with the new interaction $T1^*$ must be deeper than the RED energies. Therefore, it would be meaningless, afterwards, to use the interaction obtained for HF+BCS calculations. Within the RED+SF calculations, however, the new interaction allows one

Table 7.6. Parameters of the original T1 (Tondeur *et al.* 1984) and the re-fitted T1* (Oulne, Petkov, and Stoitsov 1989) set of Skyrme-type interaction and the associated nuclear matter characteristics.

Parameters	T1	T1*
t_0(MeV.fm^3)	-1794.0	-1799.5614
t_1(MeV.fm^5)	298.0	298.9238
t_2(MeV.fm^5)	-298.0	-298.9238
t_3(MeV.fm^4)	12812.0	12851.717
x_0	0.154	0.154
x_1	-0.5	-0.5
x_2	-0.5	-0.5
x_3	0.089	0.089
α	1/3	1/3
W_0(MeV.fm^5)	110.0	110.0
V_0(MeV.fm^3)	210.0	210.0
nuclear matter characteristics		
ρ_0 (fm^{-3})	0.161	0.160
E_{NM} /N (MeV)	-15.98	-16.1
K_{NM} (MeV)	236.1	237.6

to estimate a number of ground-state characteristics with an accuracy comparable to that of the HF+BCS method.

We show in Table 7.7 the results for the total binding energies of some even-even nuclei throughout the periodic table calculated with the renormalized T1* interaction. For comparison we also list the HF+BCS results (Tondeur *et al.* 1984) obtained with the set T1 as well as the experimental data of Jaeger, de Vries and de Vries (1974). The RED+SF results reproduce the experimental data with an accuracy comparable to that of HF+BCS theory. At the same time, it should be remembered that RED+SF calculations are extremely simple and numerically fast for all nuclei including light, heavy, and superheavy ones.

Table 7.7. Total binding energies (in MeV) for some even-even nuclei calculated in RED+SF approach (Oulne, Petkov, and Stoitsov 1989), HF+BCS theory (Tondeur *et al.* 1984), Strutinski integral (SI) and expectation-value (EVM) methods (Dutta *et al.* 1986b). Experimental data taken from Jaeger, de Vries, and de Vries (1974).

Nuclei	RED+SF	HF+BCS	Exp	SI	EVM
$^{16}O_8$	-126.6	-126.9	-127.6	-127.6	-127.0
$^{32}S_{16}$	-269.3	-269.7	-271.8	—	—
$^{40}Ca_{20}$	-341.0	-342.4	-342.1	-344.8	-343.5
$^{48}Ca_{20}$	-418.2	-416.1	-416.0	-414.9	-414.3
$^{56}Fe_{26}$	-489.4	-490.8	-492.3	—	—
$^{90}Zr_{40}$	-783.6	-783.9	-783.9	-785.7	-784.3
$^{132}Sn_{50}$	-1106.8	-1103.9	-1102.9	-1094.0	-1094.9
$^{138}Ba_{56}$	-1152.3	-1158.9	-1158.3	—	—
$^{208}Pb_{82}$	-1637.0	-1636.1	-1636.5	-1628.9	-1628.1
$^{226}Pb_{82}$	-1800.4	-1807.1	—	—	—
$^{354}X_{126}$	-2356.3	-2360.4	—	—	—

Table 7.7 also demonstrates that RED+SF energies reproduce the HF+BCS binding energies with a very good accuracy even for superheavy nuclei such as $^{226}Pb_{82}$ and $^{354}X_{126}$. This gives us confidence in the predictions of the approach for the ground-state properties of nuclei in the superheavy region, some of which are also listed in Table 7.7.

Natural introduction of collective coordinates

It is well known that the choice of collective variables $u = (u_1, u_2, u_3, \ldots)$ and their introduction in a basic collective function

$$\Psi(\mathbf{u}) = \Psi(x_1,x_2,\ldots,x_N;\mathbf{u}), \tag{7.83}$$

is a very difficult problem that occurs in essentially all theories of the nuclear collective motion. In general, the methods which solve the problem can be classified according to two levels of sophistication and accuracy:

(i) The simplest way, and the only possible for the phenomenological collective methods, is simply to choose $\Psi(\mathbf{u})$ on the basis of physical intuition. The Nilsson model (Nilsson 1955) and all phenomenological calculations (Bohr and Mottelson 1974) do that.

(ii) A better way, obviously more accurate but more complicated, is the application of the constrained HF (CHF) method as considered in Section 6.4.

Petkov (1983) and Petkov and Stoitsov (1985) have proposed another quite natural way which introduces collective variables by means of the local density distribution $\rho(\mathbf{r})$. The basic idea is to start from a local density distribution $\rho_{\mathbf{u}}(\mathbf{r}) \equiv \rho(\mathbf{r};\mathbf{u})$ parameterized to depend on a few well-chosen local density characteristics as the nuclear half-radius R or the diffuseness parameter b, some moments of the density or its deformation parameters. Then, following the procedure described in Section 7.1, one can construct a basic collective wave function according to eqn (7.1)

$$\Psi(\mathbf{u}) = \Psi(x_1,x_2,\ldots,x_N;[\rho_{\mathbf{u}}]) = \mathcal{U}_{f[\rho_{\mathbf{u}}]}\bar{\Psi}(x_1,x_2,\ldots,x_N). \tag{7.84}$$

Any known approximation to the many-body wave function may be used in eqn (7.84) as a model function $\bar{\Psi} \in \mathcal{H}^N$. For example, wave functions following from the harmonic oscillator or Woods-Saxon shell model, and even that following from static HF calculations or beyond, may be used at this point.

The density $\rho_{\mathbf{u}}(\mathbf{r})$ should be derived by minimizing the energy density functional $E[\rho]$ with respect to $\rho(\mathbf{r})$ within a density variational space in which all densities are constrained to give a definite value for $\mathbf{u}$. Within the orbit $\mathcal{O}_{SL} \subset \mathcal{H}^{N}_{HF}$, such a procedure corresponds to a specific local density CHF problem in which the densities are constrained to reproduce some given values for the collective variables.

The above procedure seems to combine the simplicity and the physical realism, which are typical of the phenomenological models on the one hand, with the realistic behavior of $\Psi(\mathbf{u})$, similar to the CHF wave functions, on the other. For example, if one considers the local density distribution (Bohr and Mottelson 1974)

$$\rho(\mathbf{r};\alpha_{\lambda\mu}) = \rho_0(r) + \sum_{\lambda=2}^{\infty} \rho_\lambda(r) \sum_{\mu=-\lambda}^{\lambda} Y^*_{\lambda\mu}(\theta,\varphi)\, \alpha_{\lambda\mu}, \qquad (7.85)$$

the deformation parameters $\alpha_{\lambda\mu}$ (all or some of them) become collective coordinates $\mathbf{u} = (\alpha_{\lambda\mu})$ for the basic collective state $\Psi(\alpha_{\lambda\mu})$ following from eqn (7.84) and eqn (7.85). This implies that the family of functions $\Psi(\alpha_{\lambda\mu})$ corresponds to various densities of the form of eqn (7.85) whose radial parts $\rho_\lambda(r)$ minimize the functional $E[\rho]$. This method has advantages compared with the use of a deformed Woods-Saxon potential, for instance. In particular, the conservation condition for the nucleon number, which is used when deforming the densities, has a much stronger physical basis than the usual volume conservation condition used for the potential well.

Among the other possible applications, the local density-dependent wave function (7.84) has been used within the generator coordinate method for a description of isoscalar giant monopole resonances in nuclei (Petkov and Stoitsov 1985, Stoitsov, Petkov, and Dimitrova 1986). Antonov, Petkov, and Stoitsov (1983, 1985) have applied eqn (7.84) in a justification of the coherent density fluctuation model, an important theoretical tool for investigating nucleon momentum distributions in nuclei (Antonov, Hodgson, and Petkov 1988). A search of generating functions associated with the optimal collective path of nuclei has been proposed on the basis of eqn (7.84) by Marinova, Petkov, and Stoitsov (1986a, 1986b).

Flexible parameterization of wave functions and densities

Let us consider a subgroup $\mathcal{F}[\sigma] \subset \mathcal{F}$ containing all local-scale transformation functions of the type of eqn (2.43)

$$\mathbf{f}_\varepsilon(\mathbf{r}) = \exp(i\varepsilon\hat{W}(\sigma))\ \mathbf{r}\ \exp(-i\varepsilon\hat{W}(\sigma)), \tag{7.86}$$

which correspond to different values of the real parameter $\varepsilon \in (-\infty,\infty)$ for a given scalar function $\sigma(\mathbf{r})$. The latter defines the operator $\hat{W}(\sigma)$ via eqn (2.39). According to eqn (2.42), both functions $f_\varepsilon(\mathbf{r}) = f_\varepsilon(r,\mathbf{r}_0)$ and $\sigma(\mathbf{r}) = \sigma(r,\mathbf{r}_0)$ are related to each other by equation

$$\sigma(r,\mathbf{r}_0) = \sigma(\mathbf{f}_\varepsilon(r,\mathbf{r}_0))\left[\frac{df_\varepsilon(r,\mathbf{r}_0)}{dr}\right]^{-1}. \tag{7.87}$$

One can easily verify that all vector functions (7.86) actually form a subgroup $\mathcal{F}[\sigma]$ of $\mathcal{F}$ when ε scans the range of values $-\infty \le \varepsilon \le \infty$. For example, the subgroup $\mathcal{F}[\sigma \equiv r] \subset \mathcal{F}$ is identical with the dilatations group $\mathcal{F}_S \subset \mathcal{F}$ because any element $\mathbf{f}_\varepsilon(\mathbf{r}) \in \mathcal{F}[\sigma \equiv r]$ is of the form $\mathbf{f}_\varepsilon(\mathbf{r}) = \exp(\varepsilon)\mathbf{r}$ which agrees with eqn (2.36).

Inserting eqn (7.86) into eqn (7.1), one obtains the following representation of the wave function within the local-scale transformation method

$$\Psi_\varepsilon[\sigma] = \Psi_\varepsilon(\mathbf{r}_1,\mathbf{r}_2,\ldots,\mathbf{r}_N;[\sigma]) = \mathcal{U}_{\mathbf{f}_\varepsilon}\bar{\Psi}(\mathbf{r}_1,\mathbf{r}_2,\ldots,\mathbf{r}_N), \tag{7.88}$$

where $\bar{\Psi} \in \mathcal{H}^N$ is a certain many-particle model wave function. The family of wave functions (7.88) for $\varepsilon \in (-\infty,\infty)$ is associated with a family of different local density distributions

$$\rho_\varepsilon(\mathbf{r}) = r^{-2} f_\varepsilon^2(\mathbf{r}) \frac{\partial f_\varepsilon(\mathbf{r})}{\partial r} \bar{\rho}(\mathbf{f}_\varepsilon(\mathbf{r})). \qquad (7.89)$$

Both $\Psi_\varepsilon[\sigma]$ and $\rho_\varepsilon(\mathbf{r})$ are completely defined once the form of the scalar function $\sigma(\mathbf{r})$ is known.

Marinova, Petkov, and Stoitsov (1989) have suggested that the function $\sigma(\mathbf{r})$, the generator of the subgroup $\mathcal{F}[\sigma]$, be determined by means of a given symmetrized-Fermi local density distribution $\rho^{SF}(r)$, eqn (7.64). At a given value $\varepsilon = \varepsilon'$, eqn (7.89) rewritten in the form

$$\rho_{\varepsilon'}(r) \equiv \rho^{SF}(r) = r^{-2} f_{\varepsilon'}^2(r)[\partial f_{\varepsilon'}(r)/\partial r]\, \bar{\rho}(\mathbf{f}_{\varepsilon'}(r)), \qquad (7.90)$$

is solved with respect to $f_{\varepsilon'}(r)$. The generator function $\sigma(r)$ then follows as a solution of eqn (7.87) with $f(r) \equiv f_{\varepsilon'}(r)$.

In the particular case of a harmonic oscillator model wave function, both functions $f_{\varepsilon'}(r)$ and $\sigma(r)$ depend on the parameters R,b of the SF local density and also on the oscillator width $\alpha = (m\omega/\hbar)^{1/2}$ coming from the model local density $\bar{\rho}(r)$. Inserting the resulting solution $\sigma(r;R,b,\alpha)$ into eqn (7.86), one defines different local-scale transformation functions $f_\varepsilon(r;R,b,\alpha)$ for different values of $\varepsilon \in (-\infty,\infty)$. In terms of these functions, one further obtains a family of many-particle wave functions

$$\Psi_\varepsilon(R,b,\alpha) = \Psi_\varepsilon(\mathbf{r}_1,\mathbf{r}_2,\ldots,\mathbf{r}_N;R,b,\alpha) \qquad (7.91)$$

with the associated family of local density distributions

$$\rho_\varepsilon(r;R,b,\alpha) = r^{-2} f_\varepsilon^2(r;R,b,\alpha)[\partial f_\varepsilon(r;R,b,\alpha)/\partial r]\bar{\rho}(f_\varepsilon(r;R,b,\alpha)). \qquad (7.92)$$

At $\varepsilon = \varepsilon'$, eqn (7.92) reproduces the smooth SF local density $\rho^{SF}(r)$, while at $\varepsilon = 0$, eqn (7.92) reproduces the harmonic oscillator model density $\bar{\rho}(r)$ which contains density shell oscillations. When $\varepsilon \in (-\infty,\infty)$, the family (7.92) contains many

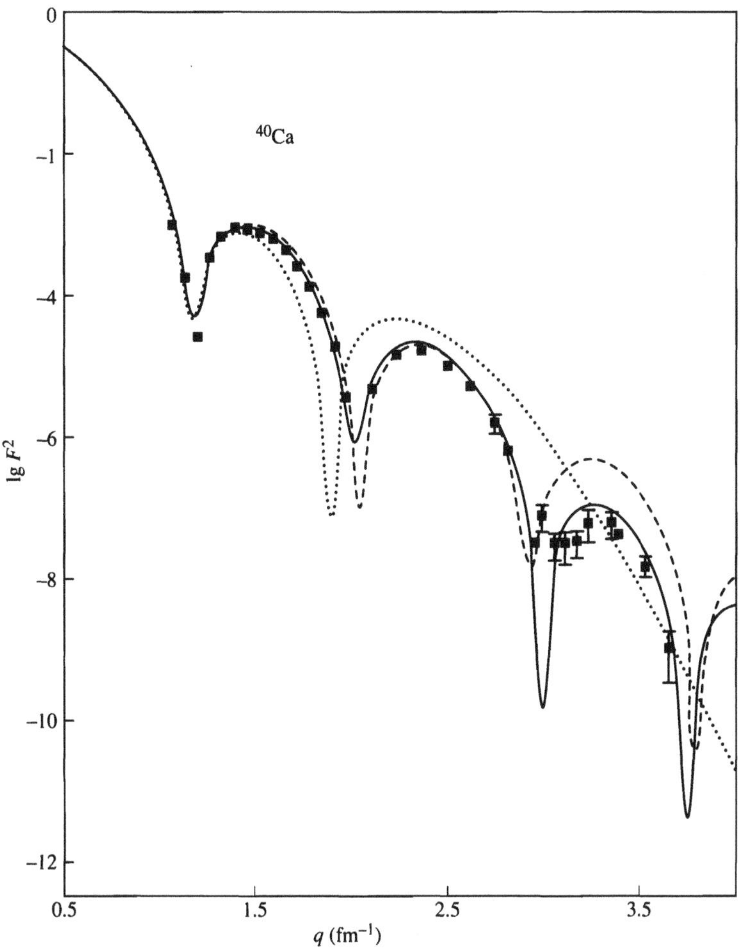

Figure 7.5. Analysis of elastic electron scattering experimental data (Bellicard *et al.* 1967, 1977, Sick *et al.* 1979) in terms of the pseudo-experimental form factor F(q) for the nucleus ^{40}Ca. The best χ^2-fit to the experimental form factor obtained within the family (7.29) (solid line) is compared with the best χ^2-fits within the harmonic oscillator shell model (dotted line) and symmetrized Fermi-type (dashed line) local density distributions.

other particular spatial forms which are well suited for simulating the actual local densities in nuclei.

Rather accurate energy density variational calculations accounting for the shell effects in nuclei have been performed within the family defined by eqn (7.92). Due to its flexibility, the parameterization $\rho_\varepsilon(r;R,b,\alpha)$ also provides a useful tool for analysis of electron elastic scattering data, whereby the many-particle function (7.91) allows the 'experimental' local density distributions to be related to some microscopic properties of nuclei. In Figure 7.5 we show an example of an analysis of elastic electron scattering data for the nucleus ^{40}Ca (Marinova,

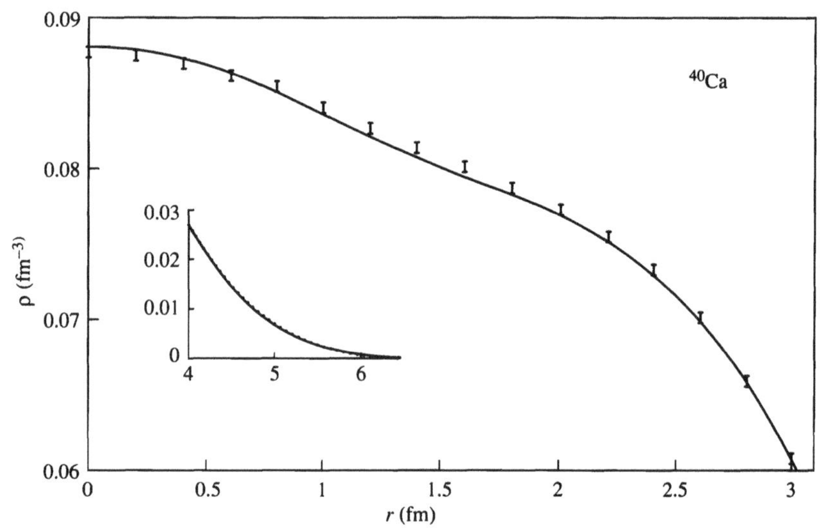

Figure 7.6. Comparison of the local density distribution (7.92) with parameters $\varepsilon = -0.225$, $\alpha = 0.595\ \text{fm}^{-1}$, $R = 3.78$ fm, and $b = 0.6$ fm extracted from the analysis of elastic electron scattering data from Figure 7.5 (solid line) with the 'experimental' charge distribution extracted within the model independent analysis of Sick *et al.* (1979) and Vries, de Jaeger, and de Vries (1987).

Petkov, and Stoitsov 1989). The extracted local density distribution is compared with the 'experimental' one in Figure 7.6.

Other applications of the approach

The RED approach has been applied also to some problems outside the nuclear theory. A rigorous derivation of approximate energy density functionals and their explicit forms have been analysed within the spin-restricted approach in quantum chemistry by Petkov, Stoitsov, and Kryachko (1986a), Kryachko, Petkov, and Stoitsov (1987a,b,c) as well as a computational procedure for reconstructing the many-electron functions on the basis of x-ray diffraction data of one-electron density (Petkov, Stoitsov, and Kryachko 1986b, Kryachko, Petkov, and Stoitsov 1987d). The RED approach has been also applied in the case of atomic systems within a single Slater determinant orbit and hydrogen model single-particle wave functions (Marinova, Petkov, and Stoitsov 1985). Other aspects of the approach have been discussed by Petkov (1988). Its applications for describing nuclear giant resonances and properties of nuclei at finite temperatures are given in Chapters 8 and 10, respectively.

8. Time-dependent processes in nuclei

In this chapter we turn to the question of how the nuclear density functional theory can be used in the description of the dynamic properties of nuclei. The problem is ill defined only in terms of the local densities because, as we have seen earlier in Section 2.3, the Hohenberg-Kohn theorem does not admit direct generalization to time-dependent processes. Density functional theory, however, is closely interlinked to nuclear fluid dynamics or the adiabatic limit of time-dependent HF theory.

8.1 Time-dependent Hartree-Fock theory

In order to describe time-dependent processes in nuclei, one is obviously forced to apply the time-dependent Schrödinger equation (1.1). If one starts at time $t = 0$ with a wave packet $\Psi(0)$, then its evolution with increasing time $\Psi(t)$ is given by the general eqn (1.2) which is too complicated for practical purposes. We have seen in Section 1.4 that in terms of density matrices, eqn (1.1) leads to an infinite system of dynamic equations which connects the one-body density matrix $\rho(\mathbf{r},\mathbf{r}',t)$ with the two-body matrix $\Gamma(\mathbf{r}_1,\mathbf{r}_2,\mathbf{r}_1',\mathbf{r}_2',t)$, the two-body with the three-body density matrices and so on. Therefore, one must seek a physically plausible truncation which specifies the higher order density matrices in terms of the lower ones. This can be done explicitly within the time-dependent HF (TDHF) theory originally proposed by Dirac (1930) and applied for the first time to nuclei by Ferrell (1957).

Time-dependent Hartree-Fock equations of motion

In TDHF theory, one assumes $\Psi(t)$ to be a single Slater determinant at all times

$$\Psi(t) \equiv \Psi(\mathbf{r}_1,\ldots,\mathbf{r}_N,t) = (N!)^{-1/2}\det|\psi_i(\mathbf{r}_j,t)| \quad (i,j=1,\ldots,N). \tag{8.1}$$

In coordinate representation this means that the time evolution affects the wave function (8.1) only by means of the time-dependent single-particle wave functions $\psi_i(\mathbf{r},t)$. The representation (8.1) is conveniently specified by defining the one-body density matrix

$$\rho(t) \equiv \rho(\mathbf{r},\mathbf{r}',t) = \sum_{i=1}^{N} \psi_i^*(\mathbf{r}',t)\psi_i(\mathbf{r},t), \tag{8.2}$$

which satisfies the conditions

$$\mathrm{tr}(\rho(t)) = N \tag{8.3}$$

and

$$\rho^2(t) = \rho(t). \tag{8.4}$$

According to Thouless (1972), eqn (8.3) and eqn (8.4) uniquely determine the wave function of the system into the form of eqn (8.1).

The remarkable property of the Slater determinant (8.1) is that all associated high-order density matrices are completely determined in terms of the one-body matrix (8.2). In particular, the two-body density matrix is of the form

$$\Gamma(\mathbf{r}_1,\mathbf{r}_2,\mathbf{r}_1',\mathbf{r}_2',t) =$$

$$\rho(\mathbf{r}_1,\mathbf{r}_1',t)\rho(\mathbf{r}_2,\mathbf{r}_2',t) - \rho(\mathbf{r}_2,\mathbf{r}_1',t)\rho(\mathbf{r}_1,\mathbf{r}_2',t). \tag{8.5}$$

Thus, the infinite system of dynamic equations (1.106) truncates in TDHF approximation after the first equation with $n = 1$ which, using eqn (8.5), transforms into the TDHF equation of motion

$$i\hbar\dot{\rho} = [\hat{W},\rho]. \tag{8.6}$$

Here, $\rho \equiv \rho(t)$ and $\dot{\rho} \equiv \partial\rho(t)/\partial t$ stand for the density matrix (8.2) and its time derivative, respectively.

In the shorthand notations introduced in Section 5.2, the TDHF Hamiltonian entering eqn (8.6)

$$\hat{W} \equiv \hat{H}_{HF} = \hat{t} + \mathrm{tr}(\hat{\omega}_A\rho), \tag{8.7}$$

is defined exactly as in the static HF case (compare with eqn (5.5)). The difference is that the TDHF Hamiltonian (8.7) contains the time-dependent density matrix $\rho(t)$ and therefore the self-consistent field is time-dependent as well.

Properties of the TDHF equations

The TDHF equation (8.6) is a nonlinear equation with respect to $\rho(t)$ and a first order partial differential equation in time. Starting with an initial guess for $\rho(0)$, it defines the solution $\rho(t)$ for all times and this solution satisfies the idempotent condition (8.4) because we have from eqn (8.6)

$$i\hbar\frac{\partial}{\partial t}(\rho^2 - \rho) = i\hbar(\dot{\rho}\rho + \rho\dot{\rho} - \dot{\rho}) = [\hat{W},\rho^2 - \rho]. \tag{8.8}$$

Therefore, the TDHF wave function (8.1) remains a single Slater determinant for all times. In other words, the density matrix $\rho(t)$, together with the supplementary conditions (8.3) and (8.4), defines a point representing the system in the space $\mathcal{H}^N_{HF}$ of all N-dimensional single Slater determinants. The TDHF dynamic equation (8.6) then defines a trajectory described by this point in the space $\mathcal{H}^N_{HF}$. Since eqn (8.6) is a first order differential equation in time, there is one and only one trajectory through each point of $\mathcal{H}^N_{HF}$.

One can verify by analogy with eqn (8.8) that the orthonormality properties of the single-particle wave functions $\psi_i(\mathbf{r},t)$, the particle number N, the mean values of symmetry operators as well as the total energy of the system

$$E_{HF} = \mathrm{tr}(\hat{t}\ \rho) + \frac{1}{2}\ \mathrm{tr}\ \mathrm{tr}\ (\rho\ \hat{\omega}_A\ \rho), \tag{8.9}$$

are conserved along each of the TDHF trajectories in $\mathcal{H}^N_{HF}$.

Considering the conservation of the energy (8.9), one can conclude that the TDHF collective motion has to take place on lines of constant energy on the multidimensional energy surface. Therefore, if the initial guess for $\rho(0)$ is the static HF solution $\rho_0(0) \equiv \rho_{HF}(\mathbf{r},\mathbf{r}')$; that is, if the system is assumed to be in the absolute ground-state minimum, then the system can never leave this ground-state point of the multidimensional energy surface. In this case, all time-dependence of $\rho(t)$ is gone and one gets the static HF equations

$$\dot{\rho} = 0, \qquad 0 = [\hat{W}_0, \rho_0]. \tag{8.10}$$

It should be stressed for further reference that the dynamic TDHF solution $\rho(t)$ is not in general time-reversal invariant because the time-reversal operator (Messiah 1966)

$$\hat{K} = -\ i\sigma_y \hat{K}_0, \tag{8.11}$$

where $\hat{K}_0$ is the complex conjugation and σ_y is the Pauli matrix, changes the solution $\rho(t)$ of eqn (8.6) and $\rho(t)$ cannot be equal to $\hat{K}\rho(t)\hat{K}^+$ (Baranger and Vénéroni 1978). The only exception is the static HF case in which the time-even single Slater determinant ($\rho_0 = K\rho_0 K^+$) transforms the TDHF equation (8.6) into the static HF equation (8.10).

Actual TDHF calculations

One can represent the TDHF equation (8.6) in coordinate space as

$$-i\hbar\, \dot{\rho}(\mathbf{r},\mathbf{r}',t) = -\frac{\hbar^2}{2m}(\Delta_r-\Delta_{r'})\rho(\mathbf{r},\mathbf{r}',t)+[U_H(\mathbf{r},t)-U_H(\mathbf{r},t)]\rho(\mathbf{r},\mathbf{r}',t)$$

$$+\int d\mathbf{r}''[U_{EX}(\mathbf{r},\mathbf{r}'',t)\rho(\mathbf{r}'',\mathbf{r}',t)-U_{EX}(\mathbf{r}'',\mathbf{r}',t)\rho(\mathbf{r},\mathbf{r}'',t)], \qquad (8.12)$$

where $U_H(\mathbf{r},t)$ and $U_{EX}(\mathbf{r},\mathbf{r}',t)$ are the time-dependent analogies of the direct and exchange single-particle HF potentials (1.137) and (1.138), respectively (the generalization to spin and isospin variables is obvious). It is clear that the solution of the TDHF equation (8.12) in terms of $\rho(\mathbf{r},\mathbf{r}',t)$ is a rather complicated task because it is a partial differential equation in seven independent variables. The same holds for eqn (8.6) in configuration space where $\rho(t)$ usually has a huge number of rows and columns.

In practical calculations, the TDHF equation is easier to handle in terms of the single-particle functions $\psi_i(\mathbf{r},t)$, whereby eqn (8.12) transforms into a system of N differential equations

$$-i\hbar\dot{\psi}_i(\mathbf{r},t)=\left[-\frac{\hbar^2}{2m}\Delta+U_H(\mathbf{r},t)\right]\psi_i(\mathbf{r},t)+\int d\mathbf{r}'\,U_{EX}(\mathbf{r},\mathbf{r}',t)\psi_i(\mathbf{r}',t). \qquad (8.13)$$

Although they require a great deal of numerical effort, the TDHF calculations have been carried out within exactly soluble models or in slab geometry by Bonche, Koonin, and Negele (1976), Yoon and Negele (1977), Krieger (1977), Nogami and Warke (1978), Kan *et al.* (1980); in two dimensions by Koonin (1976), Cusson and Maruhn (1976), Maruhn and Cusson (1976); in quasi-three dimensions by Koonin at al (1977), Maruhn-Rezwani, Davies, and Koonin (1977), Negele *et al.* (1978); in three dimensions by Cusson, Smith, and Maruhn (1976), Flocard and Weiss (1978), Flocard, Koonin, and Weiss (1978), Bonche, Grammaticos, and Koonin (1978), Cusson, Maruhn, and Meldner (1978), Stöcker *et al.* (1980). An interesting TDHF procedure in which the local density distribution is taken as a constraint was presented by Cusson *et al.* (1985). In this kind of calculation all processes such as fusion, compound nucleus formation, dissipation, strongly damped collisions, shock wave propagation or fragmentation are seen. An extensive review of TDHF theory and its extensions is given by Negele (1982), Mädler (1984) and Stöcker and Greiner (1986).

8.2 Adiabatic time-dependent Hartree-Fock theory

We have noticed that the TDHF theory, in a sense, shows classical deterministic behaviour. One has trajectories in the space of Slater determinants $\mathcal{H}^N_{HF}$ with the energy constant along a trajectory. Further, the adiabatic time-dependent Hartree-Fock (ATDHF) theory defines classical coordinates and canonical conjugate momenta and allows us to derive a classical Hamiltonian of the collective motion which could be accordingly quantized.

ATDHF equations of motion

In order to formulate the ATDHF theory, Baranger and Vénéroni (1978) decompose the TDHF density matrix $\rho(t)$ into the form

$$\rho(t) = \exp((i/\hbar)\chi(t))\ \rho_0(t)\ \exp(-(i/\hbar)\chi(t)), \tag{8.14}$$

where $\rho_0(t) \equiv \rho_0(\mathbf{r},\mathbf{r}',t)$ is a time-even one-body density matrix $(\hat{K}\rho_0(t) = \rho_0(t))$ and $\chi(t) \equiv \chi(\mathbf{r},\mathbf{r}',t)$ is a Hermitian time-odd operator $(\hat{K}\chi(t) = -\chi(t))$.

According to eqns (8.3), (8.4) and (8.14), the time-even density matrix $\rho_0(t)$ also satisfies the conditions

$$\mathrm{tr}(\rho_0(t)) = N \tag{8.15}$$

and

$$\rho_0^2(t) = \rho_0(t), \tag{8.16}$$

and it therefore corresponds to a certain time-even Slater determinant

$$\Phi(t) \equiv \Phi(\mathbf{r}_1,\ldots,\mathbf{r}_N,t) = (N!)^{-1/2}\det|\varphi_i(\mathbf{r}_j,t)|, \qquad (8.17)$$

with single-particle wave functions $\varphi_i(\mathbf{r},t)$ different from the functions $\psi_i(\mathbf{r},t) = \exp((i/\hbar)\chi(t))\varphi_i(\mathbf{r},t)$ entering eqn (8.1).

The decomposition (8.14) reveals the TDHF dynamics of the system as a collective motion along a trajectory in the space of time-even Slater determinants, any dynamic state being a point on this trajectory with coordinate $\rho_0(t)$ and conjugate momentum $\chi(t)$. The adiabatic approximation consists in the restriction of the motion only over the trajectories in the ρ_0-space which are adiabatic in the sense that the operator $\chi(t)$ must be small.

In order to derive the ATDHF equations of motion, one may start by expanding eqn (8.14) as a power series in $\chi(t)$

$$\rho(t) = \exp((i/\hbar)\chi)\rho_0\exp(-(i/\hbar)\chi) = \rho_0(t) + \rho_1(t) + \rho_2(t) + \ldots$$

$$=\rho_0(t)+(i/\hbar)[\chi(t),\rho_0(t)]+\tfrac{1}{2}(i/\hbar)^2[\chi(t),[\chi(t),\rho_0(t)]]+\ldots, \qquad (8.18)$$

where

$$\rho_1 \equiv \rho_1(t) = (i/\hbar)[\chi(t),\rho_0(t)] \qquad (8.19)$$

and

$$\rho_2 \equiv \rho_2(t) = \tfrac{1}{2}(i/\hbar)^2[\chi(t),[\chi(t),\rho_0(t)]]. \qquad (8.20)$$

Similarly, one can expand $\mathbf{W}(t)$ from eqn (8.7) and eqn (8.14)

$$\hat{W}(t) = \hat{W}_0(t) + \hat{W}_1(t) + \hat{W}_2(t) + \ldots, \qquad (8.21)$$

with

$$\hat{W}_0(t) = \hat{t} + \mathrm{tr}(\hat{\omega}_A\rho_0), \qquad (8.22)$$

$$\hat{W}_1(t) = \mathrm{tr}(\hat{\omega}_A \rho_1), \tag{8.23}$$

$$\hat{W}_2(t) = \mathrm{tr}(\hat{\omega}_A \rho_2). \tag{8.24}$$

Substituting in the TDHF equation (8.6) the expansions (8.19) and (8.21), separating the time-even from the time-odd parts and stopping at the lowest significant orders in $\chi(t)$, one obtains the following ATDHF equations (Baranger and Vénéroni 1978)

$$\text{ATDHF(I):} \quad i\hbar\dot{\rho}_0 = [\hat{W}_0, \rho_1] + [\hat{W}_1, \rho_0], \tag{8.25}$$

$$\text{ATDHF(II):} \quad i\hbar\dot{\rho}_1 = [\hat{W}_0, \rho_0] + [\hat{W}_1, \rho_1] + [\hat{W}_2, \rho_0] + [\hat{W}_0, \rho_2]. \tag{8.26}$$

The right-hand side of eqn (8.25) is of the first order in $\chi(t)$, while all terms equal to or higher than the third order in $\chi(t)$ are neglected in eqn (8.26).

The ATDHF(I) equation reflects the physical nature of the ATDHF approximation used. It shows that $\dot{\rho}_0(t)$, the velocity in the space of time-even Slater determinants, is first order in $\chi(t)$. Therefore, the adiabatic approximation is understood as a slow motion of the system, in a sense that its velocity $\dot{\rho}_0(t)$ is small. If we write A for the amplitude and ω for the angular frequency of the adiabatic motion, then from the Fourier analysis of $\rho_0(t)$ it becomes clear that the smallness of the velocity $\dot{\rho}_0(t) \sim A\omega$ means that either the frequency is small (which can be satisfied even with a large amplitude motion), or the amplitude is small (when a large frequency is possible), or both.

From the ATDHF(II) equation it is seen that in general the commutator $[\hat{W}_0, \rho_0]$ is not zero, but small, independent of the fact that $\hat{W}_0$ and ρ_0 separately could be large. In the limit when $\chi(t)$ goes to zero, both eqn (8.25) and eqn (8.26) lead to the static HF equation (8.10). However, when $\chi(t)$ is not zero the smallness of $[\hat{W}_0, \rho_0]$ means that the static HF condition is almost satisfied by

$\rho_0(t)$ at all times. Therefore, the nucleus is almost in static equilibrium at all times during the adiabatic collective motion.

Baranger and Vénéroni (1978) show that the quantity $[\hat{W}_0, \rho_0]$ is closely related to the acceleration of the representative point in $\rho_0(t)$-space. Then the smallness of $[\hat{W}_0, \rho_0]$ is a necessary and sufficient condition for the existence of adiabatic trajectories in $\rho_0(t)$-space. In order to describe the collective motion along one of the adiabatic trajectories one has to solve the ATDHF equations I and II with respect to $\rho_0(t)$ and $\chi(t)$ with appropriate initial conditions which may be either the values of $\rho_0(0)$ and $\chi(0)$ or the values of $\rho_0(0)$ and $\dot{\rho}_0(0)$.

Total energy along the collective path

In order to obtain the total energy of the system in the ATDHF approximation, we insert the expansion (8.18) into eqn (8.9). Up to second order with respect to $\chi(t)$, the result is

$$E \equiv \mathcal{H}(\chi, \rho_0) = \mathcal{K} + \mathcal{V}, \tag{8.27}$$

where

$$\mathcal{V} \equiv \mathcal{V}(\rho_0) = \mathrm{tr}(\hat{t}\rho_0) + \frac{1}{2}\,\mathrm{tr}\;\mathrm{tr}\;(\rho_0 \hat{\mathcal{V}}_A \rho_0) \tag{8.28}$$

is the zero-order term further identified with the collective potential energy of the system. It is just the expectation value of the nuclear Hamiltonian $\hat{H}$ for the time-even Slater determinant (8.17) defining $\rho_0(t)$.

The second-order term $\mathcal{K}$ in eqn (8.27) is quadratic in $\chi(t)$. According to ATDHF(I) which gives $\dot{\rho}_0 \sim \chi$, this term is quadratic with respect to the velocity $\dot{\rho}_0(t)$ and therefore it can be identified with the collective kinetic energy of the system. Its explicit form

$$\mathcal{K} \equiv \mathcal{K}(\chi, \rho_0) = \mathrm{tr}(\hat{W}_0 \rho_2) + \frac{1}{2}\,\mathrm{tr}\;\mathrm{tr}\;(\rho_1 \hat{\mathcal{V}}_A \rho_1) \tag{8.29}$$

may be simplified in the general case to give (Brink, Giannoni, and Vénéroni 1976)

$$\mathcal{K} = \frac{1}{2}\,\mathrm{tr}(\chi(t)\dot{\rho}_0(t)). \tag{8.30}$$

One thus obtains an explicit expression for the collective energy of the system which is easy for calculation if once the ATDHF equations are solved.

Reduction to a few collective variables

It is no easier to solve the ATDHF equations than to solve the full TDHF problem. For this reason, one should try to reduce somehow the degrees of freedom of the problem, which could be done for instance by imposing on the system a collective motion within a restricted set of time-even single Slater determinants. A particularly interesting case is to pick a trial family $\rho_0(u)$ depending upon the time by means of a finite number of real collective parameters $\mathbf{u} = (u_1,u_2,u_3,...)$ and to leave the operator $\chi(t)$ free to be determined as a solution of the ATDHF equation I. We neglect the equation ATDHF(II) for a moment but it may be further used to define the optimal collective path for the system in the ρ_0-space.

With the collective path $\rho_0(\mathbf{u})$ imposed on the system, the equation ATDHF(I) reads

$$[\hat{W}_0,\rho_1] + [\hat{W}_1,\rho_0] = i\hbar\sum_l(\partial\rho_0(u)/\delta u_l)\dot{u}_l, \tag{8.31}$$

where $\dot{u}_l$ stands for the time derivative $du_l(t)/dt$. This is a set of linear equations with respect to the matrix elements of $\rho_1(t)$. The operator $\chi(t)$ may be then determined by means of $\rho_1(t)$ and it will depend on the collective vector $\mathbf{u}$ and its time derivative $\dot{\mathbf{u}}$.

It should be underlined that such a procedure is not uniquely determined due to the non-uniqueness of the decomposition (8.14). The ATDHF equations I and II do not determine all matrix elements of $\chi(t)$. Both equations have only particle-hole and hole-particle matrix elements and therefore particle-particle and hole-hole matrix elements of $\chi(t)$ are completely arbitrary (Brink, Giannoni, and Vénéroni 1976). A natural choice is to set them equal to zero, although other conditions may by applied as well (Bonche and Quentin 1978). Baranger and Vénéroni (1978) use this choice to assume the supplementary conditions

$$\rho_0 \chi \rho_0 = (1 - \rho_0)\chi(1 - \rho_0) = 0 \tag{8.32}$$

and

$$-\frac{\pi}{4}\hbar < \text{eigenvalues of } \chi < +\frac{\pi}{4}\hbar, \tag{8.33}$$

which uniquely determine $\rho_0(t)$ and $\chi(t)$ from the general matrix $\rho(t)$ by means of eqn (8.14).

It is now possible from eqn (8.19) and eqn (8.32) to express $\chi(t)$ in the form

$$\chi = -i\hbar[\rho_1, \rho_0]. \tag{8.34}$$

Hence, inverting eqn (8.31), one obtains

$$\chi(t) = \sum_i K_i(u)\, \dot{u}_i, \tag{8.35}$$

where the quantities $K_i(u)$ depend only on u. Finally, inserting eqn (8.35) into eqn (8.30), one derives the classical Hamiltonian (8.27) in the form

$$\mathcal{H} = \mathcal{K}(u,\dot{u}) + \mathcal{V}(u) = \frac{1}{2}\sum_{i,j} m_{ij}(u)\, \dot{u}_i \dot{u}_j + \mathcal{V}(u), \tag{8.36}$$

where $m_{ij}(u)$ is the real collective mass tensor and $\mathcal{V}(u) \equiv \mathcal{V}(\rho_0(u))$ is the collective potential energy (8.28).

In order to get a collective Hamiltonian $\hat{\mathcal{H}}$ which allows us to calculate the collective wave functions $\Psi(u)$ and other characteristics of the system under consideration, one has to requantize eqn (8.36). For this purpose, one can apply the standard method of the Pauli quantization as done by Baranger and Vénéroni (1978), or one can use another quantization procedure taking into account the zero-point energy in nuclei as done for instance by Goeke (1976, 1977).

We do not want to go into more details of the ATDHF method because the field is too large and there are numerous attempts to develop approximations along this line as well as widespread applications in nuclear physics. Among them, we mention the formulation of ATDHF theory given by Villars (1977) as well as the so-called quantized ATDHF theory developed and widely applied by Goeke and Reinhard (1978), Goeke, Reinhard, and Rowe (1981), Goeke, Gruemmer, and Reinhard (1983), Slavov *et al.* (1986), Reinhard and Goeke (1987), where additional information is used from ATDHF(II) in order to determine the optimal collective path of the system.

8.3 Local-scale version of the ATDHF approach

We shall now demonstrate how the RED approach presented in the previous chapter allows the collective kinetic and potential energies following from the ATDHF approach to be expressed as functionals of the time-dependent local density distribution in nuclei.

ATDHF problem in coordinate space

Engel *et al.* (1975) have formulated the ATDHF approach with Skyrme-type effective forces after imposing on the system a collective path $\rho_0(t)$ and then solving ATDHF(I) in coordinate space. They start with the time-even Slater determinant (8.17) derived within the CHF method with a suitable set of time-even constraint operators $\hat{\mathbf{Q}} = (\hat{Q}_1, \hat{Q}_2, \hat{Q}_3, \ldots)$ and consider the associated Lagrange multipliers $\mathbf{u} = (u_1,u_2,u_3,\ldots)$. The resulting collective path

$$\rho_0(\mathbf{u}) \equiv \rho_0(\mathbf{r},\mathbf{r}',t) = \sum_{l=1}^{N} \varphi_l^*(\mathbf{r}',t)\varphi_l(\mathbf{r},t), \tag{8.37}$$

solves the CHF problem

$$[\hat{W}_0 - \mathbf{u}\cdot\hat{Q},\rho_0(t)] = 0, \tag{8.38}$$

and thus depends on collective variables $\mathbf{u}$.

From eqns (8.16), (8.19), and (8.32) one obtains the equality

$$\rho_1 = \rho_0\rho_1 + \rho_1\rho_0, \tag{8.39}$$

which, together with eqn (8.37), puts $\rho_1(t)$ into the form

$$\rho_1(\mathbf{r},\mathbf{r}',t) = i\sum_{l=1}^{N}\left[\varphi_l^*(\mathbf{r}',t)\eta_l(\mathbf{r},t) - \eta_l^*(\mathbf{r}',t)\varphi_l(\mathbf{r},t)\right], \tag{8.40}$$

The new single-particle functions

$$\eta_l(\mathbf{r},t) = \int d\mathbf{r}'\ \rho_1(\mathbf{r},\mathbf{r}',t)\varphi_l(\mathbf{r}',t) \tag{8.41}$$

should be derived as a solution of the ATDHF(I) equation in coordinate space (Engel *et al.* 1975)

$$(\hat{W}_0 - \varepsilon_l)\eta_l(\mathbf{r},t) = i\hbar\int d\mathbf{r}'\ \dot{\rho}_1(\mathbf{r},\mathbf{r}',t)\varphi_l(\mathbf{r}',t)$$

$$- \int d\mathbf{r}'\langle\mathbf{r}|[\hat{W}_1,\rho_0]|\mathbf{r}'\rangle\ \varphi_l(\mathbf{r}',t). \tag{8.42}$$

With the Skyrme-type effective forces, eqn (8.42) is an inhomogeneous differential equation for the unknown functions $\eta_l(\mathbf{r},t)$ which should be solved by iterations subject to the boundary condition that $\eta_l(\mathbf{r},t)$ is regular and goes to zero for $r\to\infty$ and $\langle\eta_m|\varphi_l\rangle = 0$ for all occupied states $\varphi_l(\mathbf{r},t)$.

Substituting eqn (8.37) and eqn (8.40) into eqn (8.34), one obtains

$$\chi(\mathbf{r},\mathbf{r}',t) = \sum_{i=1}^{N} \left[\eta_i^*(\mathbf{r}',t)\varphi_i(\mathbf{r},t) + \varphi_i^*(\mathbf{r}',t)\eta_i(\mathbf{r},t)\right], \qquad (8.43)$$

which together with the time derivative of eqn (8.37) gives the collective kinetic energy (8.30) in the form

$$\mathcal{K} = \frac{1}{2}\hbar \int d\mathbf{r} \sum_{i=1}^{N} \left[\eta_i^*(\mathbf{r},t)\dot{\varphi}_i(\mathbf{r},t) + \dot{\varphi}_i^*(\mathbf{r},t)\eta_i(\mathbf{r},t)\right]. \qquad (8.44)$$

In this way, the collective kinetic energy in the coordinate representation (8.44), the collective potential energy $\mathcal{V}(\mathbf{u}) = \mathcal{V}(\rho_0(\mathbf{u}))$ and, therefore, the classical Hamiltonian (8.36) are completely determined after solving the CHF problem (8.38) as well as the ATDHF(I) equation (8.42).

Generalized scaling properties

An approximate approach that avoids the solution of the ATDHF equation (8.38) is supplied by using single-particle wave functions $\varphi_i(\mathbf{r},t)$ which satisfy the generalized scaling property (Engel *et al.* 1975). This could in principle be done within the CHF method with the help of additional time-dependent external fields. It appears, however, more appealing and at the same time easier to use for this purpose the time-even Slater determinant

$$\Phi(t) \equiv \Phi[\rho_\mathbf{u}] = (N!)^{-1/2}\det|\varphi_i(\mathbf{r}_j;[\rho_\mathbf{u}])| \qquad (i,j=1,\dots,N), \qquad (8.45)$$

derived as in eqn (7.43) in terms of the density-dependent wave functions

$$\varphi_i(\mathbf{r},t) \equiv \varphi_i(\mathbf{r};[\rho_\mathbf{u}]) = \left[\frac{f^2_{[\rho_\mathbf{u}]}(r)}{r^2}\,\frac{\partial f_{[\rho_\mathbf{u}]}(r)}{\partial r}\right]^{1/2} \bar{\varphi}_i(\mathbf{f}_{[\rho_\mathbf{u}]}(\mathbf{r})). \qquad (8.46)$$

The local density distribution $\rho_\mathbf{u}(\mathbf{r})$ entering eqn (8.46) depends on the time by means of the collective variables $\mathbf{u} = \mathbf{u}(t)$ which may be naturally introduced as described in Section 7.5.

Dimitrova, Petkov, and Stoitsov (1988a) show the important fact that the single-particle wave functions (8.46) satisfy the generalized scaling properties

$$\dot{\varphi}_l^q(\mathbf{r},t) = \mathbf{v}_q \cdot \nabla\varphi_l^q(\mathbf{r},t) - \frac{1}{2}\varphi_l^q(\mathbf{r},t)\nabla \cdot \mathbf{v}_q \tag{8.47}$$

with real vector functions

$$\mathbf{v}_q \equiv \mathbf{v}_q(\mathbf{r},t) = - \mathbf{r}_0[\dot{f}_{\rho^q_{\mathbf{u}}}(\mathbf{r})/f'_{\rho^q_{\mathbf{u}}}(\mathbf{r})], \tag{8.48}$$

where $\dot{f}_{\rho^q_{\mathbf{u}}}(\mathbf{r}) = \partial f_{\rho^q_{\mathbf{u}}}(\mathbf{r})/\partial t$ and $f'_{\rho^q_{\mathbf{u}}}(\mathbf{r}) = \mathbf{r}_0 \cdot \nabla f_{\rho^q_{\mathbf{u}}}(\mathbf{r})$. Under conditions (8.47), all nucleons are forced to move in a common velocity field $\mathbf{v}_q$, q = (n,p) for each kind of nucleons.

Classical Hamiltonian as an energy density functional

Inserting eqn (8.47) into eqn (8.44) one obtains, after a partial integration and some rearrangement, the expression

$$\mathcal{K} = (\hbar/2)\int d\mathbf{r}(\mathbf{v}_n \cdot \mathbf{j}_n + \mathbf{v}_p \cdot \mathbf{j}_p), \tag{8.49}$$

where

$$\mathbf{j}_q \equiv \mathbf{j}_q(\mathbf{r},t) = \frac{\hbar}{2mi}[(\nabla_r - \nabla_{r'})(\rho_0 + \rho_1)\Big|_{r'=r} = \frac{\hbar}{2mi}[(\nabla_r - \nabla_{r'})\rho_1(t)\Big|_{r'=r}$$

$$= \frac{\hbar}{2m}\sum_{l=1}^{N_q}(\varphi_l^{q*}\nabla\eta_l^q - \eta_l^q\nabla\varphi_l^{q*} - \eta_l^{q*}\nabla\varphi_l^q + \varphi_l^q\nabla\eta_l^{q*}), \tag{8.50}$$

is the current density in the ATDHF approximation and $\mathbf{j}_{0q}(t)$ vanishes identically because $\rho_0(t)$ is time-even.

As shown by Engel *et al.* (1975), the ATDHF equation (8.42) leads to a continuity equation which, rewritten by means of eqn (8.45) and the Skyrme-type effective forces , reads

$$\dot{\rho}_u^q(\mathbf{r}) + \nabla.\mathbf{J}_q(\mathbf{r},t) = 0 \tag{8.51}$$

with a generalized current density

$$\mathbf{J}_q \equiv \mathbf{J}_q(\mathbf{r},t) = \mathbf{j}_q(\mathbf{r},t) + \beta[\rho_u^{\bar{q}}\mathbf{j}_q(\mathbf{r},t) - \rho_u^q\mathbf{j}_{\bar{q}}(\mathbf{r},t)], \tag{8.52}$$

where the index $\bar{q} = (p,n)$ is used together with the index $q = (n,p)$, $\mathbf{j}_q(\mathbf{r},t)$ is given by eqn (8.50) and the non-locality parameter

$$\beta = (m/2\hbar^2)(t_1 + t_2) \tag{8.53}$$

comes from the exchange terms of the effective forces.

On the other hand, the time-derivative $\dot{\rho}_u^q(\mathbf{r})$ in eqn (8.51) satisfies the equation

$$\dot{\rho}_u^q(\mathbf{r}) + \nabla \, . \, [\rho_u^q(\mathbf{r})\mathbf{v}_q] = 0, \tag{8.54}$$

which follows when multiplying eqn (8.47) by $\varphi_1^{q*}(\mathbf{r},t)$, adding the complex conjugate resulting equation and then summing over all occupied states.

The comparison of eqn (8.51) with eqn (8.54) leads to the relation

$$\nabla.\left\{[\rho_u^q(\mathbf{r})\mathbf{v}_q] - \left[\mathbf{j}_q + \beta[\rho_u^{\bar{q}}\mathbf{j}_q(\mathbf{r},t) - \rho_u^q\mathbf{j}_{\bar{q}}\,(\mathbf{r},t)]\right]\right\} = 0, \tag{8.55}$$

which allows the current density $\mathbf{j}_q(\mathbf{r},t)$ to be expressed by means of the local densities $\rho_u^q(\mathbf{r})$ and their time derivatives $\dot{\rho}_u^q(\mathbf{r})$, $q = (n,p)$. Assuming that the expression in the brackets of eqn (8.48) is irrotational, as for example in the case of breathing vibrations of spherical nuclei, one obtains

$$\frac{m}{\hbar}\mathbf{j}_q = \beta\,\frac{\rho_u^q\,\rho_u^{\bar{q}}}{1+\beta\rho_u}\,\mathbf{v}_q + \rho_u^q\,\frac{1+\beta\rho_u^q}{1+\beta\rho_u}\,\mathbf{v}_{\bar{q}}\,, \tag{8.56}$$

where β is given by eqn (8.53) and $\rho_{\mathbf{u}} = \rho^{n}_{\mathbf{u}} + \rho^{p}_{\mathbf{u}}$. Finally, inserting eqn (8.56) into eqn (8.49), the collective kinetic energy takes the form of a local density functional

$$\mathcal{K} \equiv \mathcal{K}[\rho_{\mathbf{u}}] = \frac{m}{2}\int d\mathbf{r}\left[\rho^{n\,2}_{\mathbf{u}}\, \frac{1+\beta\rho^{n}_{\mathbf{u}}}{1+\beta\rho_{\mathbf{u}}}\, v^{2}_{n} + \beta\, \frac{\rho^{n}_{\mathbf{u}}\,\rho^{p}_{\mathbf{u}}}{1+\beta\rho_{\mathbf{u}}}\, v_{n} v_{p} + \rho^{p\,2}_{\mathbf{u}}\, \frac{1+\beta\rho^{p}_{\mathbf{u}}}{1+\beta\rho_{\mathbf{u}}}\, v^{2}_{p}\right] d\mathbf{r}, \tag{8.57}$$

since the velocity fields $\mathbf{v}_{q}(\mathbf{r},t)$ are already determined by $\rho^{q}_{\mathbf{u}}(\mathbf{r})$ and $\dot{\rho}^{q}_{\mathbf{u}}(\mathbf{r})$ according to eqn (8.48). The same holds reasonably for the collective potential energy

$$\mathcal{V}[\rho_{\mathbf{u}}] \equiv E[\rho_{\mathbf{u}}] = \langle\Phi[\rho_{\mathbf{u}}]|\hat{H}|\Phi[\rho_{\mathbf{u}}]\rangle. \tag{8.58}$$

which in fact identifies the RED functional in $\mathcal{O}_{SL} \subset \mathcal{H}^{N}_{HF}$. In this way, eqn (8.57) and eqn (8.58) express the classical Hamiltonian (8.36) completely as a functional of the local density distribution

$$\mathcal{H}[\rho_{\mathbf{u}}] = \mathcal{K}[\rho_{\mathbf{u}}] + \mathcal{V}[\rho_{\mathbf{u}}]. \tag{8.59}$$

There exist a number of approaches which assume the collective potential energy in the dynamic case to be a static TF-like energy density functional of the time-dependent densities (see the next section). Such a procedure may be questioned due to the distortion of the Fermi sphere during the collective motion. In contrast, we see that both $\mathcal{K}[\rho_{\mathbf{u}}]$ and $\mathcal{V}[\rho_{\mathbf{u}}]$ are obtained simultaneously in eqn (8.59) and the inconsistency mentioned above is avoided due to the particular form of the energy density functional derived on the basis of a given model many-particle time-even Slater determinant.

Giant monopole resonances

The local-scale version of the ATDHF approach presented above has been applied by Dimitrova, Petkov, and Stoitsov (1986b, 1988a,b) to isoscalar (ISMR) and isovector (IVMR) giant monopole resonances in spherical nuclei. They consider the collective path $\rho_0(\mathbf{u})$ of Slater determinants

$$\Phi(R_n,R_p,b_n,b_p) \equiv \Phi(\mathbf{r}_1,\ldots,\mathbf{r}_N;[\rho_n^{SF},\rho_p^{SF}]) = \mathcal{U}_{f[\rho_{SF}]}\bar{\Phi}(\mathbf{r}_1,\ldots,\mathbf{r}_N),$$

belonging to the orbit $\mathcal{O}_{SL}^{HO} \subset \mathcal{H}_{HF}^{N}$ and associated with symmetrized Fermi-type (SF) local density distributions (7.58), that is, the choice $\rho_{\mathbf{u}}^{q}(r) \equiv \rho^{SF}(r;R_q,b_q)$ is assumed. The collective variables are thus naturally represented by SF parameters $\mathbf{u} \equiv \mathbf{u}(t) = (R_n(t),R_p(t),b_n(t),b_p(t))$. This choice of collective variables allows us to investigate nuclear giant monopole resonances as coupled bulk (R-type) and surface (b-type) oscillations of the nuclear densities. Similar collective variables have been considered with applications to monopole vibrations in nuclei by Jensen and Larsen (1981), Brack and Stöcker (1983), Shlomo and Siemens (1985), and Gleissl *et al.* (1988).

Using eqn (8.60), the collective potential energy (8.58) represents a multidimensional energy surface

$$\mathcal{V}(R_n,R_p,b_n,b_p) \equiv E(R_n,R_p,b_n,b_p) \tag{8.61}$$

with an absolute minimum at the static ground-state energy as obtained after solving eqn (7.66) of Section 7.4. The collective kinetic energy (8.57) is then given in the form

$$\mathcal{K} = \frac{m}{2}\sum_{i,j=1}^{4} m_{ij}(\mathbf{u})\dot{u}_i\dot{u}_j, \tag{8.62}$$

with mass tensor matrix elements

$$m_{ij} = \int\left[\frac{\beta\rho_{\mathbf{u}}^{q_i}\rho_{\mathbf{u}}^{q_j}(1-\delta_{q_iq_j})+\rho_{\mathbf{u}}^{q_i}(1+\beta\rho_{\mathbf{u}}^{q_i})\delta_{q_iq_j}}{(1+\beta\rho_{\mathbf{u}})}\right] v_i v_j\, d\mathbf{r}, \tag{8.63}$$

where the isospin index $q_i = n(p)$ for $i = 1,3(2,4)$ is ordered to correspond to the index $i = 1,2,3,4$ labelling the collective vector $\mathbf{u} \equiv (u_1,u_2,u_3,u_4) = (R_n,R_p,b_n,b_p)$. Taking into account eqn. (7.64) and eqn (8.48), one obtains the velocity fields (measured in units $\dot{u}_i = du_i(t)/dt$)

$$v_i(r,t) = \frac{\int_0^r (\partial \rho_{\mathbf{u}}^{q_i}/\partial u_i) r'^2 dr'}{\rho_{\mathbf{u}}^{q_i} r^2} . \tag{8.64}$$

Collecting eqn (8.61) and eqn (8.62), one gets the classical Hamiltonian (8.59) in the desired form of eqn (8.36)

$$\mathcal{H} = \frac{m}{2}\sum_{i,j=1}^{4} m_{ij}(\mathbf{u})\dot{u}_i\dot{u}_j + \mathcal{V}(\mathbf{u}). \tag{8.65}$$

Dimitrova, Petkov, and Stoitsov (1988a) solve the requantization problem in the same context as in RPA by looking for small oscillations around equilibrium of the spherical nucleus. In this way the quantized monopole vibration splits into four normal modes of monopole excitations. Two of them, lower in energy, correspond to ISMR and IVMR, respectively. The other two normal modes are of the so-called antiscaling type in which the nuclear surface (b-type vibrations) oscillates opposite in phase to the bulk (or R-type) vibrations. The antiscaling-type excitations carry relatively little strength and are pushed up in energy due to the strong coupling between the bulk and surface vibrations.

We show in Table 8.1 typical results for excitation energies, isoscalar and isovector energy-weighted sum rules (EWSR) and the amplitudes of the transition densities obtained with SkM* forces for all four normal modes of excitation. The calculated excitation energies are compared in Figures 8.1 and 8.2 with

Table 8.1. Characteristics of the four normal modes of giant monopole excitations calculated with SkM* forces by Dimitrova, Petkov and Stoitsov (1988a): excitation energies E_α (in MeV), isoscalar (EMS) and isovector (EMV) energy weighted sum rules (in percent) and coefficients $t_{1\alpha}$ (in fm) of the transition density (8.66).

Nucleus	α	E_α	EMS	EMV	$t_{1\alpha}$	$t_{2\alpha}$	$t_{3\alpha}$	$t_{4\alpha}$
^{16}O	4	58.99	0.07	20.13	0.609	-0.427	-0.117	0.076
	3	50.53	4.34	1.59	0.393	0.608	-0.092	-0.126
	2	30.59	0.08	65.01	-0.051	0.113	0.094	-0.104
	1	23.31	79.35	0.01	0.067	0.037	0.075	0.085
^{40}Ca	4	51.98	0.00	25.52	0.339	-0.293	-0.082	0.067
	3	40.46	2.80	0.94	0.277	0.337	-0.086	-0.094
	2	29.85	0.00	69.14	0.005	0.040	0.060	-0.074
	1	20.77	91.40	0.01	0.102	0.091	0.044	0.046
^{90}Zr	4	47.57	0.02	22.46	0.164	-0.229	-0.049	0.073
	3	36.10	0.46	0.74	0.195	0.158	-0.076	-0.069
	2	29.83	0.00	74.37	0.013	-0.038	0.042	-0.044
	1	17.79	97.29	0.79	0.109	0.121	0.022	0.018
^{208}Pb	4	42.21	0.00	16.23	0.086	-0.157	-0.034	0.066
	3	31.22	0.00	1.23	0.117	0.081	-0.061	-0.052
	2	27.50	0.21	79.79	0.027	-0.053	0.026	-0.028
	1	13.59	98.93	2.08	0.102	0.106	0.006	0.010

available experimental data for ISMR and IVMR, respectively. The nice agreement seen in these figures is very important because giant monopole resonances can be related to some characteristics of effective forces in nuclei. For example we show in Figure 8.3 the almost linear dependence of the ISMR excitation energies on the nuclear matter incompressibility coefficient K_{NM} (see Table 2.1).

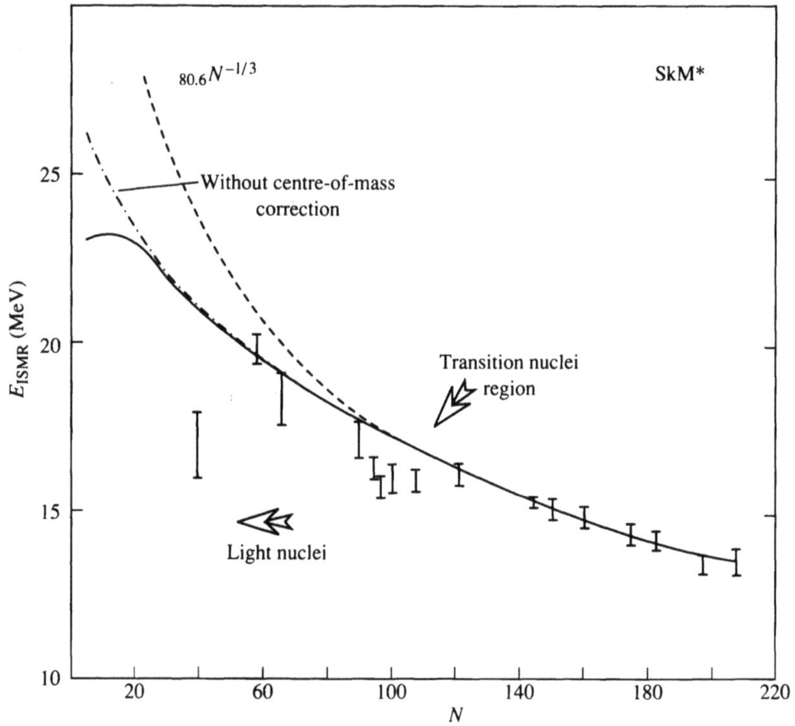

Figure 8.1. Excitation energies of the isoscalar giant monopole resonance (ISMR) of spherical nuclei obtained from the local-scale version of the adiabatic time-dependent Hartree-Fock approach using SkM* effective forces with (full line) and without (short-dashed line) centre-of-mass correction (Dimitrova, Petkov and Stoitsov 1988a). Some experimental ISMR peak energies (Bertrand 1981, Buenerd 1984) are also given. The dashed curve corresponds to the classical trend $E_{ISMR} \cong 80$ MeV $N^{-1/3}$.

It should be underlined that the transition density associated with the four normal modes of excitation have a simple analytical form

$$\rho_{tr}(r)=t_{1\alpha}[\partial\rho_{SF}/R_n]+t_{2\alpha}[\partial\rho_{SF}/R_p]+t_{3\alpha}[\partial\rho_{SF}/b_n]+t_{4\alpha}[\partial\rho_{SF}/b_p], \quad (8.66)$$

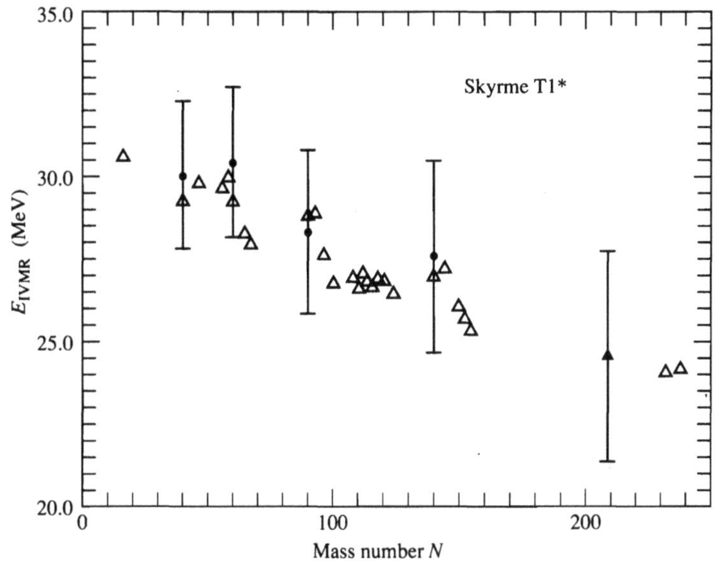

Figure 8.2. Excitation energies of the isovector giant monopole resonance (IVMR) of spherical nuclei obtained from the local-scale version of the adiabatic time-dependent Hartree-Fock approach using $\mathrm{T1}^{*}$ effective forces (see Table 7.6). Some experimental IVMR peak energies (Bowman *et al.* 1983, 1984, Erell *et al.* 1984, 1986, Irom *et al.* 1986) are also given.

where the amplitude coefficients $t_{1\alpha}$ (see Table 8.1) do not contain any phenomenological parameters and depend only on the Skyrme-type forces used (Dimitrova, Petkov, and Stoitsov 1986a). At the same time, the transition densities (8.66) are rather close to the RPA ones. This makes it possible to use eqn (8.66) in the DWBA analysis of ISMR experimental data as done by Morsch and Decowski (1979) and Morsch *et al.* (1980, 1983).

8.4 Nuclear fluid dynamics

The classical nature of TDHF theory gives the possibility of attempting a fluid dynamic description (Bertsch 1975, Wong,

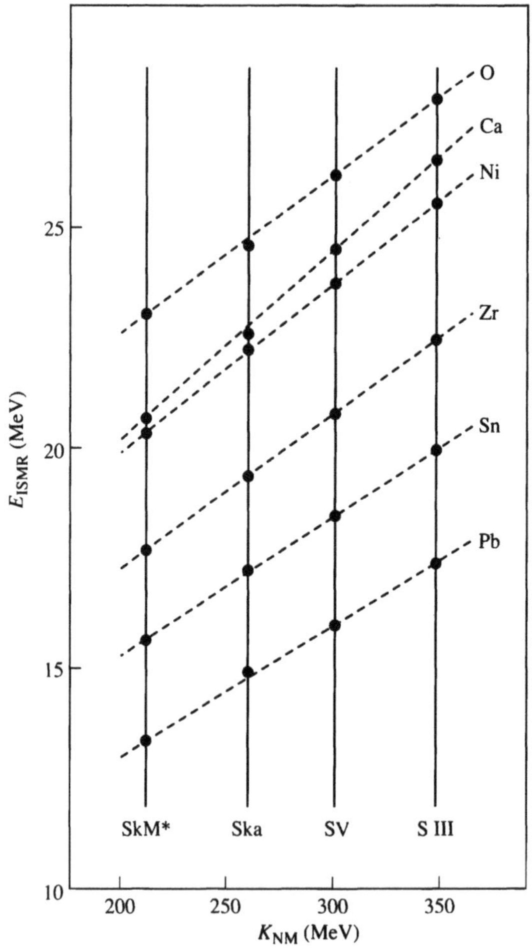

Figure 8.3. ISMR excitation energies versus nuclear matter incompressibility K_{NM} (Dimitrova, Petkov, and Stoitsov 1988a). Skyrme-type forces SIII, SV, Ska and SkM* are used.

Maruhn, and Welton 1975) whose distinctive and appealing feature is the choice of the local density distribution $\rho(\mathbf{r},t)$ and the velocity

$$\mathbf{v}(\mathbf{r},t) = \rho^{-1}(\mathbf{r},t)\frac{\hbar}{2mi}\left[(\nabla_{\mathbf{r}} - \nabla_{\mathbf{r}'})\rho(\mathbf{r},\mathbf{r}',t)\right]\Big|_{\mathbf{r}'=\mathbf{r}} \tag{8.67}$$

as basic dynamic ingredients, rather than the exact time-dependent wave function $\Psi(\mathbf{r}_1,\ldots,\mathbf{r}_N,t) \in \mathcal{H}^N$. The justification of this choice is not possible *a priori* but it may be obtained *a posteriori* by making comparisons with other methods and the actual physical situation.

Collisionless Boltzman equation

The simplest approach in deriving nuclear fluid dynamics is to take the semiclassical limit of the TDHF equation (8.6) which can easily be done in terms of its Wigner transform (Kerman and Koonin 1976)

$$\frac{\partial f(\mathbf{q},\mathbf{p},t)}{\partial t} + \frac{2}{\hbar} f(\mathbf{q},\mathbf{p},t)\sin\left[\frac{\hbar}{2}\,(\overleftarrow{\nabla}_q.\overrightarrow{\nabla}_p - \overleftarrow{\nabla}_p.\overrightarrow{\nabla}_q)\right] H(\mathbf{q},\mathbf{p}) = 0. \quad (8.68)$$

The time-dependent Wigner distribution function $f(\mathbf{q},\mathbf{p},t)$ in eqn (8.68) is obtained as the Wigner transform of the TDHF Hamiltonian (8.7) which for simplicity may be assumed local and of the form of eqn (4.15).

Equation (8.68) is known as the quantum kinetic equation and it is strictly equivalent to the TDHF equation (8.6) (Baym and Pethick 1978). If one expands it in powers of $\hbar$ and retains only terms that do not vanish for $\hbar \to 0$, one obtains from eqn (8.68) exactly the Landau kinetic equation (Landau 1957a, 1957b, 1958) with a vanishing collision term (Vlasov 1954). From this, assuming also that $\nabla_p H(\mathbf{q},\mathbf{p}) = \text{Const}$ and $\nabla_q H(\mathbf{q},\mathbf{p}) = \nabla_q U$, one obtains precisely the collisionless Boltzmann equation (Kadanoff and Baym 1962)

$$\left[\frac{\partial}{\partial t} + \frac{\mathbf{p}}{m}\cdot\nabla_q - \nabla_q U \cdot \nabla_p\right] f^{(0)}(\mathbf{q},\mathbf{p},t) = 0, \quad (8.69)$$

where the distribution function $f^{(0)}(\mathbf{q},\mathbf{p},t)$ is the classical analogue of the Wigner distribution function $f(\mathbf{q},\mathbf{p},t)$.

In general, neither eqn (8.68) nor eqn (8.69) can be solved analytically due to their complicated partial differential character in seven variables. An approach to the problem is direct solution of eqn (8.68) in phase space as done, for example, in the fluid dynamic consideration due to Brink and Di Toro (1981), Di Toro, Lombardo, and Russo (1985), or the expansion of the solution of the Vlasov equation for small intervals of time (Bertsch 1978). Another alternative approach is given by the method of moments we shall briefly outline below.

Method of moments

According to the standard procedure of the method of moments one transforms eqn (8.68) or eqn (8.69) into an infinite system of equations for the moments of $f(\mathbf{q},\mathbf{p},t)$ taken with respect to $\mathbf{p}$.

Let us define the average $\langle A\rangle$ for an arbitrary quantity $A(\mathbf{q},\mathbf{p},t)$ as

$$\langle A\rangle \equiv \langle A(\mathbf{q},t)\rangle = \frac{\int d\mathbf{p}\; A(\mathbf{q},\mathbf{p},t) f^{(0)}(\mathbf{q},\mathbf{p},t)}{\int d\mathbf{p}\; f^{(0)}(\mathbf{q},\mathbf{p},t)}. \tag{8.70}$$

Integrating first eqn (8.69) over $\mathbf{p}\,d\mathbf{p}$, one obtains just the continuity equation (conservation of mass)

$$\frac{\partial\rho(\mathbf{q},t)}{\partial t} + \nabla_{\mathbf{q}} \cdot (\rho(\mathbf{q},t)\mathbf{v}(\mathbf{q},t)) = 0, \tag{8.71}$$

where $\rho(\mathbf{q},t)$ is the local density and $\mathbf{v}(\mathbf{q},t)$ is the average velocity with components $v_i(\mathbf{q},t) = \langle p_i/m\rangle$, $i = (1,2,3)$.

The next step is to integrate eqn (8.69) over $p_i d\mathbf{p}$. One obtains the Euler equation of motion (conservation of momentum)

$$m\rho\frac{d\mathbf{v}}{dt} \equiv m\rho(\mathbf{q},t)\left(\frac{\partial v_i}{\partial t} + \mathbf{v} \cdot \nabla_{\mathbf{q}} v_i\right) = -\rho(\mathbf{q},t)\nabla_{q_i} U - \sum_{j=1}^{3} \frac{\partial P_{ij}}{\partial q_j}, \tag{8.72}$$

which contains a new quantity, the pressure tensor

$$P_{ij} = m\rho(\mathbf{q},t)\langle(p_i/m - v_i)(p_j/m - v_j)\rangle. \tag{8.73}$$

Integrating eqn (8.69) over $\mathbf{p}^2 d\mathbf{p}$, a dynamic equation for the pressure tensor results

$$m\rho(\mathbf{q},t)\frac{d}{dt}(Q/\rho) + \nabla \cdot \mathbf{W} = -\sum_{i,j=1}^{3} P_{ij} D_{ij}, \tag{8.74}$$

which contains some new quantities

$D_{ij}(\mathbf{q},t) = \frac{1}{2}(\partial v_i/\partial q_j + \partial v_j/\partial q_i)$, *rate of strain tensor*;

$u_i(\mathbf{q},\mathbf{p},t) = p_i/m - v_i$, *deviation from the average velocity*;

$W_i(\mathbf{q},t) = \frac{1}{2} m\rho\langle u_i u^2\rangle$, *heat current density*;

$Q(\mathbf{q},t) = \frac{1}{2} m\rho\langle u^2\rangle$, *thermal energy*.

The above procedure could in principle be continued to give an infinite system of equations and if it were carried out with respect to eqn (8.68) then quantum corrections (terms of higher order in $\hbar$) would be taken into account.

Any one piece of this infinite system, as it stands, is completely useless because it does not form a closed system of equations. Therefore, one must seek a physically plausible truncation which specifies the higher moments of $f^{(0)}(\mathbf{q},\mathbf{p},t)$ in terms of the lower ones. The simplest assumption (Stringari 1985) is to set

$$P_{ij} = 0 \quad (\text{for } i \neq j), \; W_i = 0, \tag{8.75}$$

whereby eqns (8.71), (8.72), and (8.74) form the closed set of equations of ordinary hydrodynamics. Such an imposition is equivalent to assuming that the distribution function $f^{(0)}(\mathbf{q},\mathbf{p},t)$ preserves the same spherical form in velocity space as at equilibrium.

In nuclear hydrodynamics, condition (8.75) can be represented by the symmetric distribution function following from the usual TF limit

$$f^{(0)}(\mathbf{q},\mathbf{p},t) \cong f_{TF}(\mathbf{q},\mathbf{p},t) = \theta\Big(\mu - (\mathbf{p}-m\mathbf{v})^2/2m - U(\mathbf{q},t)\Big), \tag{8.76}$$

which is the lowest order term (4.25) of the semiclassical expansion (4.29) but performed in the time-dependent case (Ring and Schuck 1980). The use of eqn (8.76) expresses the pressure tensor in terms of $\rho(\mathbf{r},t)$ and $\mathbf{v}(\mathbf{r},t)$, and thus truncates the infinite system to the first two equations, eqn (8.71) and eqn (8.72). Their solution determines both the density $\rho(\mathbf{r},t)$ and the velocity $\mathbf{v}(\mathbf{r},t)$.

An alternative solution of the problem may be achieved by following the natural extension of the method of moments (Bertsch 1978, Nix and Sierk 1980, 1982) due to Balbutsev *et al.* (1981), Balbutsev, Vaishvila, and Mikhailov (1982), Balbutsev, Mikhailov, and Vaishvila (1986), Balbutsev and Piperova (1988). They perform simultaneous integrations with different weights of eqn (8.68) or eqn (8.69) in the coordinate as well as in the momentum space. The resulting infinite system of virial equations for the moments of $f^{(0)}(\mathbf{q},\mathbf{p},t)$ with respect to different powers of momentum and spatial coordinates quite naturally truncates to a closed system of dynamic equations. In this case, the nuclear density functional theory may be used as an effective tool for easier calculations of the associated moments at the nuclear equilibrium (Balbutsev, Mikhailov, and Stoitsov 1989).

Variational formulation of nuclear fluid dynamics

The variational derivation of fluid dynamic equations is more general than the method of moments because in principle it is free of the assumption for single Slater determinant representation of the time-dependent wave function. Instead, the following variational ansatz for the trial wave function is assumed

$$\Psi(t) \equiv \Psi(\mathbf{r}_1,\ldots,\mathbf{r}_N,t)$$
$$= \Psi_0(\mathbf{r}_1,\ldots,\mathbf{r}_N,t)\,\exp((im/\hbar)S(\mathbf{r}_1,\ldots,\mathbf{r}_N,t)), \quad (8.77)$$

where

$$S(\mathbf{r}_1,\ldots,\mathbf{r}_N,t) = \sum_{i=1}^{N} s(\mathbf{r}_i,t). \quad (8.78)$$

This ansatz is quite general because the exact wave function of the system can always be decomposed into a real amplitude Ψ_0 and a real phase function S. The assumption (8.78), however, is already a strong restriction under which all single-particle states are forced to have the same real phase function s(**r**,t).

Let us introduce eqn (8.77) and eqn (8.78) into the action integral (1.48). The result

$$L = \int_{t_2}^{t_1} dt \; \langle\Psi(t)|i\hbar\frac{\partial}{\partial t} - \hat{H}|\Psi(t)\rangle$$

$$= \int_{t_2}^{t_1} dt\left[-m\int d\mathbf{r}\; s(\mathbf{r},t)\dot{\rho}_0(\mathbf{r},t) - \langle\Psi(t)|\hat{H}|\Psi(t)\rangle\right], \qquad (8.79)$$

contains the collective energy

$$\mathcal{H} \equiv \langle\Psi(t)|\hat{H}|\Psi(t)\rangle = \frac{m}{2}\int d\mathbf{r}\; \mathbf{v}^2(\mathbf{r},t)\rho_0(\mathbf{r},t) + \langle\Psi_0(t)|\hat{H}|\Psi_0(t)\rangle. \qquad (8.80)$$

where $\rho_0(\mathbf{r},t)$ is the local density and $\mathbf{v}(\mathbf{r},t) = \nabla s(\mathbf{r},t)$ is the velocity (8.67), both derived with the wave function (8.77) under condition (8.78).

It should be stressed that the condition of irrotationality is implicitly contained in eqn (8.78) because $\mathbf{v}(\mathbf{r},t) = \nabla s(\mathbf{r},t)$ and, therefore, rot $\mathbf{v} \equiv \nabla \times \mathbf{v}(\mathbf{r},t) = \nabla \times \nabla s(\mathbf{r},t) \equiv 0$ (Giannoni *et al.* 1976). Rotational components can be introduced in eqn (8.77) by replacing the local phase s(**r**,t) in eqn (8.78) with a more general nonlocal operator (Providencia and Holzwarth 1983).

The collective kinetic energy in eqn (8.80),

$$\mathcal{K} \equiv \mathcal{K}[\rho_0,\mathbf{v}] = \frac{m}{2}\int d\mathbf{r}\; \mathbf{v}^2(\mathbf{r},t)\rho_0(\mathbf{r},t). \qquad (8.81)$$

depends on the local density $\rho_0(\mathbf{r},t)$ and the velocity $\mathbf{v}(\mathbf{r},t)$, while the collective potential energy, the second term in eqn (8.80), depends on the many-particle wave function $\Psi_0(t)$

$$\mathcal{V} \equiv \langle\Psi_0(t)|\hat{H}|\Psi_0(t)\rangle. \tag{8.82}$$

Note that the potential energy (8.82) should be represented as a functional $E[\rho_0]$ of the local density $\rho_0(\mathbf{r},t)$ in order to achieve a hydrodynamical description. This could in principle be done on the basis of nuclear density functional theory. As mentioned in Section 2.3, however, there is no analogue of the Hohenberg-Kohn theorem in the time-dependent case. Therefore, in order to close a hydrodynamical description, one should assume that the functional $E[\rho_0] = \langle\Psi_0(t)|\hat{H}|\Psi_0(t)\rangle$ exists and then $\mathcal{V} \equiv E[\rho_0]$. The latter may be justified at least in the context of the RED approach as applied in the previous section.

Assuming that the explicit form of the functional $E[\rho_0]$ is known, one can further vary the action (8.79) with respect to $s(\mathbf{r},t)$ and $\rho_0(\mathbf{r},t)$, independently. The variation with respect to $s(\mathbf{r},t)$ gives (see eqn (8.71))

$$\frac{\partial\rho_0(\mathbf{r},t)}{\partial t} + \nabla \cdot (\rho_0(\mathbf{r},t)\mathbf{v}(\mathbf{r},t)) = 0, \tag{8.83}$$

while the variation with respect to $\rho_0(\mathbf{r},t)$ yields an Euler-like equation

$$\frac{\partial\mathbf{v}(\mathbf{r},t)}{\partial t} + (\mathbf{v}(\mathbf{r},t) \cdot \nabla)\mathbf{v}(\mathbf{r},t) = - m^{-1}\frac{\delta E[\rho_0]}{\delta\rho_0}. \tag{8.84}$$

The simultaneous solution of eqn (8.83) and eqn (8.84) allows both unknown quantities, $\rho_0(\mathbf{r},t)$ and $\mathbf{v}(\mathbf{r},t)$, to be obtained. Within the phenomenological functional (3.40), for example, the nuclear fluid dynamic equations (8.83) and (8.84) have been solved numerically as time-dependent TF equations by Németh, Dalili, and Ngô (1985) and Dalili, Németh, and Ngô (1985).

8.5 Dynamic energy density functionals

In general, it is difficult to find the functional $E[\rho_0]$ in the dynamic case and one is forced to make some approximations at this point. We shall further consider how one can derive approximate energy density functionals for the dynamical description of atomic nuclei.

Imposed nuclear dynamics

In many cases, the main dynamical properties of nuclear systems may be analysed after investigating their behaviour under an imposed collective motion, as in the case of the ATDHF approach given in the previous section. For definiteness, we consider the case when $\Psi_0(\mathbf{r}_1,\ldots,\mathbf{r}_N,t)$ entering eqn (8.77) is a time-even Slater determinant obtained for example within the CHF method with suitably chosen constraint operators. In this case, the time-even one-body density matrix associated with $\Psi_0(\mathbf{r}_1,\ldots,\mathbf{r}_N,t)$ can always be expressed into the form (Baranger and Vénéroni 1978)

$$\rho_0(\mathbf{r},\mathbf{r}',t) = \exp(-i\varphi(\mathbf{r},t)/\hbar)\ \bar{\rho}(\mathbf{r},\mathbf{r}')\ \exp(\ i\varphi(\mathbf{r},t)/\hbar), \qquad (8.85)$$

where $\varphi(\mathbf{r},t) = \varphi^{+}(\mathbf{r},t)$ is a time-odd operator and $\bar{\rho}(\mathbf{r},\mathbf{r}')$ is the one-body density matrix of the static HF problem.

Let us further analyse a particular ansatz for the operator $\varphi(\mathbf{r},t)$ leading to the expression

$$\rho_0(\mathbf{r},\mathbf{r}',t) = \exp\left\{-\frac{1}{2}\left[\nabla_r\ .\ \mathbf{Q}(\mathbf{r},t) + \nabla_{r'}\ .\ \mathbf{Q}(\mathbf{r}',t) + \mathbf{Q}(\mathbf{r},t)\ .\ \nabla_r + \mathbf{Q}(\mathbf{r}',t)\ .\ \nabla_{r'}\right]\right\}\bar{\rho}(\mathbf{r},\mathbf{r}'), \qquad (8.86)$$

where the gradient operator ∇ acts on every function to its right. In the limit of small amplitude motion, we also assume that the operator **Q**(**r**,t) factorizes

$$\mathbf{Q}(\mathbf{r},t) = \alpha(t)\mathbf{v}(\mathbf{r}), \tag{8.87}$$

with a vector function **v**(**r**) having the following particular form in the monopole, dipole, and quadrupole cases, respectively

$$L = 0 \qquad \mathbf{v}_0 \equiv \frac{\hbar^2}{m}\mathbf{r} = \frac{\hbar^2}{m}\nabla(r^2/2),$$

$$L = 1 \qquad \mathbf{v}_1 \equiv \frac{\hbar^2}{m}(0,0,1) = \frac{\hbar^2}{m}(\nabla\, z),$$

$$L = 2 \qquad \mathbf{v}_2 \equiv \frac{\hbar^2}{m}(-x,-y,2z) = \frac{\hbar^2}{m}\nabla(2z^2-x^2-y^2). \tag{8.88}$$

Under the above rather special choice, one obtains the Wigner transform (see eqn (4.16)) of the time-even one-body density matrix (8.86) in the form

$$f_0(\mathbf{q},\mathbf{p},t) = \exp(\alpha\mathbf{v}(\mathbf{q}) \,.\, \nabla_q)\; \bar{f}(\mathbf{q},\tilde{\mathbf{p}}), \tag{8.89}$$

where

$$\bar{f}(\mathbf{q},\mathbf{p}) = \int d\mathbf{s}\; \bar{\rho}(\mathbf{q} + \frac{\mathbf{s}}{2},\mathbf{q} - \frac{\mathbf{s}}{2})\; \exp(-i\frac{\mathbf{p}\cdot\mathbf{s}}{2}), \tag{8.90}$$

is the Wigner transform of the static HF density matrix $\bar{\rho}(\mathbf{r},\mathbf{r}')$ and

$$\tilde{p}_l = \sum_{k=1}^{3} (\exp(\alpha\mathbf{A}))_{lk}\; p_k. \tag{8.91}$$

The diagonal matrix $(\mathbf{A})_{ij} = \delta_{ij}A_i$ in eqn (8.91) has nonzero matrix elements

$$L = 0 \qquad A_1 = A_2 = A_3 = -1,$$

$$L = 1 \qquad A_1 = A_2 = A_3 = 1.$$

$$L = 3 \qquad A_1 = A_2 = 1\;,\; A_3 = -2. \tag{8.92}$$

The physical meaning of the present example may be now

clarified. The integration of eqn (8.89) over **p** d**p** defines the time-dependent local density distribution

$$\rho_0(\mathbf{r},t) = \exp(\alpha(t)\nabla \cdot \mathbf{v}(\mathbf{r}))\, \bar{\rho}(\mathbf{r}), \tag{8.93}$$

where $\bar{\rho}(\mathbf{r})$ is the static local density distribution. Therefore, comparing the time-derivative of eqn (8.93) with the continuity equation (8.83), it becomes clear that the vector function $\mathbf{v}(\mathbf{r})$ is proportional to the collective velocity imposed on the system $\mathbf{v}(\mathbf{r},t)) \equiv \nabla s(\mathbf{r},t) = \dot{\alpha}(t)\mathbf{v}(\mathbf{r})$. Under conditions (8.88), the associated time evolution of the density (8.93) is determined during the imposed collective motion as

$$\begin{aligned} &L = 0 && \rho_0(\mathbf{r},t) = \tilde{\alpha}^3\, \bar{\rho}(\tilde{\alpha}\mathbf{r}), \\ &L = 1 && \rho_0(\mathbf{r},t) = \bar{\rho}(\mathbf{r} - \tilde{\alpha}\hat{e}_z), \\ &L = 2 && \rho_0(x,y,z,t) = \bar{\rho}(\tilde{\alpha}x,\tilde{\alpha}y,z/\tilde{\alpha}^2). \end{aligned} \tag{8.94}$$

where $\tilde{\alpha} \equiv \tilde{\alpha}(t) = \exp(-\alpha(t))$.

Finally, the unknown time-dependent function $\alpha(t)$ follows as a solution of the second nuclear fluid dynamic equation (8.84). Examples for calculations with an imposed collective dynamics may be found in Bertsch (1975), Engel *et al.* (1975), Vautherin (1975), Bertsch and Stricker (1976), Holzwarth and Eckart (1978), Sagawa and Holzwarth (1978), Serr, Bertsch, and Borysowicz (1980), and Suzuki (1980).

Static and dynamic energy density functionals

Suppose we know the static HF one-body density matrix $\bar{\rho}(\mathbf{r},\mathbf{r}')$ or its Wigner transform $\bar{f}(\mathbf{q},\mathbf{p})$, but expressed in the form which allows the ground-state HF energy to be represented as an energy density functional $E[\bar{\rho}]$ of the static local density $\bar{\rho}(\mathbf{r})$. Then one has two possibilities to build up energy density functionals in the particular case of an imposed collective dynamics. The first one consists in replacing the static density $\bar{\rho}(\mathbf{r})$ in $E[\bar{\rho}]$ by the dynamic one $\rho_0(\mathbf{r},t)$. The resulting functional we shall denote by $E^{HD}[\rho_0] \equiv E[\rho_0]$. The second possibility is to use the Wigner transform (8.88) of $\bar{f}(\mathbf{q},\mathbf{p},t)$ which gives rise to the dynamic functional $E^D[\rho_0]$. It is important then to compare the difference of both the static functional in which the static local density is replaced by the dynamic one ($E^{HD}[\rho_0]$) and the dynamical functional $E^D[\rho_0]$.

For the purpose of demonstration, we analyse only the internal kinetic energy produced by the collective motion imposed on the system. With the TF expression for the static one-body density (3.3), one obtains the Wigner transform (8.90) in the form

$$\bar{f}(\mathbf{q},\mathbf{p}) = \theta\left(\frac{\mathbf{p}^2}{2m} - \frac{p_F^2(\mathbf{q})}{2m}\right), \tag{8.95}$$

where the Fermi momentum $p_F^2(\mathbf{q}) = \hbar k_F^2(\mathbf{q})$ is defined as in eqn (3.4). Then, inserting eqn (8.95) into eqn (8.89) and integrating the result over p^2dp, one obtains the dynamic expression for the kinetic energy density functional in the TF approximation (Holzwarth and Eckart 1978)

$$T_{TF}^D[\rho_0] = W_L(\alpha)T_{TF}[\bar{\rho}] = W_L(\alpha)\int d\mathbf{r}\bar{\rho}^{5/3}(\mathbf{r}), \tag{8.96}$$

where

$$W_{L=0}(\alpha) = \exp(2\alpha),$$
$$W_{L=1}(\alpha) = 1,$$
$$W_{L=2}(\alpha) = \frac{1}{3}\,(2\exp(-2\alpha) + \exp(4\alpha)). \tag{8.97}$$

It is obvious that due to the collective motion imposed on the system, the dynamic energy density functional $T_{TF}^D[\rho_0]$ differs from the static one $T_{TF}[\bar{\rho}]$ as well as from the functional

$$T_{TF}^{HD}[\rho_0] \equiv T_{TF}[\rho_0] = C_k\int d\mathbf{r}\ (\exp(\alpha\nabla\ .\ \mathbf{v})\ \bar{\rho}(\mathbf{r}))^{5/3}, \tag{8.98}$$

Comparing eqn (8.96) with eqn (8.98), one sees that the functional $T^{HD}_{TF}[\rho_0]$ is still identical with $T^{D}_{TF}[\rho_0]$ in the monopole and dipole cases. In the dipole case, one simply has $T^{D}_{TF}[\rho_0] = T^{HD}_{TF}[\rho_0] = T_{TF}[\bar{\rho}]$, while in the monopole case eqn (8.98) transforms into eqn (8.96) after using eqn (8.94). This is not true, however, in the quadrupole case where eqn (8.96) and eqn (8.98) are completely different. As we shall see further, the difference is due to the fact that the nucleus is a Fermi liquid rather than an ordinary fluid.

Holzwarth and Eckart (1978) have shown that the result (8.96) is also true for the ITF functional (3.40) considered in Section 3.3. In the general case, however, the derivation of the functional $E^{D}[\rho_0]$ is a difficult problem. The main steps of the procedure required are pointed out by Holzwarth and Eckart (1979).

The nucleus as a Fermi liquid

Fluid dynamic models have been successfully applied in systematic investigations of the fission mode and surface vibrations since the pioneering works on the liquid-drop model (Bohr and Kalckar 1937, Meitner and Frisch 1939, Bohr and Wheeler 1939). Later, successful description of higher energy giant monopole resonances was achieved by the models of Steinwedel and Jensen (1950), Goldhaber and Teller (1948), Bohr (1952), Tassie (1956), and Werntz and Überall (1966).

Bertsch (1974, 1975) has suggested that the nucleus can vibrate as an elastic medium, rather than as ordinary fluid. The elastic vibrations are characterized by the presence in the pressure tensor of non-diagonal elements, which are absent in the hydrodynamical picture. The presence of these components is related to the deformation of the Fermi sphere in momentum space during the nuclear motion, a typical quantum effect exhibited by Fermi systems. This permits the description of a new class of phenomena such as giant quadrupole resonances in nuclei.

In order to show the difference between elastic and hydrodynamical descriptions, we consider the distorted momenta following from eqn (8.91) and eqn (8.92)

$$\tilde{p}_{L=0} = \exp(-\alpha)p,$$
$$\tilde{p}_{L=1} = p,$$
$$\tilde{p}_{L=2} = (\exp(\alpha)p_x, \exp(\alpha)p_y, \exp(-2\alpha)p_z). \tag{8.99}$$

Obviously, the momentum sphere stays spherical in the monopole and dipole cases. In the quadrupole case, however, one obtains a highly non-isotropic Fermi surface

$$\tilde{p}^2 = p^2(\exp(2\alpha)\sin(\theta_p) + \exp(-4\alpha)\cos^2(\theta_p)) \le p_F^2(q), \tag{8.100}$$

or

$$p^2 \le p_F^2(q, \theta_p) \tag{8.101}$$

with

$$p_F^2(q, \theta_p) = p_F^2(q)(\exp(2\alpha)\sin(\theta_p) + \exp(-4\alpha)\cos^2(\theta_p))^{-1}. \tag{8.102}$$

The Fermi surface at point **q** is an ellipsoid in momentum space. In other words, instead of the isotropic equilibrium distribution $\bar{f}(\mathbf{q},\mathbf{p}) = \theta(p^2 - p_F^2(q))$, eqn (8.94), one obtains the highly anisotropic distribution $f_0(\mathbf{q},\mathbf{p}) = \theta(p^2 - p_F^2(q, \theta_p))$.

This feature, associated with zero sound, is ignored in purely hydrodynamical models where the Fermi surface is always assumed to be spherical. In particular, this is the reason for the modification of the functional (8.96). The isotropic TF distribution (8.76) (that is ordinary hydrodynamics) gives the expression (8.98) which is the static ground-state energy density functional with the static local density replaced by the dynamical one.

We have seen that the functional $E^{HD}[\rho_0]$ is identical with the dynamical one $E^D[\rho_0]$ in monopole and dipole cases, but this is not true in the quadrupole case. Therefore, the L = 0,1 modes can be treated in the usual hydrodynamic picture, but for the L ≥ 2 modes important modifications occur due to the fact that the nucleus is a Fermi liquid rather than an ordinary fluid. In particular, the hydrodynamic picture gives an $N^{-1/2}$ law (Bohr and Mottelson 1974) for the quadrupole resonance excitation energy, while accounting for the deformation of the momentum sphere, leads to the correct N-dependence, $\hbar\omega \cong 60N^{-1/3}$, supported by the experimental data (Youngblood *et al.* 1976).

Generalized scaling model

Following the suggestion of Bertsch (1974), similar approaches describing the collective oscillations as vibrations of an elastic medium have been developed by Stringari (1977), Wong and McDonald (1977), Holzwarth and Eckart (1978, 1979), and Eckart, Holzwarth, and Schwesinger (1980). They use the generalized scaling assumption (8.86) and approximate energy density functionals of TF-like forms. The hydrodynamic limit is then obtained by ignoring in this approach the terms associated with the distortion of the Fermi sphere. Stringari (1983) has discussed in detail differences among generalized scaling and hydrodynamic models in connection with giant resonances in nuclei.

The first attempts to determine the operator $\mathbf{Q}(\mathbf{r},t)$ entering the generalized scaling eqn (8.86) by a variational principle were made by Holzwarth and Eckart (1979). They use the semiclassical approximation for the kinetic energy density functional, outgoing from a deformed local Fermi surface, and derive linearized equations of motion for $\mathbf{Q}(\mathbf{r},t)$. Andō and Eckart (1984) developed continuum fluid dynamics based on generalized scaling.

In conclusion, nuclear fluid dynamics has contributed enormously to our understanding of nuclei. For example, Krivine, Treiner, and Bohigas (1980) have adopted on this ground the set SkM* of Skyrme-type effective forces. Holzwarth and Eckart (1977) have investigated non-normal parity modes in spherical nuclei and suggested an interesting twist mode with velocity field $\mathbf{v} \cong (xy, -xz, 0)$ in which the local density stays constant during the collective motion, but the momentum distribution becomes distorted.

9. The sum rule approach and nuclear density functional theory

In this chapter nuclear density functional theory is used in the context of the sum rule approach to investigate small amplitude collective motions in nuclei.

The random phase approximation (RPA) is one of the widely used methods for full microscopic description of excited states of nuclei. This method, however, requires great computational efforts and the results, although very complete, are always numerical. The sum rule approach gives less detailed information than RPA, but the calculations are simple enough to allow a large-scale investigation of the average properties of the collective spectrum. In any case, the sum rule approach used with nuclear density functional theory leads to analytic results and provides a better study of properties such as the location of collective states, the N-dependence of their excitation energies and correlations between the characteristics of the effective forces used.

9.1 Random phase approximation

The ràndom phase approximation is usually derived as a small amplitude limit of the TDHF theory (Section 8.1) where the complete description of collective motion is contained in the full TDHF one-body density matrix $\rho(t) = \rho(\mathbf{r},\mathbf{r}',t)$ which satisfies the equation of motion (8.6). Within the limit of small amplitudes, one considers harmonic vibrations of small amplitude $\delta\rho$ around some equilibrium one-body density matrix $\rho_0(\mathbf{r},\mathbf{r}')$. Then, the equation of motion for $\delta\rho$ can be obtained from the TDHF equation

(8.6) which is represented in the limit of small amplitudes by the ATDHF equations (8.25) and (8.26).

RPA equations of motion

If one expresses the full density matrix $\rho(t) = \rho(\mathbf{r},\mathbf{r}',t)$ in the form

$$\rho = \rho_0 + \delta\rho, \tag{9.1}$$

then the equation of motion for $\delta\rho(t)$ follows from the linearization of the TDHF equation (8.6). The ATDHF(I) gives immediately the response equation

$$i\hbar\,\delta\dot{\rho} = [W_0,\delta\rho] + [\delta W,\rho_0], \tag{9.2}$$

where $\delta W = \mathrm{tr}(\hat{\omega}_A\delta\rho)$ is in accordance with eqn (8.22), while the ATDHF(II) yields the HF equation $[W_0,\rho_0] = 0$ for the equilibrium density $\rho_0 \equiv \rho_0(\mathbf{r},\mathbf{r}')$. For a collective motion with frequency ω, one assumes that

$$\delta\rho(t) = \delta\bar{\rho}\,\exp(-i\omega t) + \delta\bar{\rho}^{+}\exp(i\omega t), \tag{9.3}$$

where the amplitude $\delta\bar{\rho}$ is time independent and $\delta\rho^{+}(t) = \delta\rho(t)$.

Inserting eqn (9.3) into eqn (9.2) and separating the positive and negative frequencies yields

$$\hbar\omega\,\delta\bar{\rho} = [W_0,\delta\bar{\rho}] + [\delta W,\rho_0], \tag{9.4}$$

for the terms containing $\exp(-i\omega t)$, and its Hermitian conjugate equation for the terms containing $\exp(i\omega t)$. These are the RPA equations of motion. They take the usual form (Thouless 1961) by considering separately the particle-hole $X_{ph} = \delta\bar{\rho}_{ph}$ and hole-particle $Y_{ph} = \delta\bar{\rho}_{hp}$ matrix elements of $\delta\bar{\rho}$. This yields

$$\begin{pmatrix} A & B \\ -B^{*} & -A^{*} \end{pmatrix} \begin{pmatrix} X_{ph} \\ Y_{ph} \end{pmatrix} = \hbar\omega \begin{pmatrix} X_{ph} \\ Y_{ph} \end{pmatrix}, \tag{9.5}$$

where matrices A and B are defined by their matrix elements

$$A_{php'h'} = (\varepsilon_p - \varepsilon_h)\,\delta_{ph}\,\delta_{p'h'} + (\hat{\upsilon}_A)_{ph'hp'},$$

$$B_{php'h'} = (\hat{\upsilon}_A)_{pp'hh'}. \tag{9.6}$$

In this section, the letters h,h′ characterize occupied levels in the HF ground state (hole states), while p,p′ are empty levels (particle states), ε_p and ε_h being particle and hole single-particle energies, respectively.

1p–1h RPA model

Let us denote the eigenfunctions of eqn (9.5) by $|\Psi_n\rangle$, (n = 0,1,2...), where $|\Psi_0\rangle$ indicates the correlated RPA ground state. If one introduces the (quasi boson) creation operator

$$\Omega_n^+ = \sum_{ph} \left[X_{ph}^{(n)}\, a_p^+ a_h - Y_{ph}^{(n)}\, a_h^+ a_p \right], \tag{9.7}$$

then one can write for all RPA wave functions

$$|\Psi_n\rangle = \Omega_n^+ \, |\Psi_0\rangle, \tag{9.8}$$

where the ground state $|\Psi_0\rangle$ appears as a vacuum $\Omega_n|\Psi_0\rangle = 0$ for all $n > 0$, and the RPA amplitudes are given by

$$X_{ph}^{(n)} = \langle\Psi_n|a_p^+ a_h|\Psi_0\rangle,$$

$$Y_{ph}^{(n)} = \langle\Psi_n|a_h^+ a_p|\Psi_0\rangle. \tag{9.9}$$

The matrix elements of a transition operator

$$\hat{Q} = \sum_{ij} Q_{ij}\, a_i^+ a_j \tag{9.10}$$

between the RPA states $|\Psi_n\rangle$ and $|\Psi_0\rangle$ now read

$$|\langle\Psi_n|\hat{Q}|\Psi_0\rangle|^2 = \sum_{ph} |Q_{ph}(X_{ph}^{(n)} + Y_{ph}^{(n)})|^2. \qquad (9.11)$$

This 1p–1h RPA model is obviously suitable for the description of those excited states that differ from the ground state mainly by 1p–1h components such as for instance giant monopole resonances.

Shell-model and self-consistent RPA

There are basically two different approaches to the actual solution of the RPA equations. The shell-model RPA approach usually starts from a phenomenological mean field and then diagonalizes an appropriately chosen ph-residual interaction. Good qualitative results were obtained in this way by Migdal (1967), Soloviev (1976, 1978), Speth, Werner, and Wild (1977).

The self-consistent RPA approach is more fundamental in the sense that both calculations use the same effective nucleon-nucleon interaction as done by Bertsch and Tsai (1975), Müther, Waghmore, and Faessler (1976), Liu and Brown (1976), Blaizot, Gogny, and Grammaticos (1976), Blaizot and Gogny (1977) and Blaizot (1980).

It is well known that the empirical single-particle levels are more compressed at the Fermi surface than the HF ones. Bertsch and Kuo (1968) and Hamamoto and Siemens (1976) showed that the coupling of the single-particle (hole) states to nuclear vibrations gives rise to a compression of the HF single-particle spectrum. Therefore, the empirical single-particle spectrum contains a definite part coming from the phonon contribution. The coupling is stronger in low-lying states and therefore, as far as low-lying collective resonances and non-collective states are concerned, the use of the empirical spectrum or the shell-model RPA is necessary. For conventional Skyrme-type effective forces, for example, the low-lying collective states are either too high in energy or too low in collectivity (Krewald *et al.* 1977).

In the case of giant resonances, however, where the effect on the single-particle spectrum due to phonons is much smaller because of the high frequency of the collective motion (Wambach, Mishra, and Li 1982), one has to use the HF single-particle spectra which contains no phonon contributions, whereas the shell-model RPA calculation always gives results lower than the experimental excitation energies.

In general, a unified self-consistent description of both the low-lying states and giant resonances should take into account 2p-2h and higher components of the wave functions.

2p-2h RPA model

The 1p-1h RPA model is expected to be good approximation for centroid energies and total transition probabilities. As far as the width of the resonances in the continuum is concerned, the 1p-1h RPA accounts only for the width due to the particles in the continuum states (escape width). The residual interaction, however, mixes the 1p-1h components into the 2p-2h ones, which spreads the transition strength. This spreading of the 1p-1h transition strength due to a coupling to more complex configurations like 2p-2h states (spreading width) is not included in the 1p-1h RPA model. Therefore, the width and the fine structure of the resonances can be understood only if 2p-2h configurations are taken into account.

Extensive 2p-2h RPA calculations were first performed by Soloviev (1976), Malov, Nesterenko, and Soloviev (1976), Soloviev, Stoyanov, Vdovin (1977), and Kyrchev *et al.* (1978) by means of their quasiparticle-phonon model (Voronov and Soloviev 1983, Soloviev and Shirikova 1986). The inclusion of 2p-2h configurations in the framework of a shell model diagonalization was done by Hoshino and Arima (1976) and Knüpfer and Huber (1976a, 1976b). Other methods for calculating the giant resonance decay

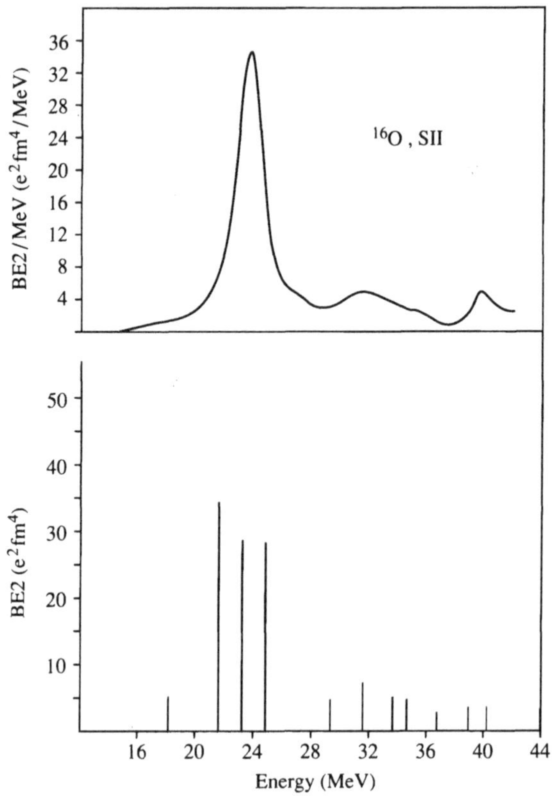

Figure 9.1. Continuum (upper part) and discretized (lower part) RPA calculations for the BE2-strength distribution of the giant quadrupole resonance in the nucleus ^{16}O performed by Krewald *et al.* (1977) with SII effective forces.

were proposed by Dehesa *et al.* (1977), Bertsch *et al.* (1979), Wambach, Mishra, and Li (1982), Bortignon and Broglia (1981). A review of RPA results obtained by several groups has been made by Speth and Van der Woode (1981), Goeke and Speth (1982) and Bertsch, Bortignon, and Broglia (1983).

Continuum RPA calculations

In the above kind of calculations, one usually makes the continuum discrete by diagonalizing the mean field in a finite basis of harmonic oscillators (Ring and Speth 1974) or by putting the nucleus into a box (Bertsch and Tsai 1975, Liu and Brown 1976). As a result, the theoretical resonances do not have any width, as illustrated in Figure 9.1. Very often, the widths are artificially included by assuming an arbitrary width for each discrete level in order to simulate the decay width of the resonances (Liu and Brown 1976, Wambach, Mishra, and Li 1982, Bortignon and Broglia 1981).

The continuum RPA approach (Krewald *et al.* 1974, 1977) takes into account the width of the single-particle resonances correctly within the ansatz

$$|\Psi_\lambda(E)\rangle = \Omega_\lambda^+(E)|\Psi_0\rangle, \tag{9.12}$$

where a generalization of the operator (9.7) is used. Thus, the escape width is taken into account correctly in 1p–1h RPA calculations (Figure 9.1) but it is, as mentioned above, only a part of the larger experimental width.

Complete self-consistent continuum RPA calculations based on the solution of the response equation in coordinate space have been made by Bertsch (1973), Liu and Van Giai (1976), Krewald *et al.* (1977), Cavinato *et al.* (1982), Auerbach and Klein (1983), Auerbach, Klein, and Siciliano (1985), and Starodubsky (1985).

9.2 Sum rule approach

The response of the ground state of the nucleus to the action of an arbitrary physical operator $\hat{Q}$ may be completely characterized by its associated strength function defined as (Bohigas, Lane, and Martorell 1979)

$$S(E) = \sum_n |\langle n|\hat{Q}|0\rangle|^2 \, \delta(E_n - E), \tag{9.13}$$

where $|0\rangle$ is ground state and $|n\rangle$ are excited states of the system with Hamiltonian $\hat{H}$ and associated energies E_n $(n = 0,1,2,\ldots)$. The strength function can be characterized by its moments (sum rules)

$$m_k = \int_0^\infty dE\ S(E)(E - E_0)^k = \sum_{n=0}^{\infty} (E_n - E_0)^k \langle n|\hat{Q}|0\rangle, \tag{9.14}$$

where k is an arbitrary integer.

Exact sum rules

When $|n\rangle$ and E_n are the exact eigenfunctions and energies of $\hat{H}$, respectively, the following energy-weighted sum rules are trivially satisfied for any integer k ($c = \hbar = 1$ units are used)

$$m_k = \sum_{n=0}^{\infty} (E_n - E_0)^k |\langle n|\hat{Q}|0\rangle|^2 = (-1)^t (i)^k \langle 0|\hat{Q}_s \hat{Q}_t|0\rangle, \tag{9.15}$$

which for odd moments transforms into

$$m_k = \frac{1}{2}(-1)^t (i)^k \langle 0|[\hat{Q}_s, \hat{Q}_t]|0\rangle, \tag{9.16}$$

where

$$\hat{Q}_s = \underbrace{[i\hat{H},[i\hat{H}, \ldots [i\hat{H}}_{s\ \text{times}},\hat{Q}]\ldots]], \tag{9.17}$$

and the s,t are arbitrary integers such that $s + t = k$. It is clear that all moments (9.14) of S(E) can be calculated only in terms of the ground-state wave function without explicitly evaluating the strength function itself. For example, we have from eqn (9.15) the moments

$$m_0 = \langle 0|\hat{Q}^2|0\rangle, \tag{9.18}$$

$$m_1 = \langle 0|\hat{Q}\ [\hat{H},\hat{Q}]|0\rangle, \tag{9.19}$$

$$m_2 = -\ \langle 0|\hat{Q}\ [\hat{H},\hat{Q}]\ [\hat{H},\hat{Q}]|0\rangle. \tag{9.20}$$

The actual calculation of the higher order moments quickly becomes a rather complicated problem but in some cases this can be done in an implicit way. For example, the $k = -1$ moment, usually called polarizability or zero-frequency response, can be derived by the so-called dielectric theorem.

Suppose that the Hamiltonian $\hat{H}$ is perturbed by a weak external field $-\lambda\hat{F}$. Assuming λ sufficiently small, the changes of the expectation values of $\hat{Q}$ and $\hat{H}$ are given by perturbation theory as

$$\delta\langle\hat{Q}\rangle = -\lambda \sum_{n=0}^{\infty} \left\{ \frac{\langle 0|\hat{F}^{+}|n\rangle\langle n|\hat{Q}|0\rangle}{(E_n - E_0)} + \frac{\langle 0|\hat{Q}|n\rangle\langle n|\hat{F}|0\rangle}{(E_n - E_0)} \right\} + O(\lambda^2),$$

$$\delta\langle\hat{H}\rangle = +\lambda \sum_{n=0}^{\infty} \left\{ \frac{\langle 0|\hat{F}^{+}|n\rangle\langle n|\hat{Q}|0\rangle}{(E_n - E_0)} \right\} + O(\lambda^3). \quad (9.21)$$

When $\hat{F} \equiv \hat{Q}$, the right-hand sides of equations (9.21) reduce to the m_{-1} moment and one obtains the dielectric theorem

$$m_{-1} = \sum_{n=0}^{\infty} \left\{ \frac{|\langle n|\hat{Q}|0\rangle|^2}{(E_0 - E_n)} \right\} = -\frac{1}{2}\left[\frac{\partial\langle\hat{Q}\rangle}{\partial\lambda}\right]_{\lambda=0} = \frac{1}{2}\left[\frac{\partial^2\langle\hat{H}\rangle}{\partial\lambda^2}\right]_{\lambda=0}. \quad (9.22)$$

The moments (9.15) are helpful to characterize how the transition strength is distributed as a function of the energy. If one defines a set of average energies as moment ratios

$$\tilde{E}_k = (m_k/m_{k-2})^{1/2} \quad (9.23)$$

and

$$\bar{E}_k = m_k/m_{k-1}, \quad (9.24)$$

then it follows from eqn (9.15) that both $\tilde{E}_k$ and $\bar{E}_k$ are equal if the strength distribution is concentrated on a single energy, that is, if S(E) is a single δ-function. For example, if this δ-function is $\delta(E_n - E_0 - E)$ then $\tilde{E}_k \equiv \bar{E}_k = E_n - E_0$; that is, eqn (9.23) and (9.24) give the excitation energy $\hbar\omega = E_n - E_0$.

The extent to which $\bar{E}_k$ and $\tilde{E}_k$ are different gives information about the spread of the strength, whereby the variance of the strength can be defined as

$$\sigma^2 = m_2/m_0 - (m_1/m_0)^2 > 0. \qquad (9.25)$$

For a Gaussian strength distribution $S(E) \cong \exp(-(E - E_0)^2/\Gamma^2)$, for example, the full width at half maximum is roughly $0.83\Gamma \cong 2.35\sigma$.

RPA sum rules

Thouless (1961) proves the important result

$$m_1(\text{RPA}) = \sum_{n=0}^{\infty}(E_n - E_0)|\langle\Psi_n|\hat{Q}|\Psi_0\rangle|^2 = \langle\Phi|\hat{Q}[\hat{H},\hat{Q}]|\Phi\rangle, \qquad (9.26)$$

which means that the k = 1 moment m_1(RPA), defined in the basis of correlated RPA states $|\Psi_n\rangle$ and energies E_n (see eqn (9.8)) can be calculated only from the HF ground state $|\Phi\rangle$. If $\hat{Q}$ is an isoscalar and velocity independent operator, then $\hat{Q}_1 = i[\hat{H},\hat{Q}]$ is also a one-body operator and, substituting it in eqn (9.26), the Thouless result (9.26) extends trivially to the k = 3 moment

$$m_3(\text{RPA}) = \sum_{n=0}^{\infty}(E_n - E_0)^3|\langle\Psi_n|\hat{Q}|\Psi_0\rangle|^2 = \langle\Phi|\hat{Q}_1[\hat{H},\hat{Q}_1]|\Phi\rangle. \qquad (9.27)$$

The Thouless theorem has been generalized also to all odd moments m_k(RPA) by Goeke and Reinhard (1980) who show that all odd RPA moments are related to each other. On this ground, one can use the HF ground state $|\Phi\rangle$ and calculate directly all odd moments m_k(RPA). The results must be strictly equivalent to the moments obtained from the self-consistent RPA strength distribution which is often used to check the completeness of the single-particle basis in the RPA calculations.

There exists also the dielectric theorem (Brown 1970, Marshalek and Da Providencia 1973) which allows the RPA polarizability to be calculated by means of CHF function $\Phi(\lambda)$:

$$m_{-1}(\text{RPA}) = -\frac{1}{2}\left[\frac{\partial}{\partial\lambda}\langle\Phi(\lambda)|\hat{Q}|\Phi(\lambda)\rangle\right]_{\lambda=0} = \frac{1}{2}\left[\frac{\partial^2}{\partial\lambda^2}\langle\Phi(\lambda)|\hat{Q}|\Phi(\lambda)\rangle\right]_{\lambda=0}, \qquad (9.28)$$

where $\Phi(\lambda)$ is the solution of the constrained HF problem with a Hamiltonian $\hat{H}_\lambda = \hat{H} + \lambda\hat{Q}$. Its generalization due to Goeke and Reinhard (1980) allows all odd RPA moments m_k(RPA) to be derived by performing CHF calculations with Hamiltonian $\hat{H}_\lambda = \hat{H} + \lambda\hat{Q}_s$, $k = 2s - 1$, where $\hat{Q}_s$ are properly chosen constraint operators.

In conclusion, there exist at least two apparently different ways to calculate the RPA sum rules: one being a generalization of the Thouless theorem, the other of the dielectric theorem. The sum rule approach (SRA) consists in the use of RPA moments when extracting important information for averaged properties of the collective excitation spectrum without explicit use of RPA results for the strength distribution function in nuclei.

9.3 Sum rule approach and giant resonances in nuclei

There are experimental evidences showing that nuclear giant multipole resonances almost exhaust the strength of the corresponding multipole operators (Bertrand 1981, Speth and Van der Woode 1981, Goeke and Speth 1982, Buenerd 1984). Therefore, it should be justified to avoid the lengthy and expensive detailed RPA calculations by just guessing the collective excitation operator

$$\hat{Q} \equiv Q(L,T = 0) = \frac{1}{2}\sum_{i=1}^{N} f_L(\mathbf{r}_i), \qquad (9.29)$$

or

$$\hat{Q} \equiv Q(L,T = 1) = \frac{1}{2}\sum_{i=1}^{N} f_L(\mathbf{r}_i)\tau_i, \qquad (9.30)$$

where $\tau_i = 1$ for neutrons and $\tau_i = -1$ for protons, and then applying the SRA.

The different multipole operators Q(L,T) entering eqn (9.29) and eqn (9.30) correspond to a special choice of the functions

$f_L(\mathbf{r})$. Assuming that only one dominant mode exists in each (L,T)-channel (the doorway hypothesis of Fallieros (1972) and Dean and Fallieros (1973)), a reliable SRA description may be expected for monopole, dipole, and quadrupole resonances, whereby a sufficient choice of functions $f_L(\mathbf{r})$ is the following (Brink and Leonardi 1976)

$$f_{L=0}(\mathbf{r}) = r^2, \qquad (\nabla f_{L=0}) = \mathbf{r},$$

$$f_{L=1}(\mathbf{r}) = z, \qquad (\nabla f_{L=1}) = (0,0,1),$$

$$f_{L=2}(\mathbf{r}) = 2z^2-x^2-y^2, \qquad (\nabla f_{L=2}) = (-x,-y,2z). \tag{9.31}$$

For higher multipolarities ($L \geq 3$), the collective strength is distributed over at least two dominant peaks (Stringari 1979).

The resulting operator $Q(L,T = 0)$ is an isoscalar operator and produces a reliable description of isoscalar ($T = 0$) giant multipole resonances. When the number of neutrons differs from the number of protons, the operator $Q(L,T = 1)$ is not exactly an isovector operator due to the neutron excess in nuclei. Nevertheless, $Q(L,T = 1)$ is still a meaningful operator in the description of isovector giant multipole resonances in nuclei (Liu and Brown 1976). We shall assume that the isoscalar and isovector modes do not couple, although a coupling in fact appears (Stringari and Lipparini 1986). In the particular case of giant monopole resonances, for instance, the coupling of both isoscalar and isovector modes was shown to be too weak (Dimitrova, Petkov, and Stoitsov 1988a; see also Table 8.1).

Scaling and constrained excitation energies

Let us consider first the excitation energies associated with eqn (9.23) at $\mathbf{k} = 3$

$$\tilde{E}_3(L,T) = (m_3(L,T)/m_1(L,T))^{1/2}. \tag{9.32}$$

In the particular case of Hamiltonian (1.114) with Skyrme-type effective forces, one obtains from eqns (9.26), (9.29), and (9.30)

$$m_1(L,T{=}0) = \langle\Phi|Q(L,T{=}0)[\hat{H},Q(L,T{=}0)]|\Phi\rangle = \frac{\hbar^2}{4m}\int d\mathbf{r}\ \rho(\mathbf{r})(\nabla f_L)^2 \quad (9.33)$$

and

$$m_1(L,T{=}1) = \langle\Phi|Q(L,T{=}1)[\hat{H},Q(L,T{=}1)]|\Phi\rangle = \frac{\hbar^2(N-Z)Z}{4m\ N^2}\int d\mathbf{r}\ \rho(\mathbf{r})(\nabla f_L)^2$$

$$+\ \frac{\hbar^2}{m}\ \beta\int d\mathbf{r}\ \rho_n(\mathbf{r})\rho_p(\mathbf{r})(\nabla f_L)^2. \quad (9.34)$$

Here, $\rho_n(\mathbf{r})$, $\rho_p(\mathbf{r})$ and $\rho(\mathbf{r}) = \rho_n(\mathbf{r}) + \rho_p(\mathbf{r})$ are local densities associated with the HF ground state $\Phi \in \mathcal{H}^N_{HF}$, and β is the nonlocal parameter defined by eqn (8.53). As before, N stands for the mass number and Z for the number of protons, while the number of neutrons is (N − Z). In deriving eqn (9.33), we use the relation that $[\hat{H},Q(L,T = 0)] \equiv [\hat{T},Q(L,T = 0]$ and therefore only the kinetic energy operator $\hat{T}$ contributes to $m_1(L,T = 0)$. For the isovector eqn (9.34), we also find terms coming from the commutator with the exchange operator P_σ in eqn (1.114).

In the isoscalar case (T = 0), eqn (9.27) should be used to evaluate the moments $m_3(L,T = 0)$ entering eqn (9.32). One can realize, however, that the k = 3 moment (9.27) can be easily calculated on the basis of the following wave function

$$|\Phi_\alpha\rangle = \exp(-i\alpha\hat{Q}_1)\ |\Phi\rangle, \quad (9.35)$$

which transforms eqn (9.27) into the form

$$m_3(\mathrm{RPA}) = \frac{1}{2}\left[\frac{\partial^2}{\partial\alpha^2}\langle\Phi_\alpha|\hat{H}|\Phi_\alpha\rangle\right]_{\alpha=0}. \quad (9.36)$$

For any isoscalar and velocity independent operator $\hat{Q}$, one has

$$\hat{Q}_1 = i[\hat{H},\hat{Q}] = i[\hat{T},\hat{Q}]. \quad (9.37)$$

Therefore, the operator $\hat{Q}_1$ associated with eqn (9.29) is of the form

$$\hat{Q}_1 = -\frac{i}{2}\sum_{i=1}^{N}(\nabla_i \cdot \mathbf{v}(\mathbf{r}_i) + \mathbf{v}(\mathbf{r}_i) \cdot \nabla_i), \tag{9.38}$$

and eqn (9.35) is just the generalized scaling ansatz

$$|\Phi_\alpha\rangle = \exp\left[-\frac{1}{2}\alpha\sum_{i=1}^{N}(\nabla_i \cdot \mathbf{v}(\mathbf{r}_i) + \mathbf{v}(\mathbf{r}_i) \cdot \nabla_i)\right]|\Phi\rangle, \tag{9.39}$$

where $\mathbf{v}(\mathbf{r}) = (\hbar^2/m)\nabla f_L(\mathbf{r})$ is the velocity associated with the collective motion. Notice that eqn (9.39) is precisely eqn (8.86) with eqns (8.87) and (8.88). Therefore, one can use eqn (9.36) with the matrix element $\langle\Phi_\alpha|\hat{H}|\Phi_\alpha\rangle$ simply evaluated by scaling the HF ground-state wave function $\Phi \in \mathcal{H}^N_{HF}$.

In order to simplify the notation, we consider the case of equal numbers of protons and neutrons and express the Skyrme-HF ground-state energy as (Bohigas, Lane, and Martorell 1979)

$$E_{HF} = \int d\mathbf{r}\mathcal{H}(\rho,\tau,\mathbf{J}) = E_{kin} + E(t_0) + E(t_1,t_2) + E(t_3) + E(W_0) + E_C, \tag{9.40}$$

where E_{kin} is the kinetic energy, $E(t_0)$ is the zero range energy originating from the terms proportional to t_0, $E(t_1,t_2)$ is the finite-range energy originating from the terms $\sim t_1, t_2$, and so on.

In the monopole case ($L = 0$), after performing the scaling (9.39), the functions $\rho(\mathbf{r})$, $\tau(\mathbf{r})$, $\mathbf{J}(\mathbf{r})$ entering eqn (9.40) transform into the form (compare with eqn (8.94))

$$\begin{aligned}\rho(\mathbf{r},\tilde{\alpha}) &= \tilde{\alpha}^3\rho(\tilde{\alpha}\mathbf{r}),\\ \tau(\mathbf{r},\tilde{\alpha}) &= \tilde{\alpha}^5\rho(\tilde{\alpha}\mathbf{r}),\\ \mathbf{J}(\mathbf{r},\tilde{\alpha}) &= \tilde{\alpha}^5\rho(\tilde{\alpha}\mathbf{r}),\end{aligned} \tag{9.41}$$

where $\tilde{\alpha} = \exp(-2\hbar^2\alpha/m)$. The scaling expectation value of the

Hamiltonian $\hat{H}$ then reads

$$E(\alpha) = \langle\Phi_\alpha|\hat{H}|\Phi_\alpha\rangle = \int d\mathbf{r}\mathcal{H}(\rho(\mathbf{r},\tilde{\alpha}),\tau(\mathbf{r},\tilde{\alpha}),\mathbf{J}(\mathbf{r},\tilde{\alpha}))$$

$$= \tilde{\alpha}^2 E_{kin} + \tilde{\alpha}^3 E(t_0) + \tilde{\alpha}^5(E(t_1,t_2)+E(W_0)) + \tilde{\alpha}^{3\gamma+3}E(t_3) + \tilde{\alpha}E_C, \quad (9.42)$$

and finally from eqn (9.36) we have

$$m_3(L,T = 0) = 4\left(\frac{\hbar^2}{2m}\right)^2 (E_{kin} + E(t_1,t_2) + \frac{1}{4}E(W_0) - \frac{1}{5}E_C). \quad (9.43)$$

For other multipolarities one may proceed in the same manner in order to obtain the scaling excitation energies $\tilde{E}_3(L,T = 0)$ (Barranco, Polls, and Martorell 1985). The situation is more complicated in the case of isovector multipole resonances. All expressions needed for the actual determination of the scaling energies $\tilde{E}_3(L,T = 1)$, eqn (9.32), are given by Reinhard and Friedrich (1985).

Alternatively, performing CHF calculations with constraint operators defined from eqns (9.29), (9.30) and (9.31), one can derive the moments $m_{-1}(L,T)$ and $m_{-3}(L,T)$ and then calculate the constrained excitation energies $\tilde{E}_1(L,T)$ and $\tilde{E}_{-1}(L,T)$, respectively.

Many authors have investigated giant multipole resonances within the RPA sum rule approach (Martorell *et al.* 1976, Bohigas, Lane, and Martorell 1979, Lipparini, Orlandini, and Leonardi 1976, Stringari *et al.* 1978a,b, Kohno and Andō 1979, Goeke and Castel 1979, Goeke, Castel, and Reinhard 1980, Suzuki 1981). Some of the SRA results are given in Table 9.1. It is seen that the energies $\tilde{E}_{-1},\tilde{E}_1,\tilde{E}_3$ are quite different from each other, especially in the case of isovector resonances in light nuclei. The reason is the increasing variance σ, eqn (9.25), of the RPA strength with decreasing mass number.

Table 9.1. Excitation energies for isoscalar and isovector giant multipole resonances (in MeV) obtained from the RPA sum rule approach with SIII effective forces.

	L = 0, T = 0[1]				L = 0, T = 1[2]			
	$\tilde{E}_{-1}$	$\tilde{E}_1$	$\tilde{E}_3$	σ	$\tilde{E}_{-1}$	$\tilde{E}_1$	$\tilde{E}_3$	σ
^{16}O	23.1	28.2	34.5	9.9	25.1	32.1	40.6	12.3
^{40}Ca	22.4	25.6	28.5	6.3	25.0	31.8	38.6	10.6
^{90}Zr	21.0	22.1	23.4	3.8	27.6	32.1	35.8	7.9
^{208}Pb	16.1	16.9	17.9	2.9	19.3	26.8	30.5	7.1

	L = 1, T = 1[3]		L = 2, T = 0[3]	L = 2, T = 1[3]	
	$\tilde{E}_3^{CHF}$	$\tilde{E}_3^{SCL}$	$\tilde{E}_3^{CHF} = \tilde{E}_3^{SCL}$	$\tilde{E}_3^{CHF}$	$\tilde{E}_3^{SCL}$
^{16}O	28	25	22	41	37
^{40}Ca	24	28	19	37	34
^{90}Zr	21	19	15	32	31
^{208}Pb	18	17	12	28	28

The values are compiled from: [1]Bohigas, Lane, and Martorell (1979); [2]Goeke, Castel, and Reinhard (1980); [3]Stringari *et al.* (1978a,b).

Connection with nuclear density functional theory

The sum rule approach is intimately related to nuclear density functional theory because HF or CHF total energies are used as a starting point in the actual derivation of the RPA sum rules. Most calculations of RPA sum rules may be thus performed by using the density functional theory rather than the HF theory because the approximate energy density functionals and equilibrium local densities compare satisfactorily with the HF ones.

The scaling excitation energies, for example, can be expressed analogously to eqn (9.43) in the framework of the nuclear density functional theory by representing eqn (9.40) as an approximate energy density functional and then using the scaled local densities (8.94). In this way, isoscalar and isovector giant monopole resonances have been investigated within the pure TF approximation by Brink and Leonardi (1976). For the same purpose, Treiner *et al.* (1981) use the modified TF functional, while the improved TF or ETF functionals have been applied to the problem at zero and finite temperatures by Krivine, Schmit, and Treiner (1982), Meyer, Quentin, and Brack (1983), Barranco, Marcos, and Treiner (1984), Barranco *et al.* (1985), Barranco, Polls, and Martorell (1985), and Gleissl *et al.* (1988).

More generally, the sum rule approach together with the nuclear density functional theory may be used to represent other collective theories completely in terms of the local density distribution, at least in their small amplitude limit. This is possible because Goeke, Lane, and Martorell (1978) have demonstrated that collective approaches such as the cranking model, the generator coordinate method or ATDHF theory can be related to RPA by means of sum rule techniques. The same was also shown in the case of nuclear fluid dynamics by Stringari (1979). A complete review of collective behaviour of nuclear systems emerging from the sum rule approach was made by Lipparini and Stringari (1989).

Nuclear incompressibility

In order to illustrate the usefulness of the density functional theory ideas combined with the sum rule approach, we consider the case of isoscalar giant monopole resonances which provide unique information on the compressibility of the nuclear medium.

If one defines the compression modulus of a finite nucleus by the expression

$$K_N = \langle r^2 \rangle_0 \left(\frac{\partial^2 (E/N)}{\partial (\langle r^2 \rangle)^2} \right)_0 , \tag{9.44}$$

where $\langle r^2\rangle_0 = ((4m)/(\hbar^2 N))\, m_1(L = 0, T = 0)$ is the ground-state rms radius, then one obtains (Jennings and Jackson 1980)

$$\tilde{E}_1 = \left[\frac{\hbar^2 K_N^{CHF}}{m\langle r^2\rangle_0}\right]^{1/2} \tag{9.45}$$

and

$$\tilde{E}_3 = \left[\frac{\hbar^2 K_N^{SCL}}{m\langle r^2\rangle_0}\right]^{1/2}, \tag{9.46}$$

depending on whether the change of the rms radius $\langle r^2\rangle$ entering eqn (9.44) is evaluated by CHF or by scaling (see eqn (9.36) and eqn (9.28), respectively).

In general, both quantities K_N^{CHF} and K_N^{SCL} will not be the same because the monopole strength function is spread out for actual heavy nuclei. According to some definite inequalities between moments, it follows that $K_N^{CHF} \leq K_N^{SCL}$. In the limit of large systems ($N \rightarrow \infty$), Jennings and Jackson (1980) show the following comparison with the nuclear matter incompressibility modulus

$$K_{NM} = \lim_{N\rightarrow\infty} (K_N^{SCL}) = \frac{10}{7} \lim_{N\rightarrow\infty} (K_N^{CHF}). \tag{9.47}$$

It has been shown in Section 8.3 (see Figure 8.3) that the ISMR excitation energy is almost linearly correlated to the nuclear matter incompressibility K_{NM}. However, the extrapolation from finite nuclei to nuclear matter is not straightforward because surface, Coulomb and asymmetry effects significantly modify the nuclear matter value K_{NM} to its value K_N at finite nuclei.

Nuclear density functional theory justifies a leptodermous $N^{-1/3}$ expansion for K_N similar to that of droplet model given in Section 4.5. Starting from eqn (9.43) expressed by means of the modified TF approximate energy density functional and assuming Fermi-type local density distributions, Treiner *et al.* (1981) find the expression

$$K_N^{SCL} = K_{NM} + K_S\, N^{-1/3} + K_\delta \left(\frac{N-2Z}{N}\right)^2 + K_C\, Z^2\, N^{-4/3}, \qquad (9.48)$$

where K_S, K_δ and K_C stand for the surface, asymmetry, and Coulomb incompressibility coefficients, respectively.

When fitting the experimental ISMR excitation energy data within their uncertainties by the parameterization following from eqn (9.46) and eqn (9.48), the following 'experimental' values appear (Treiner *et al.* 1981)

$$\begin{aligned} K_{NM} &= 220 \pm 20 \text{ MeV}, \\ K_S &= -240 \pm 70 \text{ MeV}, \\ K_\delta &= -300 \pm 100 \text{ MeV}, \end{aligned} \qquad (9.49)$$

This procedure is similar to that of extracting nuclear matter parameters from the knowledge of nuclear binding energies within the droplet model. The major difference is, however, that in this latter case more than thousand experimental data are known for about 1900 nuclei (Myers and Swiatecki 1982), whereas in the former case only about fifty experimental data are known (Bertrand 1981, Buenerd 1984).

10. Nuclear systems at finite temperature

One of the most challenging possibilities opened by heavy-ion reactions is the study of the thermodynamic behaviour of excited compound nuclei formed in the course of a heavy-ion collision. The properties of many-particle systems under high pressure and extreme temperature are of current interest in nuclear, atomic and solid-state physics, astrophysics and other branches of many-particle theory. Therefore, a rigorous generalization of the density functional theory to account for finite temperatures is highly desirable.

In this section we present a natural extension of previous developments to the case of many-particle systems at finite temperatures. The field is still under rapid investigation. There have been numerous attempts to develop approximations to the energy density functionals at finite temperatures. Several new features and problems arise in the case of finite temperatures that are just at the initial stage of intensive study.

10.1 Nuclear density functional theory at finite temperature

A description of nuclear systems at finite temperatures requires the apparatus of statistical physics (Balescu 1975, Reichl 1980, Bogolubov and Bogolubov 1984). In particular, the nuclear grand potential, the Gibbs free energy Ω, is of crucial importance because an equilibrium state of the system at fixed temperature T and chemical potential μ is the state of minimum Ω. It is also important to define the full space $\mathcal{D}$ of Hermitian and positive definite density matrix operators with unit trace

$$\mathcal{D} = \{\hat{\rho}^{+} = \hat{\rho},\ \mathrm{Tr}(\hat{\rho}) = 1\}, \tag{10.1}$$

The operators $\hat{\rho} \in \mathcal{D}$ represent the quantum statistical states of the system. The trace in eqn (10.1) and in what follows implies an ensemble average of the operators over all states with any number of particles (Bogolubov and Bogolubov 1984).

The average value of an operator $\hat{A}$ associated with a physical quantity A is defined in the state $\hat{\rho} \in \mathcal{D}$ as

$$A = \mathrm{tr}(\hat{\rho}\ \hat{A}). \tag{10.2}$$

Therefore, in order to define the equilibrium values of any physical quantity, one needs the actual density matrix operator $\hat{\rho}_0 \in \mathcal{D}$ associated with the equilibrium state of the system.

Quantum statistical variational principle and method

According to Mermin (1963), the quantum statistical variational principle for an N-particle system with Hamiltonian $\hat{H}$ and number operator $\hat{N}$ at chemical potential μ and temperature T states that the grand canonical density matrix operator

$$\hat{\rho}_0 = \exp(-\ \beta[\hat{H} - \mu\hat{N}])/\mathrm{tr}(\exp(-\ \beta[\hat{H} - \mu\hat{N}])), \tag{10.3}$$

minimizes the Gibbs free energy

$$\Omega[\hat{\rho}] = \mathrm{tr}(\hat{\rho}[\hat{H} - \mu\hat{N} + \beta^{-1}\ln\hat{\rho}]) \tag{10.4}$$

over the whole density matrix operator space $\mathcal{D}$. Here β is the inverse temperature $\beta = (k_B T)^{-1}$ and we put the Boltzmann constant k_B equal to unity in order to measure the temperature T in MeV. The solution of equation $\delta\Omega[\hat{\rho}] = 0$ is then precisely eqn (10.3), and the fact that $\hat{\rho}_0 \in \mathcal{D}$ gives the absolute minimum

$$\Omega_0 \equiv \min_{\hat{\rho}\in\mathcal{D}} \{\Omega[\hat{\rho}]\} = \Omega[\hat{\rho}_0] = -\mathrm{tr}(\exp(-\ \beta[\hat{H} - \mu\hat{N}])), \tag{10.5}$$

follows from the second variation of $\Omega[\hat{\rho}]$ for which $\delta^2\Omega[\hat{\rho}] > 0$ at $\hat{\rho} \equiv \hat{\rho}_0$ (Mermin 1963).

On one hand, the quantum-statistical variational principle leads to the following unequality

$$\Omega[\hat{\rho}] \geq \Omega[\hat{\rho}_0], \tag{10.6}$$

which means that the Gibbs free energy $\Omega[\hat{\rho}]$ attains its absolute minimum when it is evaluated with the correct density matrix operator $\hat{\rho}_0$ from eqn (10.3). On the other hand, the quantum-statistical variational principle enables one to apply a variational method to find an upper bound to the exact grand canonical potential value (10.5). According to eqn (10.6), the variational method consists in minimizing the Gibbs free energy $\Omega[\hat{\rho}]$ over a more restricted class $\mathcal{D}_{tr} \subset \mathcal{D}$ of trial density matrix operators $\hat{\rho} \subset \mathcal{D}_{tr}$ at each T and μ.

Actually, all variational quantum statistical techniques are obtained solely by means of different trial density matrix operator classes on which the grand potential (10.4) is minimized. In particular, the temperature-dependent Hartree-Fock (THF) theory consists of using the THF density matrix operator $\hat{\rho}_{THF} \in \mathcal{D}$ which minimizes eqn (10.4) among all candidates of the form

$$\hat{\rho} = \exp\left(-\sum_{ij} a_i^{+}\gamma_{ij}a_j\right) / \mathrm{tr}\left(\exp\left(-\sum_{ij} a_i^{+}\gamma_{ij}a_j\right)\right), \tag{10.7}$$

where $a_i^{+}(a_i)$ are particle creation (annihilation) operators and $\gamma = (\gamma_{ij})$ is an arbitrary Hermitian matrix with unit trace (Mermin 1963).

Reformulation in terms of the local density distribution

Nuclear density functional theory at finite temperatures must reformulate exactly the quantum-statistical variational principle, and thus the associated variational methods, completely by means of the nuclear local density distribution

$$\rho(\mathbf{r}) = \mathrm{tr}(\hat{\rho} \sum_{i=1}^{N} \delta(\mathbf{r}-\mathbf{r}_i)). \tag{10.8}$$

This could in principle be done by a procedure due to Mermin (1965). The resulting Hohenberg-Kohn-Mermin theorem has been presented in Section 2.3. However, one may also proceed as in Chapter 7 within the rigorous energy density functional approach extended to the case of finite temperatures. The complete procedure has been given by Stoitsov, Petkov, and Kryachko (1987) and Stoitsov and Petkov (1988). It allows us also in the case of finite temperatures to derive a set of local density functionals which are practically applicable and overcome some difficulties in the treatment due to Mermin (1965).

Let us consider the following trial density matrix operators

$$\hat{\rho}[\rho] = \mathcal{U}^{+}_{f[\rho]}\, \hat{\bar{\rho}}\, \mathcal{U}_{f[\rho]}, \tag{10.9}$$

where $\hat{\bar{\rho}} \in \mathcal{D}$ is a given (say model or prototypical) density matrix operator and the unitary operators $\mathcal{U}_{f[\rho]}$ belong to the representation $\mathcal{U}_{\mathcal{F}}$ of the local-scale point transformation group $\mathcal{F}$ introduced in Section 2.4. The scalar local-scale transformation function $f[\rho] \equiv f_{[\rho]}(\mathbf{r})$ entering eqn (10.9) is obtained in such a way that the local-scale transformation applied to the model local density

$$\bar{\rho}(\mathbf{r}) = \mathrm{tr}\left[\hat{\bar{\rho}} \sum_{i=1}^{N} \delta(\mathbf{r}_i-\mathbf{r})\right] \tag{10.10}$$

transforms $\bar{\rho}(\mathbf{r})$ exactly into a given local density distribution belonging to the space $\bar{D}_N$ (see Section 2.4)

$$\rho(\mathbf{r}) = r^{-2}\, f^2_{[\rho]}(\mathbf{r})\, \frac{\partial f_{[\rho]}(\mathbf{r})}{\partial r}\, \bar{\rho}(\mathbf{f}_{[\rho]}(\mathbf{r})). \tag{10.11}$$

This equation follows after substituting $\hat{\rho}[\rho]$ from eqn (10.9) into eqn (10.8). As in the ground-state case, its solution $f_{[\rho]}(\mathbf{r})$ always exists (Stoitsov and Petkov 1988).

It is clear now that eqn (10.9) defines a density matrix operator $\hat{\rho}[\rho] \in \mathcal{D}$ associated with a given local density distribution $\rho(\mathbf{r}) \in \overline{D}_N$. If with every element $\rho(\mathbf{r}) \in \overline{D}_N$ we associate a density matrix operator $\hat{\rho}[\rho] \in \mathcal{D}$ according to eqn (10.9), then the following set of density matrix operators results:

$$\mathcal{C} = \left\{ \hat{\rho}[\rho] = \mathcal{U}^{+}_{f[\rho]} \, \hat{\bar{\rho}} \, \mathcal{U}_{f[\rho]} \; ; \; \rho(\mathbf{r}) \in \overline{D}_N \right\}. \tag{10.12}$$

From the group theory point of view (Naimark 1976), the set (10.12) forms the so-called conjugate class of the space $\mathcal{D}$ corresponding to the prototypical density matrix operator $\hat{\bar{\rho}} \in \mathcal{D}$. Hence, the full density matrix operator space $\mathcal{D}$ subdivides into a set of noninteracting or disjoint conjugate classes of the type of eqn (10.12) with all density matrix operators in a given class being related via local-scale point transformations (10.9).

One can realize the close analogy that exists between the full Hilbert space $\mathcal{H}^N$ and its orbits $\mathcal{O} \subset \mathcal{H}^N$, on one hand, and the full density matrix operator space $\mathcal{D}$ and its conjugate classes $\mathcal{C} \subset \mathcal{D}$, on the other. In particular, one can define in $\mathcal{D}$ the so-called proper or 'eigen' conjugate class $\mathcal{C}_H \subset \mathcal{D}$ which involves the exact grand canonical density matrix operator (10.3).

In formulating the quantum-statistical variational principle by means of the local density $\rho(\mathbf{r})$ one then proceeds as in Chapter 7 by restricting the many-body problem to a given conjugate class $\mathcal{C} \subset \mathcal{D}$. The Gibbs free energy (10.4) calculated with respect to all density matrix operators $\hat{\rho}[\rho] \in \mathcal{C}$ transforms into a functional of the local density distribution

$$\Omega[\rho] \equiv \Omega[\hat{\rho}] \Big|_{\mathcal{C} \subset \mathcal{D}} = \mathrm{tr}\left[\hat{\rho}[\rho](\hat{H} - \mu\hat{N} + \beta^{-1}\ln \hat{\rho}[\rho])\right]. \tag{10.13}$$

Actually, eqn (10.13) defines a set of local density functionals $\Omega[\rho]$ associated with different conjugate classes of the space $\mathcal{D}$. Since the full space $\mathcal{D}$ is a union of the disjoint conjugate classes $\mathcal{C} \subset \mathcal{D}$, the minimization of $\Omega[\hat{\rho}]$ over the whole space $\mathcal{D}$ is equivalent to the procedure that consists in determining the minima of the functionals (10.13) defined over all different conjugate classes $\mathcal{C} \subset \mathcal{D}$ and further comparing them; that is,

$$\Omega_0 = \min_{\hat{\rho}\in\mathcal{D}} \left\{ \Omega[\hat{\rho}] \right\} = \min_{\mathcal{C}\subset\mathcal{D}} \left\{ \min_{\rho\in\bar{D}_N} \left\{ \Omega[\rho] \right\}\right\}. \qquad (10.14)$$

In this way it becomes clear that a complete reformulation of the quantum-statistical variational principle is achieved in terms of the local density distribution $\rho(\mathbf{r})$. This statement is close in its context to the Hohenberg-Kohn-Mermin theorem as demonstrated by Stoitsov and Petkov (1988). In fact, the exact Hohenberg-Kohn-Mermin functional is reproduced by eqn (10.13) when it is defined within the eigen class $\mathcal{C}_H \subset \mathcal{D}$. Therefore, the problem of determining the exact functional $\Omega[\rho]$ is again equivalent to the exact solution of the many-body problem.

In contrast to the Hohenberg-Kohn-Mermin theorem, however, the functional eqn (10.13) is valid within each definite trial conjugate class $\mathcal{C} \subset \mathcal{D}$ where a constructive procedure for handling its explicit form is obvious. This constructive procedure of handling the quantum-statistical many-body problem within a given conjugate class $\mathcal{C} \subset \mathcal{D}$ completely in terms of the local density distribution can be viewed as a natural generalization of the RED approach (Chapter 7) to the case of nuclear ensembles at finite temperatures.

Local density functionals at finite temperature

We begin by separating the functional (10.13) defined within a given conjugate class $\mathcal{C} \subset \mathcal{D}$ into the following parts:

$$\Omega[\rho] \equiv F[\rho] - \mu N[\rho] = E[\rho] - T\,S[\rho] - \mu N[\rho], \qquad (10.15)$$

where

$$S[\rho] = -\mathrm{tr}(\hat{\rho}[\rho]\ \ln\ \hat{\rho}[\rho]) \tag{10.16}$$

is the entropy of the system as a functional of the local density $\rho(\mathbf{r})$ and $N[\rho]$ is the particle number

$$N[\rho] = \mathrm{tr}(\hat{\rho}[\rho]\ \hat{N}) = \int d\mathbf{r}\ \rho(\mathbf{r}). \tag{10.17}$$

The Helmholtz free energy

$$F[\rho] = E[\rho] - TS[\rho] \tag{10.18}$$

contains the total internal energy at constant temperature

$$E[\rho] = T[\rho] + V[\rho], \tag{10.19}$$

where the internal kinetic $T[\rho]$ and potential $V[\rho]$ energies are associated with the particular structure of the Hamiltonian $\hat{H}$ as

$$T[\rho] = \mathrm{tr}(\hat{\rho}[\rho]\ \hat{T}) \tag{10.20}$$

and

$$V[\rho] = \mathrm{tr}(\hat{\rho}[\rho]\ \hat{V}), \tag{10.21}$$

respectively.

In order to determine the explicit form of $E[\rho]$ within a trial class $\mathcal{C} \subset \mathcal{D}$, one should start with the local-density representation of one- and two-body density matrices (see Section 1.4). The model density matrix operator $\hat{\bar{\rho}} \in \mathcal{D}$ uniquely defines both the first and the second order model density matrices

$$\bar{\rho}^{(1)} \equiv \bar{\rho}^{(1)}(\mathbf{r};\mathbf{r}') = \mathrm{tr}(\hat{\bar{\rho}}\ \psi^{+}(\mathbf{r})\psi(\mathbf{r}')) = \sum_{\alpha,\beta} n_{\alpha\beta}\bar{\varphi}^{*}_{\alpha}(\mathbf{r})\bar{\varphi}_{\beta}(\mathbf{r}'), \tag{10.22}$$

and

$$\bar{\Gamma} \equiv \bar{\Gamma}(r_1,r_2;r_1',r_2') = \mathrm{tr}(\hat{\bar{\rho}}\, \psi^{+}(r_1)\psi^{+}(r_2)\psi(r_1')\psi(r_2'))$$

$$= \sum_{\alpha,\beta,\gamma,\delta} \Gamma_{\alpha\beta\gamma\delta}\, \bar{\varphi}^{*}_{\alpha}(r_1)\bar{\varphi}^{*}_{\beta}(r_2)\bar{\varphi}_{\gamma}(r_1')\bar{\varphi}_{\delta}(r_2'), \tag{10.23}$$

where $\psi^{+}(r)$ and $\psi(r)$ are particle field creation and annihilation operators, respectively. All states $(\alpha,\beta,\gamma,\delta)$ belong to an orthonormal set of model single-particle wave functions $\{\bar{\varphi}_{\alpha}(r)\}$. The model set of coefficients $n_{\alpha\beta}$ and $\Gamma_{\alpha\beta\gamma\delta}$ is also known once the trial class $\mathscr{C} \subset \mathscr{D}$ is given.

After performing the local-scale transformation (10.9), one obtains the local-scale transformed first and second order density matrices

$$\rho^{(1)}[\rho] \equiv \rho^{(1)}(r,r';[\rho]) = \mathrm{tr}(\hat{\rho}[\rho]\psi^{+}(r)\psi(r'))$$

$$= D^{1/2}[f_{[\rho]}(r);r]\; D^{1/2}[f_{[\rho]}(r');r']\bar{\rho}^{(1)}(f_{[\rho]}(r);f_{[\rho]}(r'))$$

$$= \sum_{\alpha,\beta} n_{\alpha\beta}\varphi^{*}_{\alpha}(r;[\rho])\varphi_{\beta}(r';[\rho]) \tag{10.24}$$

and

$$\Gamma[\rho] = \Gamma(r_1,r_2;r_1',r_2';[\rho]) = \mathrm{tr}(\hat{\rho}[\rho]\psi^{+}(r_1)\psi^{+}(r_2)\psi(r_1')\psi(r_2'))$$

$$= \sum_{\alpha,\beta,\gamma,\delta} \Gamma_{\alpha\beta\gamma\delta}\, \varphi^{*}_{\alpha}(r_1;[\rho])\varphi^{*}_{\beta}(r_2;[\rho])\varphi_{\gamma}(r_1';[\rho])\varphi_{\delta}(r_2';[\rho]), \tag{10.25}$$

respectively. Here, $D(f_{[\rho]}(r);r)$ is the Jacobian (2.32) and the density-dependent single-particle wave functions

$$\varphi_\alpha(\mathbf{r}) \equiv \varphi_\alpha(\mathbf{r};[\rho]) = D^{1/2}[\mathbf{f}_{[\rho]}(\mathbf{r});\mathbf{r}]\ \bar{\varphi}_\alpha(\mathbf{f}_{[\rho]}(\mathbf{r})) \qquad (10.26)$$

are the local-scale transformed model wave functions $\bar{\varphi}_\alpha(\mathbf{r})$.

Further, using eqn (10.20), eqn (10.21) and the density-dependent representations (10.24) and (10.25), one obtains the energy density functional (10.19) in the form

$$E[\rho] = \sum_{\alpha,\beta} n_{\alpha\beta}\ \langle\varphi_\alpha[\rho]|\hat{t}|\varphi_\beta[\rho]\rangle$$

$$+ \sum_{\alpha,\beta,\gamma,\delta} \Gamma_{\alpha\beta\gamma\delta}\ \langle\varphi_\alpha[\rho]\varphi_\beta[\rho]|\hat{\omega}|\varphi_\gamma[\rho]\varphi_\delta[\rho]\rangle_A. \qquad (10.27)$$

It should be pointed out that carrying out elementary calculations like the calculations performed in Section 7.2, then one obtains the kinetic energy density functional in the form (Stoitsov and Petkov 1988)

$$\tau[\rho] = \tau_W[\rho] + \tau_\gamma[\rho], \qquad (10.28)$$

which has a structure similar to the ground-state kinetic energy density functional (7.35) calculated earlier at zero temperature. The first term $\tau_W[\rho]$ is just the well known von Weizsäcker term (7.24) but now represented in terms of the temperature-dependent local density distribution. This term emerges independently of the conjugate class chosen and therefore it is a piece of the exact kinetic energy density functional. The second term in eqn (10.28) is also a temperature-dependent analogue of the nonlocal term (7.33). Using the same technique, one can derive the actual form of the entropy functional $S[\rho]$, the potential energy functional $V[\rho]$, and therefore the complete functional $\Omega[\rho]$ within $\mathcal{C} \subset \mathcal{D}$.

Euler-Lagrange equation at finite temperature

A minimization of the Gibbs free energy (10.13) with respect to $\rho(\mathbf{r})$ leads to the variational equation

$$\frac{\delta}{\delta\rho(\mathbf{r})}\int d\mathbf{r}\left\{\frac{\hbar^2}{2m}\tau[\rho] + \upsilon[\rho] - T\sigma[\rho] - \mu\rho + P_0\right\} = 0, \qquad (10.29)$$

where $\tau[\rho]$, $\upsilon[\rho]$ and $\sigma[\rho]$ are densities associated with the functionals $T[\rho]$, $V[\rho]$ and $S[\rho]$, respectively. The normalization condition

$$N[\rho] = \int d\mathbf{r}\ \rho(\mathbf{r}) = N, \qquad (10.30)$$

coupled to the problem via the Lagrange multiplier μ elucidates the meaning of μ as the chemical potential of the system, $\mu = \delta\Omega[\rho]/\delta N$. The other Lagrange multiplier $T = \beta^{-1}$ is the temperature $T = \delta\Omega[\rho]/\delta S$, where δS is the change of the entropy (10.16).

The variation (10.29) at the particular kinetic energy density functional (10.28) leads to the following Euler-Lagrange equation

$$-\frac{\hbar^2}{4m}\nabla^2\rho(\mathbf{r}) + \frac{\hbar^2}{8m}\frac{(\nabla\rho)^2}{\rho(\mathbf{r})} + \rho(\mathbf{r})\left[\nabla_{\mathbf{r}}.\nabla_{\mathbf{r}'}\bar{\gamma}(f_{[\rho]}(\mathbf{r});f_{[\rho]}(\mathbf{r}'))\right]_{\mathbf{r}'=\mathbf{r}}$$

$$+\ \rho^2\frac{\delta}{\delta\rho}\left[\nabla_{\mathbf{r}}.\nabla_{\mathbf{r}'}\bar{\gamma}(f_{[\rho]}(\mathbf{r});f_{[\rho]}(\mathbf{r}'))\right]_{\mathbf{r}'=\mathbf{r}} + \upsilon(\mathbf{r};[\rho]) = \mu\rho(\mathbf{r}), \qquad (10.31)$$

where

$$\upsilon(\mathbf{r};[\rho]) = \delta\upsilon[\rho]/\delta\rho(\mathbf{r}) - T\sigma[\rho]/\delta\rho(\mathbf{r}). \qquad (10.32)$$

This is a highly nonlinear second-order integro-differential equation for the density $\rho(\mathbf{r})$ which should be solved with appropriate boundary conditions.

A particular set of boundary conditions may approximately be reached locally in a collapsing massive star, where condensed nuclei exist in phase equilibrium with a surrounding gas of nucleons and leptons (Lattimer 1981, Bonche and Vautherin 1981, 1982). The quantity P_0 entering eqn (10.29) is the finite external pressure necessary to maintain this phase equilibrium (Bartel, Brack, and Durand 1985). The appropriate boundary conditions are such that for r = 0 the local density should be constant, $\rho(0) = \rho_0$, and that far outside the system the gas has a constant density ρ_g.

The equilibrium situation, however, does not apply for a hot but isolated system which is not stable. The nucleus can emit particles, and therefore cannot be described in static variational calculations. Assuming, however, the evaporation time to be much larger than the equilibration time, one may consider the system as being in a metastable phase. In this case, the variation in eqn (10.29) has to be done at zero external pressure $P_0 = 0$. The variation leads to the same Euler-Lagrange equation (10.31), but with new boundary conditions, $\rho(r\to 0) = \rho_0$ and $\rho(r\to\infty) = 0$.

In both cases, $P_0 \neq 0$ and $P_0 = 0$, one obtains the local density distribution $\rho(\mathbf{r};\mu,T)$ as a solution of eqn (10.31). With this solution $\rho(\mathbf{r};\mu,T)$, one then obtains the equilibrium density matrix operator (10.9) and, hence, all other physical characteristics of the system under study. According to eqn (10.6) and eqn (10.13), the following inequality holds:

$$\Omega[\bar{\rho}] \geq \Omega[\rho(\mathbf{r};\mu,T)] \geq \Omega_0, \tag{10.33}$$

where $\Omega[\bar{\rho}]$ is the Gibbs free energy associated with the model local density distribution (10.10), $\Omega[\rho(\mathbf{r};\mu,T)]$ is the actual minimum within the conjugate class $\mathcal{C} \subset \mathcal{D}$ and Ω_0 is the exact grand potential value (10.5).

10.2 Density functional theory in THF approximation

In view of the tradition that the density functional theory is usually formulated in the framework of the Hartree-Fock theory we shall consider how a rigorous formulation of the temperature-dependent HF theory can be obtained in terms of the local density in nuclei.

Temperature-dependent mean-field approximation

Let us separate in $\mathcal{D}$ a subset $\mathcal{D}_{THF} \in \mathcal{D}$ which contains all density matrix operators of the type of eqn (10.7). The Hermitian matrix $\gamma = (\gamma_{ij})$ entering eqn (10.7) can always be diagonalized. We denote its diagonal matrix elements by γ_i. The orthonormalized single- particle basis in which $\gamma_{ij} = \gamma_i \delta_{ij}$ is denoted by $\{\varphi_i(\mathbf{r})\}$. Further, one may consider the following mean-field (MF) trial density matrix operators (des Cloizeaux 1968)

$$\hat{\rho}_{MF} = \exp\left[-\beta\sum_i \gamma_i a_i^+ a_i\right] / \mathrm{tr}(\exp\left[-\beta\sum_i \gamma_i a_i^+ a_i\right]), \tag{10.34}$$

where γ_i are real c-numbers and $a_i^+ (a_i)$ is the particle creation (annihilation) operator in the single-particle state $\bar{\varphi}_i(\mathbf{r})$.

The grand partition function entering eqn (10.34) can be calculated in the Fock space,

$$Z_{MF} = \mathrm{tr}(\exp\left[-\beta\sum_i \gamma_i a_i^+ a_i\right] = \sum_{\tilde{n}_1 \tilde{n}_2 \ldots = 0,1} \langle \tilde{n}_1 \tilde{n}_2 \ldots | \exp\left[-\beta\sum_i \gamma_i \tilde{n}_i\right] | \tilde{n}_1 \tilde{n}_2 \ldots \rangle$$

$$= \sum_{\tilde{n}_1 \tilde{n}_2 \ldots = 0,1} \prod_i \exp\left(-\beta\gamma_i \tilde{n}_i\right) = \prod_i \sum_{\tilde{n}_i = 0,1} \exp\left(-\beta\gamma_i \tilde{n}_i\right) = \prod_i \exp\left(-\beta\gamma_i\right). \tag{10.35}$$

It leads to the following representation of the trial density matrix operator

$$\hat{\rho}_{MF} = \prod_i \left(n_i \, a_i^+ a_i + (1 - n_i)(1 - a_i^+ a_i)\right), \tag{10.36}$$

where $n_i = [1 + \exp(\beta\gamma_i)]^{-1}$ are particle occupation probabilities (obviously $0 \le n_i \le 1$) entering the one-body density matrix

$$n_{ij} \equiv \mathrm{tr}(\hat{\rho}_{MF} a_j^+ a_i) = n_i \delta_{ij}. \tag{10.37}$$

According to eqn (10.36), one obtains also the two-body density matrix

$$\Gamma_{klmn} = \mathrm{tr}(\hat{\rho}_{MF} a_k^+ a_l^+ a_m a_n) = n_k n_l (\delta_{km}\delta_{ln} - \delta_{lm}\delta_{kn}). \tag{10.38}$$

Hence, the expectation value of the nuclear Hamiltonian $\hat{H}$ in the trial state (10.36) is

$$E_{MF} = \mathrm{tr}(\hat{\rho}_{MF}\hat{H}) = \sum_i n_i \langle\varphi_i|\hat{T}|\varphi_i\rangle + \frac{1}{2}\sum_{i,j} n_i n_j \langle\varphi_i\varphi_j|\hat{\omega}|\varphi_i\varphi_j\rangle_A. \tag{10.39}$$

Using eqn (10.36) and eqn (10.37), the entropy reads

$$S_{MF} = -\mathrm{tr}(\hat{\rho}_{MF} \ln \hat{\rho}_{MF}) = -\sum_i \Big[n_i \ln n_i + (1 - n_i)\ln(1 - n_i) \Big]. \tag{10.40}$$

Finally, from eqns (10.4), (10.39) and (10.40), one obtains the Gibbs free energy in the MF approximation

$$\Omega[\hat{\rho}_{MF}] = \sum_i n_i \langle\varphi_i|\hat{t}|\varphi_i\rangle + \frac{1}{2}\sum_{i,j} n_i n_j \langle\varphi_i\varphi_j|\hat{\omega}|\varphi_i\varphi_j\rangle_A$$

$$+ T\sum_i \Big[n_i \ln n_i + (1-n_i)\ln(1-n_i) \Big] - \mu\sum_i n_i. \tag{10.41}$$

According to the quantum-statistical variational principle, eqn (10.41) has to be minimized with respect to all parameters entering the trial density matrix operator $\hat{\rho}_{MF}$; that is, with respect to the occupation numbers n_i and the single-particle wave functions $\varphi_i(\mathbf{r})$. The result gives rise to the self-consistent equation of the THF theory, while any other choice for n_i and $\varphi_i(\mathbf{r})$ represents a mean-field approximation to the quantum-statistical many-body problem.

Temperature-dependent Hartree-Fock theory

Let us first minimize eqn (10.41) with respect to $\varphi_i(\mathbf{r})$ by including additional constraints $\langle\varphi_i|\varphi_i\rangle = 1$. This could be done by introducing Lagrange multipliers λ_i and minimizing the quantity $\tilde{\Omega} = \Omega[\hat{\rho}_{MF}] + \sum\lambda_i\langle\varphi_i|\varphi_i\rangle$. One finds the result (des Cloizeaux 1968)

$$\hat{H}_{THF}\varphi_i(\mathbf{r}) = \varepsilon_i\, \varphi_i(\mathbf{r}), \tag{10.42}$$

where the THF single-particle Hamiltonian is of the form

$$\langle\varphi_i|\hat{H}_{THF}|\varphi_k\rangle = \langle\varphi_i|\hat{T}|\varphi_k\rangle + \sum_j n_j\langle\varphi_i\varphi_j|\hat{v}|\varphi_k\varphi_j\rangle_A. \tag{10.43}$$

and the single-particle energies ε_i are related to λ_i as $\varepsilon_i = \lambda_i/n_i$.

The minimization of eqn (10.41) with respect to n_i gives

$$\delta\Omega[\hat{\rho}_{MF}]/\delta n_i \equiv \varepsilon_i - T\,\ln[n_i/(1 - n_i)] - \mu = 0, \tag{10.44}$$

where the single-particle energies

$$\varepsilon_i \equiv \delta E_{MF}/\delta n_i = \langle\varphi_i|\hat{H}_{THF}|\varphi_i\rangle \tag{10.45}$$

are precisely the same as those entering eqn (10.42). Equation (10.44) can be solved exactly and gives the Fermi-Dirac distribution

$$n_i = \frac{1}{1 + \exp[(\varepsilon_i - \mu)/T]}, \tag{10.46}$$

which confirms the meaning of the Lagrange multipliers μ and T as the Fermi surface and temperature, respectively.

Equation (10.42) together with eqn (10.43) and eqn (10.46) form the infinite set of the self-consistent THF equations whose solution determines the single-particle wave functions $\varphi_i(\mathbf{r})$ and occupation probabilities n_i at any given temperature T and chemical potential μ. Therefore, after performing the self-consistent THF calculations, one determines all nuclear characteristics in terms of the optimal THF density matrix operator

$$\hat{\rho}_{THF} = \exp(-\beta[\hat{H}_{THF} - \mu\hat{N}])/\mathrm{Tr}(\exp(-\beta[\hat{H}_{THF} - \mu\hat{N}])). \qquad (10.47)$$

In other words, the THF theory represents a natural generalization of the HF ground-state problem. The difference is only in the fact that the summation in eqn (10.43) extends over all single-particle states, each weighted with its occupation probability (10.46). For the case of Skyrme-type effective forces, for example, the THF formalism leads to exactly the same expressions as those following from the ground-state Skyrme-HF theory (Section 1.7). One has only to replace the ground-state densities $\rho(\mathbf{r})$, $\tau(\mathbf{r})$ and $\mathbf{J}(\mathbf{r})$ by their THF analogues

$$\rho(\mathbf{r}) = \sum_{l=1}^{\infty} n_l \, |\varphi_l(\mathbf{r})|^2,$$

$$\tau(\mathbf{r}) = \sum_{l=1}^{\infty} n_l \, |\nabla\varphi_l(\mathbf{r})|^2,$$

$$\mathbf{J}(\mathbf{r}) = \sum_{l=1}^{\infty} n_l \, (\varphi_l^*(\mathbf{r}) \, \mathbf{l}.\mathbf{s} \, \varphi_l(\mathbf{r})), \qquad (10.48)$$

where the occupation numbers are defined by eqn (10.46) with the chemical potential μ adjusted to satisfy the particle number condition $\sum n_l = N$.

Actual THF calculations have been performed with Skyrme-type forces by Brack and Quentin (1974a,b, 1981), Mosel, Zint, and Passler (1974), Sauer, Chandra, and Mosel (1976), Bonche and Vautherin (1981, 1982), Bonche, Levit, and Vautherin (1984, 1985), Sagawa and Toki (1986). In these calculations the parameters of the effective forces are kept constant at their values extracted at zero temperature.

The temperature-dependence of the effective forces in nuclei may appear by means of the density matrix expansion (Negele and Vautherin 1975) in which the force parameters themselves depend on the density and thus also on the temperature. The temperature-dependence of the Skyrme forces has been recently investigated in nuclear matter calculations by Baldo *et al.* (1986). It has been shown that the temperature affects mainly the chemical potential and the pressure, while the energy is essentially unaffected. The temperature-dependence of the effective interaction following from the Brückner theory has been investigated by Lejeune *et al.* (1986) and Cugnon *et al.* (1989).

RED approach in THF approximation

On the basis of the RED approach extended to finite temperature, Stoitsov (1987) and Stoitsov and Petkov (1988) derive an approximate Gibbs free energy density functional defined within a conjugate class $\mathcal{C} \subset \mathcal{D}_{THF}$ with prototypical density matrix operator of the form of eqn (10.36). The result

$$\Omega[\rho] = \sum_i n_i[\rho]\langle\varphi_i[\rho]|\hat{t}|\varphi_i[\rho]\rangle$$

$$+ \frac{1}{2}\sum_{i,j} n_i[\rho]n_j[\rho]\langle\varphi_i[\rho]\varphi_j[\rho]|\hat{\omega}|\varphi_i[\rho]\varphi_j[\rho]\rangle_A$$

$$+ T\sum_i\Big[n_i[\rho]\ln n_i[\rho] + (1-n_i[\rho])\ln(1-n_i[\rho])\Big] - \mu\sum_i n_i[\rho], \qquad (10.49)$$

where

$$n_i[\rho] = \frac{1}{1 + \exp[(\varepsilon_i[\rho] - \mu)/T]} \qquad (10.50)$$

and

$$\varepsilon_i[\rho]=\langle\varphi_i[\rho]|\hat{T}|\varphi_i[\rho]\rangle+\sum_j n_j[\rho]\langle\varphi_i[\rho]\varphi_j[\rho]|\hat{\omega}|\varphi_k[\rho]\varphi_j[\rho]\rangle_A, \qquad (10.51)$$

contains density-dependent single-particle wave functions

$\varphi_1[\rho] \equiv \varphi_1(\mathbf{r};[\rho])$ defined as in eqn (10.26) on the basis of a given set of model single-particle functions $\bar{\varphi}_1(\mathbf{r})$.

The functional (10.49) is an exact one within the THF framework if the model single-particle set $\bar{\varphi}_1(\mathbf{r})$ follows as a solution of the THF equations. In this case, $\Omega[\rho]$ attains its absolute minimum $\Omega_{THF} = \Omega[\rho_{THF}]$ when eqn (10.49) is evaluated with the exact THF local density $\rho_{THF}(\mathbf{r})$. Any other choice for the set $\{\bar{\varphi}_1(\mathbf{r})\}$, different from the THF one, creates a well defined approximate functional (10.49) which could be used as a sufficient approximation to the more involved THF problem. One has only to solve the Euler-Lagrange equation associated with eqn (10.49) in order to obtain the density $\rho(\mathbf{r})$ and other nuclear characteristic of interest.

Stoitsov (1987) applied the RED approach at different temperatures within the particular choice of the model set $\{\bar{\varphi}_1(\mathbf{r})\}$ given by the harmonic-oscillator shell model wave functions (7.57). The resulting functional (10.49) is then minimized within the restricted space of symmetrized Fermi-type (SF) local density distributions (7.64). In Figure 10.1, some results of these RED+SF calculations obtained at different temperatures with SkM effective forces for the mass rms radius of the nucleus ^{208}Pb are compared with the self-consistent THF results of Bonche, Levit, and Vautherin (1984, 1985). The same comparison is made in Figure 10.2 with respect to the neutron single-particle spectrum. It is seen that the RED+SF approach reproduces the average as well as the microscopic self-consistent THF results of Bonche, Levit and Vautherin (1984, 1985) with satisfactory accuracy.

It should be pointed out that all single-particle states of the first nine major shells ($N_{sh} = 8$) have been included in the RED+SF calculations (Stoitsov 1987). This configuration space, however, is not large enough to estimate the convergence of the entropy and excitation energy above T = 5 MeV. As an illustration, we plot in Figure 10.3 the excitation energy $E^*(T) = E(T) - E(0)$ versus temperature T. It is obvious that significant differences

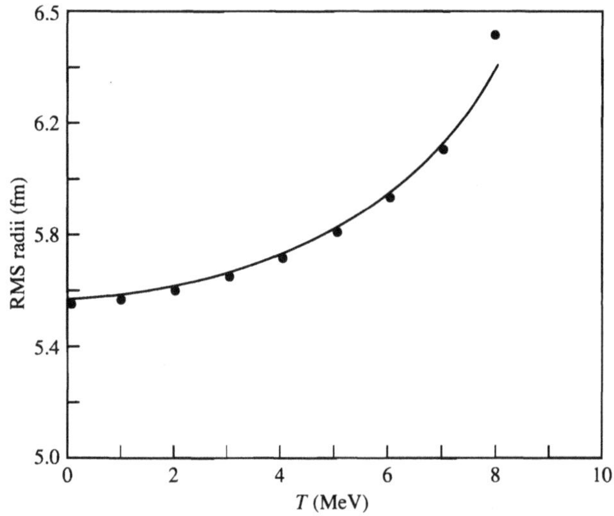

Figure 10.1. Nuclear mass rms radius as a function of temperature T for the nucleus ^{208}Pb calculated with SkM forces. The results obtained from the rigorous energy density approach (dots) applied within $N_{sh} = 8$ major shells (Stoitsov 1987) are compared with the temperature-dependent Hartree-Fock results (full line) of Bonche, Levit, and Vautherin (1984,1985).

occur for temperatures above T = 5 MeV when including in a perturbative way the contributions of all single-particle states of the first 21 major shells. Sagawa and Toki (1986) obtained similar results, which are also given in Figure 10.3. We shall see below that the contributions of the high-lying single-particle states are of special importance when including them within the self-consistent procedure.

Accounting for the continuum states

The physical meaning of the single-particle states with energies above the Fermi level is not completely clear within the THF theory. Nevertheless, they are usually invoked to simulate those quasibound or even unbound single-particle states which are

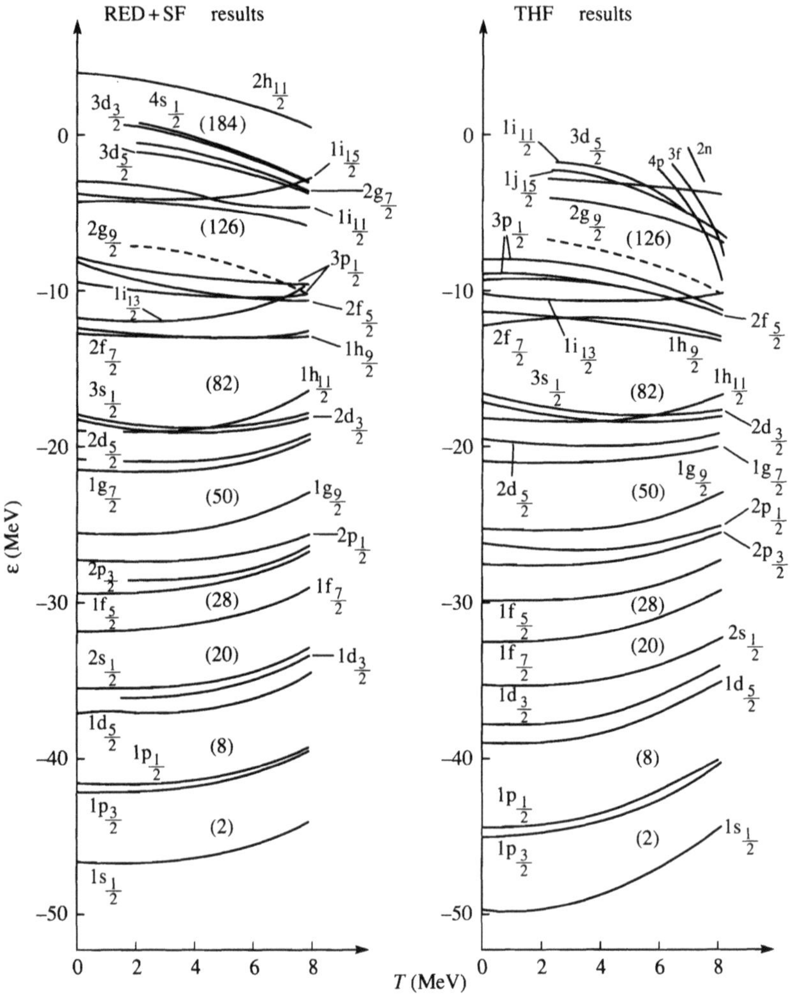

Figure 10.2. Comparison between RED+SF (left part) and THF (right part) neutron single-particle spectra for the nucleus ^{208}Pb calculated with SkM forces (Stoitsov 1987).

occupied with nonzero occupation probabilities in nuclei at high temperatures. As a result, the local density may not vanish at large distances and some convergence problems appear when calculating the expectation values of observable quantities. Therefore, in order to extract the properties of the hot nucleus

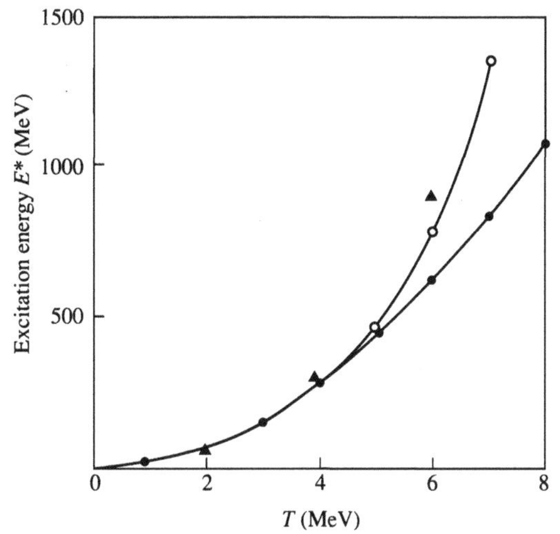

Figure 10.3. Excitation energy $E(T)^* = E(T) - E(0)$ versus temperature T for the nucleus ^{208}Pb calculated with SkM forces within $N_{sh} = 8$ (dots) and $N_{sh} = 21$ (circles) major shells (Stoitsov 1987). For comparison, the temperature-dependent constrained Hartree-Fock results derived by Sagawa and Toki (1986) are shown by triangles.

from a static THF calculation, one should adopt a kind of subtraction or truncation procedure identifying unambiguously the contribution of the continuum states (the external nucleon vapour).

According to the truncation procedure applied by Brack and Quentin (1974a,b), the nucleus is treated statistically as a metastable system totally neglecting the external gas phase. Thus, one takes into account only those single-particle states which are bound and quasibound by their Coulomb and centrifugal barriers. This truncation procedure obviously ignores continuum states and, therefore, is presumably a reasonable approximation at small temperatures. Some results for the entropy as a function of T are shown in Figure 10.4.

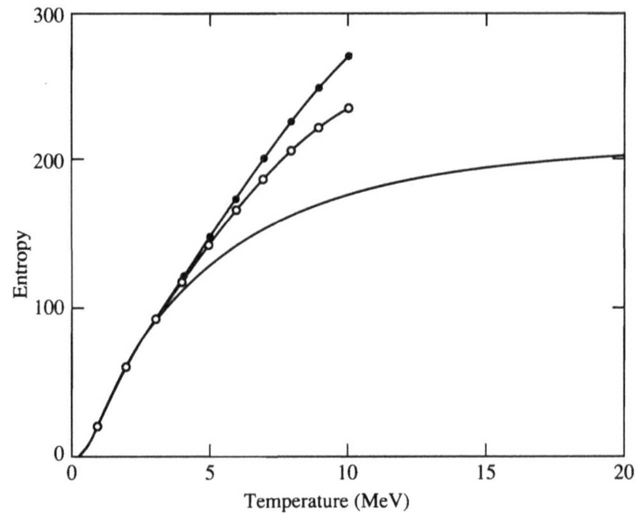

Figure 10.4. Variation of the entropy with temperature calculated in the nucleus ^{208}Pb with SII forces. The truncation procedure of Bonche, Levit, and Vautherin (1984,1985) leads to the results denoted by dots. The circles indicate results obtained when including bound and quasibound orbits only. The results obtained when including continuum effects but neglecting changes in the mean field with temperature are also shown (Bonche, Levit, and Vautherin 1984).

Another procedure (Mosel, Zint, and Passler 1974, Brack and Quentin 1974a,b, Sauer, Chandra, and Mosel 1976) works with a truncated oscillator basis and includes all unbound single-particle states obtained when diagonalizing the THF Hamiltonian (10.43) in configuration space. This procedure is still unsatisfactory due to the arbitrary cut-off used. It is difficult to include a large configuration space within the self-consistent THF diagonalization procedure.

The RED+SF procedure is close by its context to the procedure of Mosel, Zint and Passler (1974). When defining the functional (10.49) with the help of harmonic oscillator shell model wave functions $\bar{\varphi}_1(\mathbf{r})$, the states $\varphi_1(\mathbf{r};[\rho])$ accordingly describe all bound, quasibound and continuum THF single-particle states (see Figure 10.2). On the other hand, the RED+SF procedure is sufficiently simple and numerically fast to take into account a large number of single-particle states.

Stoitsov (1988) has included within the self-consistent RED+SF procedure all states of the first 21 major shells. With increasing temperature T, it has been found that beyond a critical value T_{cr} the nucleus no longer exists. In numerical calculations this effect is detected by looking at the Helmholtz free energy whose minimum at lower temperatures (Figure 10.5) disappears at temperatures $T \geq T_{cr}$ (Figure 10.6). For SkM* effective forces, T_{cr} takes values 5.5 MeV, 6.0 MeV and 6.5 MeV for nuclei ^{40}Ca, ^{90}Zr and ^{208}Pb, respectively. These results agree with recently reported experimental data giving $T_{cr} = 6.2 \pm 0.4$ MeV (Fabris *et al.* 1987).

In the astrophysical context, finite nuclei may coexist with a gas of nucleons in local thermodynamic equilibrium under external pressure. Although it may not be the appropriate picture for an isolated nucleus in a nuclear laboratory, the equilibrium between both phases is used in nuclear THF calculations. The subtraction procedure of Bonche, Levit, and Vautherin (1984, 1985), for example, is based on the existence of two solutions of the THF equations: one corresponding to the hot nucleus plus the surrounding vapour of nucleons, and the other to the vapour phase alone. The nucleus+vapour solution is obtained when starting the THF iterations with a Saxon-Woods initial guess for the THF mean-field, while a null initial potential leads to the vapour solution. The asymptotic coincidence of the nucleus+vapour and the vapour solutions at large distances is illustrated in Figure 10.7. The grand potential of the nucleus $\Delta\Omega$ is then defined as a difference $\Delta\Omega = \Omega(\text{nucleus+vapour}) - \Omega(\text{vapour})$. In this kind of calculation, the compound nucleus is stable up to critical temperature $T_c > T_{cr}$ which is of the order of 8 MeV and 11 MeV for

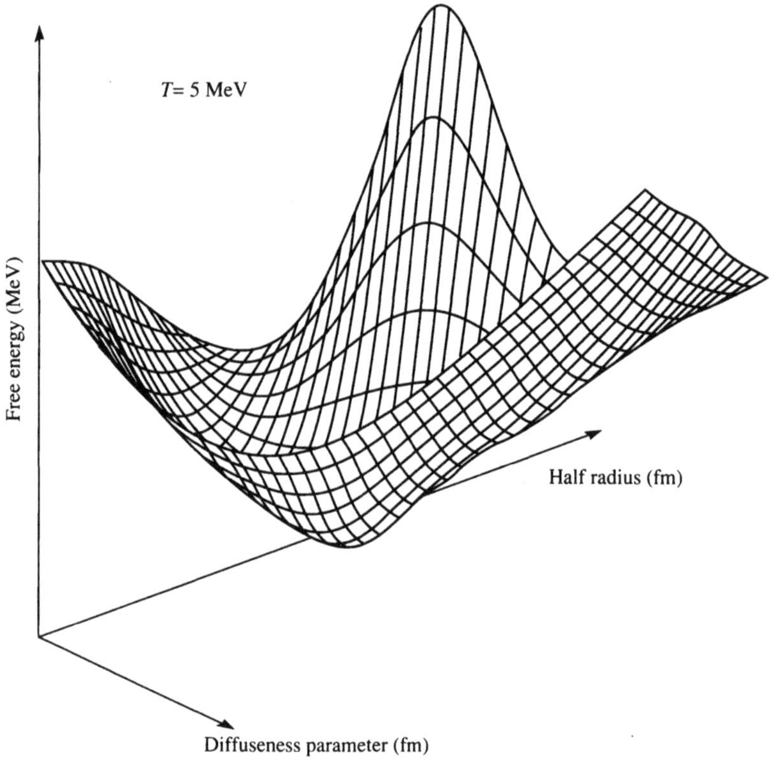

Figure 10.5. The Helmholtz free energy surface F(R,b) for the nucleus ^{40}Ca calculated with SkM* forces at temperature T = 5 MeV. The absolute minimum at the equilibrium point is evident.

SkM and SIII forces, respectively. Beyond this temperature the so-called Coulomb instability of nuclei appears (Bonche, Levit, and Vautherin 1985).

The subtraction procedure of Bonche, Levit, and Vautherin (1984, 1985) is not free of difficulties because the vapour phase of protons causes a divergence problem for the Coulomb energy. It must be assumed that the vapour nucleons do not contribute to the Coulomb repulsion. Of course, other subtraction or truncation procedures are also possible. For example, Dean and Mosel (1985) take the continuum into account via the Levinson theorem.

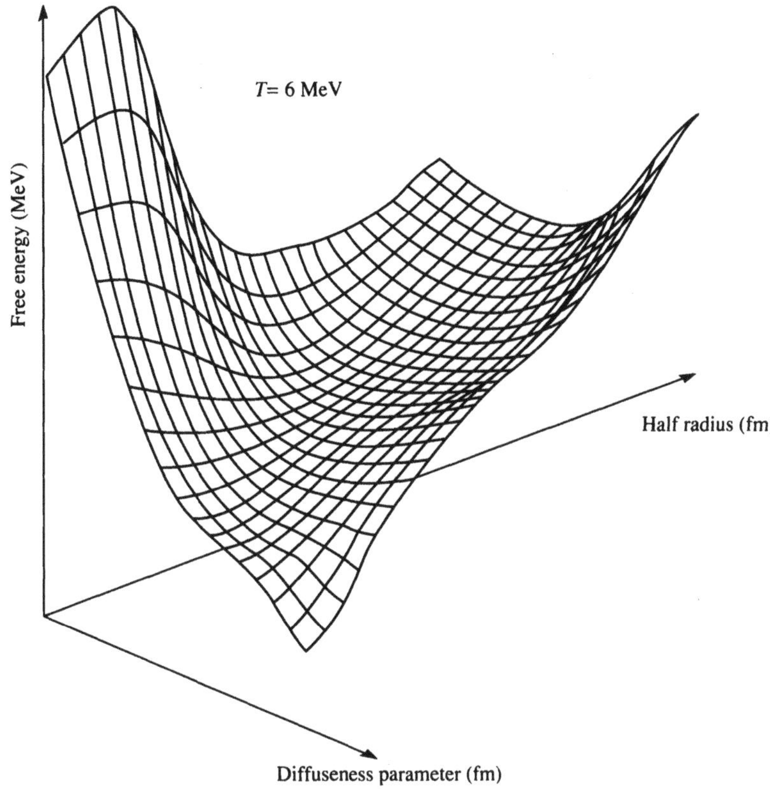

Figure 10.6. The Helmholtz free energy surface F(R,b) for the nucleus Ca^{40} calculated with SkM^* forces at temperature T = 6 MeV. The absolute minimum seen at temperature T = 6 MeV disappears at temperatures above the critical one, T_{cr} = 5.5 MeV.

10.3 Semiclassical functionals at finite temperature

It is well known from earlier THF calculations (Brack and Quentin 1974b) that the local density oscillations due to nuclear shell effects completely disappear beyond a critical temperature $T_{sh} \cong$ 2.5–3 MeV which is roughly the same for all nuclei. Systems at temperatures $T \geq T_{sh}$ are thus appropriate objects for studies within the semiclassical density functional theory because the

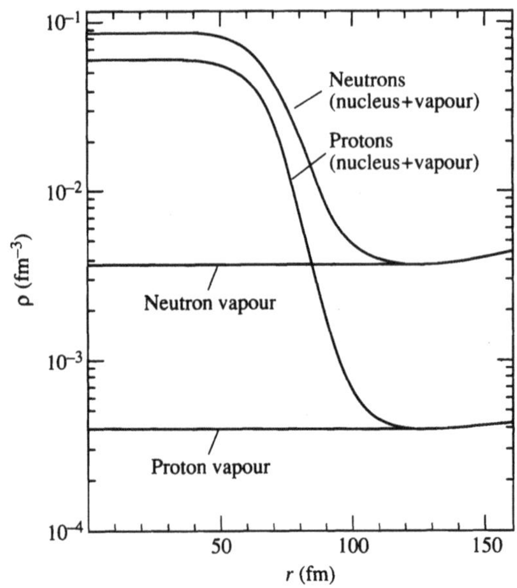

Figure 10.7. Neutron and proton densities of the nucleus+vapour and vapour solutions of the temperature-dependent Hartree-Fock equations for the nucleus ^{208}Pb at temperature T = 7 MeV with SkM forces and Coulomb interaction switched off (Bonche, Levit, and Vautherin 1985).

semiclassical description is suitable for reproducing only the averaged properties of nuclei.

Thomas-Fermi approximation at finite temperature

In order to define the TF approximation at finite temperatures, one may start with the Gibbs free energy density functional (10.49) defined within the conjugate class $\mathcal{C}_{PW} \subset \mathcal{D}_{THF}$ built upon model single-particle states $\bar{\varphi}_i(\mathbf{r})$ taken just as plane waves (PW) $\bar{\varphi}_i(\mathbf{r}) \equiv \bar{\varphi}_{\mathbf{k}_i}(\mathbf{r}) = V^{-1/2}\exp(i\mathbf{k}_i \cdot \mathbf{r})$ of momentum $\mathbf{k}_i$ and normalized in the volume V (Stoitsov and Petkov 1988). In this case, the local-scale transformed states (10.26) take the form

$$\varphi_{\mathbf{k}_1}[\rho] \equiv \varphi_{\mathbf{k}_1}(\mathbf{r},[\rho]) = \left[\frac{\rho(\mathbf{r})}{N}\right]^{1/2} \exp\left[i\mathbf{k}_1 \cdot \mathbf{f}_{[\rho]}(\mathbf{r})\right], \qquad (10.52)$$

where

$$f_{[\rho]}(\mathbf{r}) \equiv f^{PW}_{[\rho]}(\mathbf{r}) = \left[3\rho_0^{-1}\int_0^r dr' \; \rho(r',\theta,\varphi)\; r'^2\right]^{1/3} \qquad (10.53)$$

is the solution of eqn (10.11) defined in $\mathcal{C}_{PW}$ with $\rho(\mathbf{r}) \in \overline{D}_N$.

From eqn (10.51) and eqn (10.52), one obtains the following expression for the single-particle energies

$$\varepsilon_1[\rho] = \mathcal{A}[\rho]\; k_1^2 + \mathcal{B}[\rho], \qquad (10.54)$$

where both functionals $\mathcal{A}[\rho]$ and $\mathcal{B}[\rho]$ depend on the particular form of the nuclear Hamiltonian. For Skyrme-type effective forces

$$\mathcal{A}[\rho] = \int dr \; \frac{\hbar^2}{2m^*[\rho]} \; \tau_{\gamma}[\rho] \qquad (10.55)$$

and

$$\mathcal{B}[\rho] = \int dr \left\{U[\rho]\rho + \frac{\hbar^2}{2m^*[\rho]} \; \tau_W[\rho]\right\}, \qquad (10.56)$$

where $U[\rho]$ and $\hbar^2/2m^*[\rho]$ are given by eqn (1.152) and eqn (1.154), respectively. The expression

$$\tau_{\gamma}[\rho] = \rho(r)\left[\frac{2}{3}\left\{f_{[\rho]}(r)/r\right\}^2 + \frac{1}{3}\left\{\nabla f_{[\rho]}(r)\right\}^2\right] \qquad (10.57)$$

is completely determined by eqn (10.53).

Inserting eqn (10.54) into eqn (10.50), one obtains the single-particle occupancies

$$n_1[\rho] = [1 + \exp(x - \eta)]^{-1}, \qquad (10.58)$$

where

$$x = \mathcal{A}[\rho]k_i^2/T \quad \text{and} \quad \eta = (\mu - \mathcal{B}[\rho])/T. \tag{10.59}$$

Using eqn (10.58) and replacing as usual the direct sum by an integral over **k**, one obtains the model local density

$$\rho_0 \equiv \sum_i n_i[\rho]|\bar{\varphi}_{\mathbf{k}_i}(\mathbf{r})|^2 = \frac{1}{2\pi^2}\left\{\frac{T}{\mathcal{A}[\rho]}\right\}^{3/2} J_{1/2}(\eta), \tag{10.60}$$

where $J_\nu(\eta)$ are the Fermi integrals (Stoner 1939)

$$J_\nu(\eta) = \int_0^\infty dy\, \frac{y^\nu}{1+\exp(y-\eta)} \qquad (\nu = -1/2,\ 1/2,\ 3/2, \dots). \tag{10.61}$$

Finally, the kinetic energy density $\tau[\rho]$ in $\mathcal{E}_{PW}$ is represented by eqn (10.28) with the nonlocal term

$$\frac{\hbar^2}{2m}\tau_\gamma[\rho] \equiv \frac{\hbar^2}{2m}\tau_{PW}[\rho] = \frac{1}{2\pi^2\rho_0[\rho]}\left\{\frac{T}{\mathcal{A}[\rho]}\right\}^{5/2} J_{3/2}(\eta)\,\tau_{\gamma'}[\rho]. \tag{10.62}$$

In the case of local densities $\rho(\mathbf{r})$, slowly varying with distance, one can integrate eqn (10.53) to obtain the solution

$$f_{[\rho]}(\mathbf{r}) = [\rho(\mathbf{r})/\rho_0]^{1/3}\mathbf{r}. \tag{10.63}$$

Inserting this into eqn (10.62) and dropping the gradient terms in the same approximation of slowly varying densities, one obtains the following TF-like expression:

$$\frac{\hbar^2}{2m}\tau_{PW}[\rho] = \frac{1}{2\pi^2}\left\{\frac{2m^*T}{\hbar^2}\right\}^{5/2}[\rho(\mathbf{r})/\rho_0]^{5/3}J_{3/2}(\eta). \tag{10.64}$$

For infinite nuclear matter, where $\rho(\mathbf{r}) \equiv \rho_0$ = Const, the Weizsäcker term $\tau_W[\rho]$ in eqn (10.28) vanishes and eqn (10.64) becomes exactly the well known TF expression for the kinetic energy at finite temperature. It is extensively used in hot nuclear matter calculations, whereby new phenomena appear, such as the liquid-vapour phase coexistence, phase transitions and bubble or droplet formation in hot and dense nuclear matter (Küpper, Wegmann, and Hilf 1974, Barranco and Buchler 1980, Lamb *et al.* 1981, Reinhard and Schulz 1985).

In finite nuclei, one can assume that the equality $\rho(\mathbf{r}) = \rho_0$ is locally satisfied under the condition of slowly varying densities. Then, eqns (10.60) and (10.64) lead to the familiar TF expression (Stocker and Burzlaff 1975, Barranco and Treiner 1981, Ravenhall, Pethick, and Lattimer 1983, Suraud 1984)

$$\tau^{T}_{TF}[\rho] = \frac{2m}{\hbar^2} A_T \, T \, J_{3/2}(\eta), \tag{10.65}$$

where $\eta = \eta(\rho, T)$ is an implicit function of $\rho(\mathbf{r})$ and T, the solution of the equation

$$\rho(\mathbf{r}) = A_T \, J_{1/2}(\eta) \tag{10.66}$$

with

$$A_T = \frac{1}{2\pi^2}\left\{\frac{2m^* T}{\hbar^2}\right\}^{3/2}. \tag{10.67}$$

In the same TF approximation, the entropy density is of the form

$$\sigma^{T}_{TF}[\rho] = \frac{5}{3} A_T T \, J_{3/2}(\eta) - \eta \, \rho. \tag{10.68}$$

There are approaches in which $\tau_W[\rho]$-like terms are added, in an *ad hoc* manner, as 'cold' gradient corrections to the 'hot' kinetic energy density usually taken in the thermal TF form (Barranco and Treiner 1981, Bartel *et al.* 1984). One can see from eqn (10.28) that such a procedure, proposed originally by Barranco and Treiner (1981), is justified only in the case when the complete nonlocal functional $\tau_\gamma[\rho]$ is taken into account. When approximating $\tau_\gamma[\rho]$ by its TF expression (10.65), however, a number of gradient terms of the order of the Weizsäcker one are obviously dropped. A systematic derivation of higher order gradient corrections to the TF approximation (10.65) can be obtained by the semiclassical method extended to finite temperatures.

Semiclassical density functionals at finite temperature

In order to generalize the semiclassical method presented in Chapter 4 to the case of nuclear systems at finite temperatures, Brack and Quentin (1981) consider the following derivative of the Fermi-Dirac occupation numbers (10.46):

$$f_T(\varepsilon-\varepsilon_i) \equiv (dn_i/d\mu)_{\mu=\varepsilon} = \frac{1}{4T}\left[\cosh\left(\frac{\varepsilon-\varepsilon_i}{2T}\right)\right]^{-2}. \tag{10.69}$$

Here, T is the temperature and ε_i are the temperature-dependent single-particle energies defined according to eqn (10.45). Brack and Quentin (1981) show that the THF density matrix

$$\rho_T(\mathbf{r},\mathbf{r}') = \sum_i n_i(\mu,T)\, \varphi_i(\mathbf{r})\varphi_i^*(\mathbf{r}') \tag{10.70}$$

can be derived formally from the 'cold' one, eqn (4.8), via convolution of the T = 0 spectral density (4.6) with the function $f_T(E)$ defined by eqn (10.69).

Due to the convolution theorem, the Bloch density matrix $C_T(\mathbf{r},\mathbf{r}';\beta)$ at T > 0 is a product (Brack, Guet, and Håkansson 1985)

$$C_T^\beta \equiv C_T(\mathbf{r},\mathbf{r}';\beta) = C_0(\mathbf{r},\mathbf{r}';\beta)\, \tilde{f}_T(\beta), \tag{10.71}$$

where C_0^β is the Bloch matrix (4.5) at zero temperature and $\tilde{f}_T(\beta)$ is the two-sided Laplace transform of eqn (10.69):

$$\tilde{f}_T(\beta) = \mathcal{L}_{\varepsilon\to\beta}[f_T(\varepsilon)] = \int_{-\infty}^{+\infty} dE\, \exp(-\beta E) f_T(E) = \frac{\pi\beta T}{\sin(\pi\beta T)}. \tag{10.72}$$

It is not difficult to verify that eqn (10.70) results from the inverse Laplace transform of eqn (10.71) divided by β (compare with the zero temperature case of eqn (4.8))

$$\rho_T(\mathbf{r},\mathbf{r}') = \frac{1}{2\pi i}\int_{c-i\infty}^{c+i\infty} d\beta\ \exp(\mu\beta)\ C_T(\mathbf{r},\mathbf{r}';\beta)/\beta$$

$$= \mathcal{L}^{-1}_{\beta\to\mu}[C^{\beta}_T/\beta] = \mathcal{L}^{-1}_{\beta\to\mu}[C^{\beta}_0\tilde{f}_T(\beta)/\beta]. \qquad (10.73)$$

One can also define at T> 0 the local density

$$\rho(\mathbf{r}) = \mathcal{L}^{-1}_{\beta\to\mu}[C_T(\mathbf{r};\beta)/\beta], \qquad (10.74)$$

and the entropy density

$$\sigma(\mathbf{r}) = \sum_i|\varphi_i(\mathbf{r})|^2\ [n_i \ln n_i + (1 - n_i) \ln (1 - n_i)]$$

$$= \mathcal{L}^{-1}_{\beta\to\mu}[\beta^{-2}\partial C_T(\mathbf{r};\beta)/\partial T], \qquad (10.75)$$

where $C_T(\mathbf{r};\beta) = C_T(\mathbf{r},\mathbf{r};\beta)$ is the diagonal part of eqn (10.71). For the complete treatment of the finite-temperature problem with Skyrme-type effective forces, one also needs the kinetic energy density $\tau(\mathbf{r})$ and the spin-density $\mathbf{J}(\mathbf{r})$ which can be derived from eqn (10.73) as

$$\tau(\mathbf{r}) = [\nabla_r \cdot \nabla_{r'}\rho_T(\mathbf{r},\mathbf{r}')]_{\mathbf{r}'=\mathbf{r}} \qquad (10.76)$$

and

$$\mathbf{J}(\mathbf{r}) = -i[\nabla\times\sigma\ \rho_T(\mathbf{r},\mathbf{r}')]_{\mathbf{r}'=\mathbf{r}}, \qquad (10.77)$$

respectively.

The semiclassical derivation of the density functionals at finite temperatures now resorts to the known procedure. One has to replace the exact Bloch density matrix $C_0(\mathbf{r},\mathbf{r}';\beta)$ in eqn (10.71) by its Wigner-Kirkwood expansion associated with eqn (4.28). Inserting the result into eqns (10.74), (10.75), (10.76), and (10.77), a semiclassical gradient expansion in powers of $\hbar$ follows for quantities $\rho_{sc}(\mathbf{r})$, $\sigma_{sc}(\mathbf{r})$, $\tau_{sc}(\mathbf{r})$ and $\mathbf{J}_{sc}(\mathbf{r})$, respectively. It should be stressed that the factor $\tilde{f}_T(\beta)$ is left untouched in eqn (10.71) when the $\hbar$-expansion is applied. If $\tilde{f}_T(\beta)$ were expanded in powers of β, one would obtain the low-temperature expansion used by Barranco and Treiner (1981), Campi and Stringari (1983), Barranco, Pi, and Viñas (1983).

Extended Thomas-Fermi model at finite temperature

One may eliminate the potential and its gradients (as well as the chemical potential μ) from the semiclassical expressions $\rho_{sc}(\mathbf{r})$, $\sigma_{sc}(\mathbf{r})$, $\tau_{sc}(\mathbf{r})$ and $\mathbf{J}_{sc}(\mathbf{r})$ in the same way as explained in Section 4.4 in order to obtain the associated local density functionals

$$\tau^T[\rho] = \tau^T_{TF}[\rho] + \tau^T_2[\rho] + \tau^T_4[\rho] + \dots, \tag{10.78}$$

$$\sigma^T[\rho] = \sigma^T_{TF}[\rho] + \sigma^T_2[\rho] + \sigma^T_4[\rho] + \dots, \tag{10.79}$$

$$\mathbf{J}^T[\rho] = \mathbf{J}^T_2[\rho] + \mathbf{J}^T_4[\rho] + \dots . \tag{10.80}$$

Here, the notation is the same as in Section 4.4 but all quantities now depend on the temperature-dependent local density distribution.

The lowest order terms give the temperature-dependent TF approximation for $\tau^T_{TF}[\rho]$ and $\sigma^T_{TF}[\rho]$ already given by eqn (10.65) and eqn (10.68), respectively. The spin-orbit density has no contribution of zero order in $\hbar$. The terms of the second order in $\hbar$ were derived for the first time by Brack (1984a,b, 1985a). The results are

$$\tau^T_2[\rho] = \xi(\eta)\frac{(\nabla\rho)^2}{\rho} + \frac{1}{3}\Delta\rho + [3\gamma(\eta)-\frac{5}{12}]\ \frac{(\nabla\rho.\nabla f)}{f}$$

$$+ [\frac{9}{4}\gamma(\eta) - \frac{7}{42}]\rho\left(\frac{\nabla f}{f}\right)^2 + \frac{1}{2}\left(\frac{2m}{\hbar^2}\right)^2\rho(\mathbf{W}/f)^2, \tag{10.81}$$

$$\sigma_2^T[\rho] = -T^{-1}\frac{\hbar^2}{2m}\,\nu(\eta)\left\{f\frac{(\nabla\rho)^2}{\rho} + \frac{9}{4}\rho\frac{(\nabla f)^2}{f} + 3(\nabla\rho\,.\,\nabla f)\right\}, \tag{10.82}$$

$$\mathbf{J}_2^T[\rho] = -\left(\frac{2m^*}{\hbar^2}\right)\rho\mathbf{W}, \tag{10.83}$$

and contain the following functions

$$\gamma(\eta) = \xi(\eta) - \nu(\eta), \tag{10.84}$$

$$\nu(\eta) = -\frac{3}{2}\xi(\eta) + 36\xi^2(\eta) - \frac{3}{8}J^2_{1/2}(\eta)J_{-5/2}(\eta)J^{-3}_{-1/2}(\eta), \tag{10.85}$$

$$\xi(\eta) = -\frac{1}{12}J_{1/2}(\eta)J_{-3/2}(\eta)J^{-2}_{-1/2}(\eta), \tag{10.86}$$

where the Fermi integrals $J_\nu(\eta)$ are defined for $\nu>-1$ from eqn (10.61). Their analytical continuation to values $\nu<-1$, for which the integral (10.61) is not defined, is given by Fernández Velicia (1984) and Bartel, Brack, and Durand (1985). The Helmholtz free energy density functional up to second order in $\hbar$ has also been derived by Perrot (1979) for a local potential by a completely different method.

To carry out corrections to fourth order terms by hand requires a great amount of work. For the case of purely local potential, the fourth order gradient corrections $\tau_4^T[\rho]$ and $\sigma_4^T[\rho]$ have been given by Bartel, Brack, and Durand (1985). The resulting temperature-dependent ETF (TETF) functional with Skyrme-type effective forces is then represented in the form

$$E_{TETF}[\rho] = \int d\mathbf{r}\;\mathcal{E}(\rho,\tau^T[\rho],\mathbf{J}^T[\rho]). \tag{10.87}$$

The functional (10.87) is implicitly expressed in terms of $\rho(\mathbf{r})$ due to the quantity $\eta = \eta(\rho,T)$. As earlier mentioned, this quantity must be calculated from eqn (10.66) taking into account the ρ-dependence of the effective mass entering eqn (10.67). This

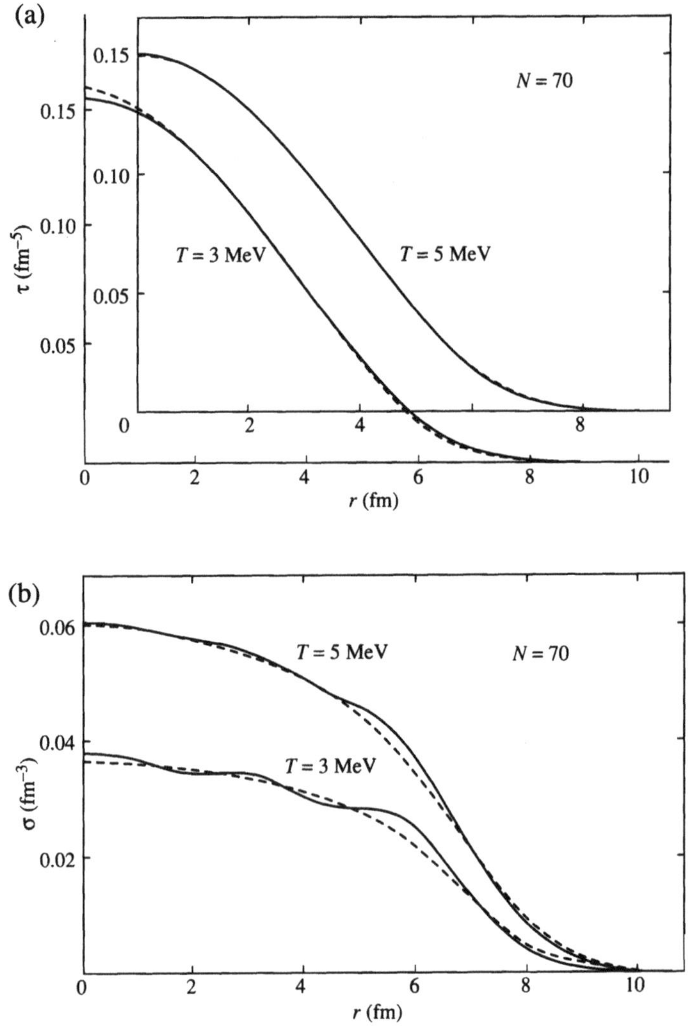

Figure 10.8. Kinetic energy density (a) and entropy density (b) for an axially deformed harmonic oscillator shell model at two different temperatures. The exact quantum-mechanical densities (full lines) are compared with the semiclassical ones (dashed lines) as explained in the text (Bartel, Brack, and Durand 1985).

can be done by inverting eqn (10.66) numerically or by adopting some approximate techniques (Bartel, Brack, and Durand 1985).

Bartel, Brack, and Durand (1985) test the functionals (10.78) and (10.79) by evaluating these quantities with the exact quantum-mechanical densities $\rho^T_{HO}(\mathbf{r})$ associated with the axially deformed harmonic oscillator shell model potential and comparing them with corresponding quantum-mechanical results for $\tau^T_{HO}(\mathbf{r})$ and $\sigma^T_{HO}(\mathbf{r})$, respectively. The kinetic energy density $\tau^T[\rho^T_{HO}] - \frac{1}{4}\rho^T_{HO}$ and the entropy density $\sigma^T[\rho^T_{HO}]$ are compared for two different temperatures T = 3 MeV and T = 5 MeV in Figure 10.8a,b with the quantum-mechanical values of $\tau^T_{HO}(\mathbf{r}) - \frac{1}{4}\rho^T_{HO}$ and $\sigma^T_{HO}(\mathbf{r})$, respectively. The total TETF kinetic energy E_{kin} and the entropy S as functions of the deformation parameter $q = \omega_\perp/\omega_z$ are compared with the exact results obtained by integrating $\tau^T_{HO}(\mathbf{r})$ and $\sigma^T_{HO}(\mathbf{r})$, in Figure 10.9a,b respectively. These results show that the TETF functionals, going up to fourth order corrections, reproduce the exact entropy to within less than 1% and the agreement is even better for the kinetic energy. The complete functionals are seen to give a correct dependence on the deformation parameter as well. This is, for instance, not the case either for the pure temperature-dependent TF approximation (the top curve in Figure 10.9b) or for the approximation in which the functional $\tau^{app}[\rho] = \tau^T_{TF}[\rho] + \tau^T_2[\rho] + \tau^{T=0}_4[\rho]$ (the top curve in Figure 10.9a) is used.

Similar comparison of the complete functional (10.87) against THF calculations shows (Brack, Guet, and Håkansson 1985) that the TETF model is an excellent approximation to the more costly THF method.

Solution of the classical turning point problem

As mentioned in Chapter 4, the semiclassical densities $\rho_{SC}(\mathbf{r})$ and $\tau_{SC}(\mathbf{r})$ diverge at the classical turning points for temperature T = 0. The ETF functional $E_{ETF}[\rho]$, eqn (4.34), has been nevertheless used in the classically forbidden region under the

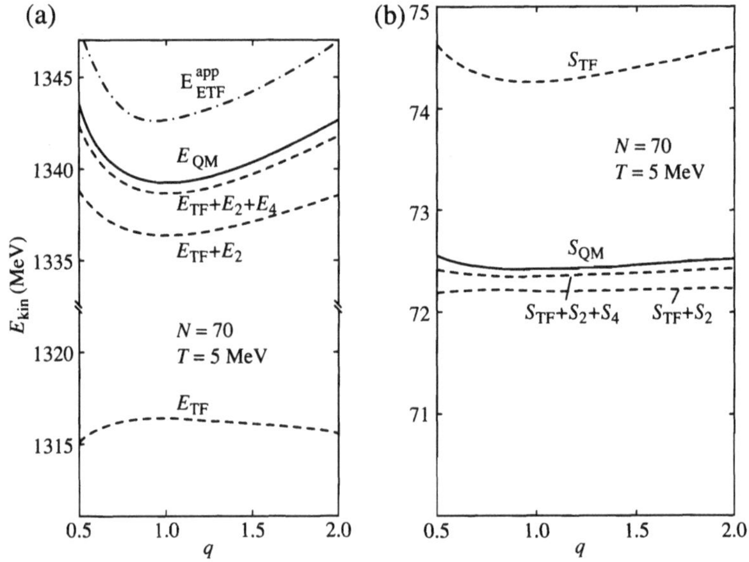

Figure 10.9. Kinetic energy (a) and entropy (b) as functions of the deformation parameter $q = \omega_{\perp}/\omega_z$ at fixed particle number N = 70 and temperature T = 5 MeV. The exact quantum-mechanical results (QM) for an axially deformed harmonic oscillator shell model are compared with different contributions emerging from the temperature-dependent Thomas-Fermi method (Bartel, Brack, and Durand 1985).

assumption that it could be analytically continued to the whole space R^3.

Brack, Guet, and Håkansson (1985) and Bartel, Brack and Durand (1985) have shown that in the limit $T \to 0$ the entropy density (10.79) goes to zero like T and the kinetic energy density (10.78) recovers the zero-temperature expression (4.35). Since the limit $T \to 0$ can be taken locally at any point in R^3 where $\rho(\mathbf{r}) > 0$, this constitutes a rigorous proof that the zero temperature ETF kinetic energy density functional is also valid beyond the classically allowed region.

It should be stressed that the term $\tau_6^T[\rho]$ and the higher order terms entering the kinetic energy density (10.78) give finite contributions to the Helmholtz free energy at $T > 0$. The limit $T \to 0$, however, shows that the term $\tau_6^T[\rho]$ at $T = 0$ diverges at large distances for any realistic density which goes exponentially to zero. The functional $\tau^{T=0}[\rho]$ must therefore be truncated after the fourth order term as discussed in Section 4.4.

10.4 Applications to hot isolated nuclei

Liquid drop parameters for hot nuclei

In many cases it is desirable to perform a 'leptodermous' (or simply LDM-type) expansion of the Helmholtz free energy

$$F \equiv E-TS = a_v^T N + a_s^T N^{2/3} + a_c^T N^{1/3} + \ldots, \qquad (10.88)$$

assuming that the nuclear density is essentially flat in the interior and varies only in a relatively narrow surface region. Such expansions have been applied, for example, in calculations of supernova formation (Pethick, Ravenhall, and Lattimer 1983) or the multi-fragmentation of hot compound nuclei (Bondorf *et al.* 1985a,b), whereby the temperature dependence of the LDM parameters is required as an input.

Whereas at zero temperatures the LDM parameters are obtained by fitting the available experimental data as discussed in Section 4.5, the only reliable way to derive their temperature dependence is to use theoretically calculated values of the free energy at finite temperature T. This is precisely what can be derived systematically from THF calculations. It is easier, however, to proceed in the framework of the TETF model by using exactly the same procedure with respect to the Helmholtz free energy $F_{TETF}[\rho] = E_{TETF}[\rho] - T\, S_{TETF}[\rho]$ as that explained in Section 4.5.

Guet, Strumberger, and Brack (1988) derive the LDM coefficients of hot nuclear systems within the TETF model and Skyrme-type effective force SkM* by expanding them up to second order in T; that is,

$$a_i(T) = a_i(0)\,(1 + x_i T^2). \tag{10.89}$$

The quadratic approximation (10.89) applied to the coefficients entering eqn (10.88) leads to errors in the total energy $F_{TETF}[\rho]$ which are smaller than 1 MeV up to temperature T = 4 MeV. As in the zero temperature case, the asymmetry energy is represented in the form

$$F_{as}(N,I) = J\,N\,I^2/[1 + (9J/4Q)\,N^{-1/3}], \tag{10.90}$$

where I = (N − 2Z)/N. Both the volume asymmetry energy J and the surface stiffness coefficient Q, and also the Coulomb energy coefficients entering the formula

$$F_{Coul}(N,I) = e_1(T)Z^2N^{-1/3} + e_2(T)Z^2N^{-1}, \tag{10.91}$$

are again fitted to the quadratic T-dependence of the form of eqn (10.89).

We give in Table 10.1 some values of the LDM parameters derived with SkM* forces by Guet, Strumberger, and Brack (1988). Their quadratic T-dependence holds up to temperatures of about

Table 10.1. Droplet model coefficients $a_i(T) = a_i(0)(1-x_iT^2)$ obtained from TETF calculations with SkM* forces (Guet, Strumberger, and Brack 1988). The first row contains the coefficients at zero temperature $a_i(0)$ (in MeV), while the second row gives the coefficients x_i (in MeV^{-2}).

	a_V	a_S	a_C	a_0	c_1	c_2	J	Q	K
$a_i(0)$	−15.776	17.22	10.24	−8.0	0.737	−1.28	30.03	354	216.7
x_i	−3360	5530	5430	−1230	763	−13200	1040	12800	5990

8 MeV. Only the constant a_0 (Table 10.1), added to eqn (10.88) in order to adjust F(N,I) to the variational TETF energy, is quadratic up to T = 4 MeV. Similar calculations with other Skyrme-type forces have been reported by Brack, Guet, and Håkansson (1985) and Bartel, Brack, and Durand (1985). The surface energy coefficient a_s has also been investigated in the pure temperature-dependent TF approximation by Ravenhall, Pethick, and Lattimer (1983) and Stocker (1984).

Temperature dependence of fission barriers

The possible decrease of the barrier heights with increasing temperature is of crucial importance in understanding nuclear fission processes. The use of the microscopic THF method in order to study how the fission barriers of excited nuclei depend on the temperature is too time-consuming for systematic investigations (Mosel, Zint, and Passler 1974, Brack and Quentin 1974a,b, Sauer, Chandra, and Mosel 1976). Nuclear density functional theory at finite temperature gives a solid basis for this goal.

Campi and Stringari (1983) and Barranco, Pi, and Vinas (1983) use the pure temperature-dependent TF kinetic energy density functional in the so-called 'low-temperature expansion'

$$\tau^{T}_{TF}[\rho] = \tau^{T=0}_{TF}[\rho] + \frac{2m}{\hbar^2}\, a(\rho) T^2, \tag{10.92}$$

where

$$a(\rho) = \frac{1}{12}\, (3\pi^2)^{1/3}\, \frac{2m}{\hbar^2}\, \rho^{1/3}. \tag{10.93}$$

Due to the low-temperature approximation a drastic overestimate of the temperature effects on the barrier heights appears.

A systematic investigation of the fission barriers at various temperatures may be obtained within the TETF model or by using the associated LDM-type expression with temperature-

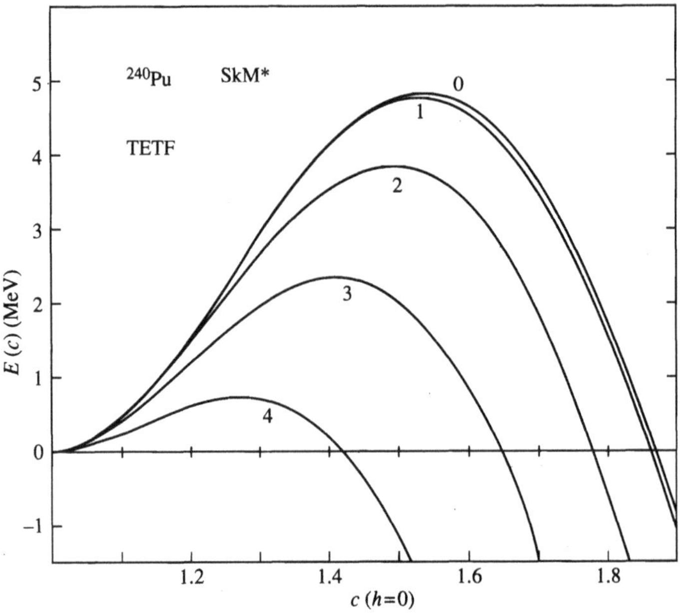

Figure 10.10. Fission barriers of the nucleus ^{240}Pu obtained with SkM^* forces at various temperatures (indicated by the numbers in MeV) using the temperature-dependent Thomas-Fermi method (Brack, Guet, and Håkansson 1985).

dependent coefficients in combination with the Strutinski-averaging procedure extended to finite temperatures by Brack and Quentin (1981). For temperatures larger than 2-3 MeV, nuclear shell effects disappear and the semiclassical or LDM barriers are meaningful.

Actual calculations have been performed by Brack, Guet, and Håkansson (1985) with 'cold' fourth order gradient corrections to the TETF functional. We illustrate the TETF results for the fission barriers of the nucleus ^{240}Pu in Figure 10.10. These calculations have been repeated by Guet, Strumberger, and Brack (1988) incorporating also the correct temperature-dependent fourth order corrections to TETF kinetic energy density.

Giant multipole resonances on excited states

The study of collective nuclear motion is currently of challenging interest because in recent heavy-ion experiments evidence for giant resonances which are built upon excited states has emerged (Newton *et al.* 1981, Draper *et al.* 1982, Barette and Sandorfi 1983, Gaardhoje *et al.* 1984, Snover 1986). This has motivated many investigations within the RPA method generalized to the case of thermally excited nuclear states (Faber, Egido, and Ring 1983, Vautherin and Vinh Mau 1983, 1984, Ring (1984), Besold, Reinhard, and Toepffer 1984, Civitarese, Broglia, and Dasso 1984, Sagawa and Bertsch 1984, Bar-Touv 1985).

Nuclear density functional theory has been applied within the RPA sum rule approach (see Section 9.4) accordingly generalized to finite temperatures by Vautherin and Vinh Mau (1983, 1984) and Barranco, Polls, and Martorell (1985). Most of the calculations of the sum rules have been performed using the TETF type functionals (Meyer, Quentin, and Brack 1983, Barranco, Marcos, and Treiner 1984, Barranco *et al.* 1985, Barranco, Polls, and Martorell 1985).

The TF-like calculations within the RPA sum rule approach lead to some spurious changes in the entropy of the system, because the two separate types of independent variations with respect to the occupation numbers and the single-particle wave functions in the THF method are lumped together in a single variation with respect to the local density. To overcome this difficulty some alternatives have been considered by Barranco, Polls, and Martorell (1985). The problem disappears when using the RED functionals, as shown by Stoitsov (1987).

Additional difficulties appear at temperatures higher than 2 MeV due to continuum effects which have been observed in the dynamic time-dependent TF calculations extended to finite temperatures by Dalili, Németh, and Ngô (1985) and Barranco *et al.* (1987). We show in Figure 10.11 the isoscalar strength function S(E) derived at two temperatures T = 0 MeV and T = 3 MeV. One can

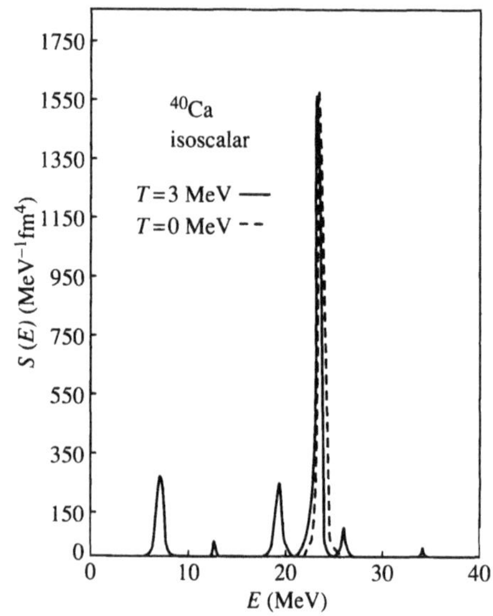

Figure 10.11. Isoscalar strength function S(E) at temperatures T = 3 MeV (full line) and T = 0 MeV (dashed line) derived for the nucleus ^{40}Ca by Barranco *et al.* (1987).

see the low-intensity peaks whose origin is related to the external nucleon gas (Barranco *et al.* 1987), whereas the big peak associated with the nucleus is only shifted downwards very little. As a consequence, it is very likely that the average energy defined as $\bar{E}_3 = (m_3/m_1)^{1/2}$ would be reliable but not $\bar{E}_1 = (m_{-1}/m_1)^{1/2}$ because the m_{-1} sum rule, containing integration of $E^{-1}S(E)$, would be most sensitive to the external gas problem. The results reported by Suraud, Barranco, and Treiner (1987) confirm this conclusion. For example, in the case of isoscalar giant monopole resonances, illustrated in Figure 10.12, the nonsubtracted value of $\bar{E}_1$ collapses with increasing temperature. Similar spurious behaviour appears for resonances with higher multipolarities. At the same time, the use of the subtraction procedure of Bonche, Levit, and Vautherin (1984, 1985) makes the temperature-dependence of the energy $\bar{E}_1$ very weak.

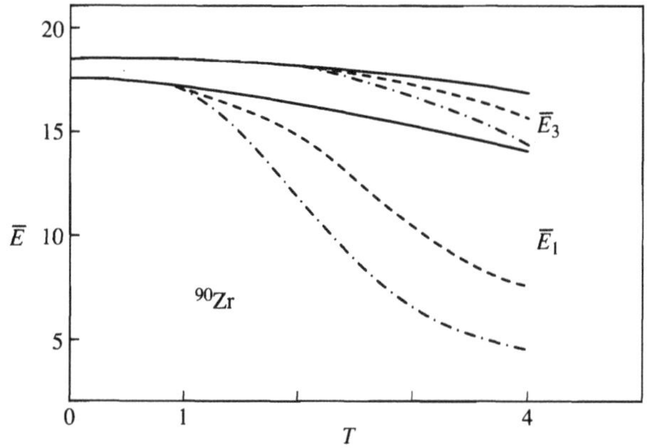

Figure 10.12 Isoscalar giant monopole resonance excitation energies $\overline{E}_3 = (m_3/m_1)^{1/2}$ and $\overline{E}_1 = (m_{-1}/m_1)^{1/2}$ for the nucleus ^{90}Zr (in MeV) as functions of the temperature T (Suraud, Barranco, and Treiner 1987). The non-subtracted values for box sizes R = 14 fm (dashed lines) and R = 12 fm (dashed-dotted lines) are compared with the subtracted values (full lines).

Bubble formation in hot nuclei

The equation of state of hot nuclear matter exhibits van der Waals behaviour and one might expect that phase transitions occur in hot nuclei. Their existence has been conjectured in experiments with high energy protons on nuclei and in heavy-ion collisions (Curtin, Toki, and Scott 1983, Panagiotou *et al.* 1984).

Theoretically, the role of statistical and quantum fluctuations which lead to phase transitions in hot nuclear matter has been first investigated by Reinhardt and Schulz (1985) on the base of the relativistic mean-field model of Walecka (1974, 1975) and Chin (1977) applying the Euclidean space path integral techniques of Coleman (1977) and Reinhardt (1981). A new phase may be formed in small volumes which are stable if they have a certain size. The picture one has particularly in mind is that bubbles in a nuclear matter environment or droplets in a gas background may be formed as they are energetically favourable in the interplay of volume and surface energies.

A semiclassical study of bubble formation in hot nuclei has been presented by Blin *et al.* (1986) and Blin, Brack, and Hiller (1986). They appropriately parameterize the bubble density profiles and calculate the grand potential in TETF approximation by using the truncation procedure of Bonche, Levit, and Vautherin (1984, 1985). In a heavy-ion collision, if the hot nucleus reaches a metastable density below a certain value, the model predicts a probability of the order of unity to form bubbles due to statistical fluctuations.

References

Note

Works with less than four authors are cited in the text by author's names in full, followed by the date, and these are listed here in strict alphabetical order. Works with four or more authors are cited by quoting the first author only, followed by 'et al' and the date, and these are listed here in date order, following those with 1–3 authors.

Abramowitz, M. and Stegun, I. A. (1965). *Handbook of mathematical functions*, Dover, New York.

Acharya, P. K. (1983). *Journal of Chemical Physics*, **78**, 2101.

Aerts, A. Th., Mulders, P. J., and de Swart, J. J. (1980). *Physical Review*, **D21**, 2653.

Aleksandrov, L. (1971). *Journal of Computational Mathematics and Mathematical Physics*, **11**, 36.

Aleksandrov, L. (1973). *Communication of JINR*, **P5–7259**, Dubna.

Aleksandrov, L., Drenska, M., and Karadjov, D. (1981). *Program system REGN for solution of nonlinear equation*, **PSR–RSIC–31**. Oak Ridge.

Aleksandrov, L., Drenska, M., and Karadjov, D. (1982a). *Journal of Computational Mathematics and Mathematical Physics*, **22**, 375.

Aleksandrov, L., Drenska, M., and Karadjov, D. (1982b). *Communication of JINR*, **B1–11–82–767**. Dubna.

Aleksandrov, L. and Karadjov, D. (1980). *Journal of Computational Mathematics and Mathematical Physics*, **20**, 923.

Anastasio, M. R., Celenza, L. S., Pong, W. S., and Shakin, C. M. (1983). *Physics Reports*, **100C**, 327.

Andō, K. and Eckart, G. (1984). *Nuclear Physics*, **A420**, 9.

Antonov, A. N., Hodgson, P.E., and Petkov, I. Zh. (1988). *Nucleon momentum and density distributions in nuclei*, Clarendon press, Oxford.

Antonov, A. N., Petkov, I. Zh., and Stoitsov, M. V. (1983). *Comptes Rendus de l'Académie Bulgare des Sciences*, **36**, 1159.

Antonov, A. N., Petkov, I. Zh., and Stoitsov, M. V. (1985). *Preprint ICTP*, **IC/85/44**, Trieste.

Argyres, P. (1965). *Physics*, **2**, 131.

Arnold, M. (1980). In *6th international conference on atomic masses and fundamental constants, East Lansing, 1979*, p.375, Plenum, New York.

Auerbach, N. and Klein, A. (1983). *Physical Review*, **C28**, 2075.

Auerbach, N., Klein, A., and Siciliano, E. R. (1985). *Physical Review*, **C31**, 682.

Balazs, N.L. and Jennings, B.K. (1984). *Physics Reports*, **104**, 347.

Balbutsev, E. B. and Piperova, I. (1988). *Preprint JINR*, **E4-88-886**, Dubna.

Balbutsev, E. B., Mikhailov, I. N., and Stoitsov, M. V. (1989). *Soviet Journal of Nuclear Physics*, **50**, 35.

Balbutsev, E. B., Mikhailov, I. N., and Vaishvila, Z. (1986). *Nuclear Physics*, **A457**, 222.

Balbutsev, E. B., Vaishvila, Z., and Mikhailov, I. N. (1982). *Soviet Journal of Nuclear Physics*, **35**, 836.

Balbutsev, E. B., Dymarz, R., Mikhailov, I. N., and Vaishvila, Z. (1981). *Physics Letters*, **B105**, 84.

Baldo, M., Di Toro, M., Giausiracusa, G., and Lombardo, U. (1986). *Physics Letters*, **B168**, 8.

Balescu, R. (1975). *Equilibrium and nonequilibrium statistical mechanics*, Vol. 1, Wiley, New York.

Baranger, M. and Vénéroni, M. (1978). *Annals of Physics*, **114**, 123.

Bardeen, J., Cooper, L. N., and Schrieffer, J. R. (1957). *Physical Review*, **108**, 1175.

Barette, J. and Sandorfi, A. M. (1983). *Communication in Nuclear and Particle Physics*, **A12**, 57.

Barranco, M. and Buchler, J.R. (1980). *Physical Review*, **C22**, 1729.

Barranco, M., Marcos, S., and Treiner, J. (1984). *Physics Letters*, **B143**, 314.

Barranco,M., Pi ,M. and Viñas, X. (1983). *Physics Letters*, **B124**, 131.

Barranco, M., Polls, A., and Martorell, J. (1985). *Nuclear Physics*, **A444**, 445.

Barranco, M. and Treiner, J. (1981). *Nuclear Physics*, **A351**. 269.

Barranco, M., Polls, A., Marcos, S., Navarro, J., and Treiner, J. (1985). *Physics Letters*, **B154**, 96.

Barranco, M., Pi, M., Németh, J., Ngô. C., and Tomasi, E. (1987). *Nuclear Physics*, **A464**, 29.

Bartel, J., Brack, M., and Durand, M. (1985). *Nuclear Physics*, **A445**, 236.

Bartel, J., Durand, M., and Brack, M. (1984). *Zeitschrift für Physik*, **A315**, 341.

Bartel, J. and Vallieres, M. (1982). *Physics Letters*, **B114**, 303.

Bartel, J., Quentin, P., Brack, M., Guet, C., and Håkansson, H.-B. (1982). *Nuclear Physics*, **A386**, 79.

Bartel, J., Brack, M., Guet, C., and Håkansson, H.-B. (1984). *Physics Letters*, **B139**, 1.

Barth, U. Von (1979). *Physical Review*, **A20**, 1693.

Barth, U. Von, and Hedin, L. (1972). *Journal of Physics*, **C5**, 1629.

Bar-Touv, J. (1985). *Physical Review*, **C32**, 1369.

Baym, G. and Pethick, C. (1978). In *The physics of liquid and solid helium*, Vol. 2 (ed. K.H.Beunemann *et al.*), p.1.

Beiner, M. (1970). Theory of nuclear structure. In *Trieste Lecture 1969*, IAEA, Vienna.

Beiner, M. and Lombard, R. J. (1974). *Annals of Physics*, **86**, 262.

Beiner, M., Lombard, R. J., and Mass, D. (1975). *Nuclear Physics*, **A249**, 1.

Beiner, M., Lombard, R. J., and Mass, D. (1976). *Atomic Data and Nuclear Data Tables*, **17**, 450.

Beiner, M., Flocard, H., Van Giai, N., and Quentin, P. (1975). *Nuclear Physics*, **A238**, 29.

Bell, L. S. and Skyrme, T. H. R. (1956). *Philosophical Magazine*, 1, 1055.

Bellicard, J.B. *et al.* (1967). *Physical Review Letters*, 19, 527.

Bellicard, J.B. *et al.* (1977). *Physical Review*, C16, 1262.

Bellicard, J.B. *et al.* (1981). *Saclay Progress Report*, CEA-N-2207, 81.

Berezin, F. A. (1986). *Method of second quantization*, Nauka, Moscow.

Berg, R. A. and Wilets, L. (1956). *Physical Review*, 101, 201.

Bertrand, F. E. (1981). *Nuclear Physics*, A354, 129.

Bertsch, G. F. (1973). *Physical Review Letters*, 31, 121.

Bertsch, G. F. (1974). *Annals of Physics*, 86, 138.

Bertsch, G. F. (1975). *Nuclear Physics*, A249, 253.

Bertsch, G. F. (1978). In *Nuclear physics with heavy ions and mesons, Les houches lectures 1977*, Vol. 1 (ed. R. Balian, M.Rho and G.Ripka), p.1174, North Holland, Amsterdam.

Bertsch, G. F., Bortignon, D. F., and Broglia, R. A. (1983). *Review of Modern Physics*, 55, 287.

Bertsch, G.F. and Kuo, T.T.S. (1968). *Nuclear Physics*, A112, 204.

Bertsch, G.F. and Stricker, K. (1976). *Physical Review*, C13, 1312.

Bertsch, G.F. and Tsai, S.F. (1975). *Physics Reports*, 18C, 125.

Bertsch, G. F., Bortignon, D. F., Broglia, R. A., and Dasso, C. H. (1979). *Physics Letters*, B80, 161.

Besold, W., Reinhard, P.-G., and Toepffer, C. (1984). *Nuclear Physics*, A431, 1.

Bethe, H. A. (1968). *Physical Review*, 167, 879.

Bethe, H. A. (1971). Theory of nuclear matter. In *Annual review of nuclear science*, Vol. 21, p. 93. Palo Alto, California.

Bethe, H. A. and Bacher, F. (1936). *Reviews of Modern Physics*, 8, 82.

Bhaduri, R. K. (1977). *Physical Review Letters*, 39, 329.

Bhaduri, R. K. and Ross, C. K. (1971). *Physical Review Letters*, 27, 606.

Blaizot, J. P. (1980). *Physics Reports*, **64C**, 171.

Blaizot, J. P. and Gogny, D. (1977). *Nuclear Physics*, **A284**, 429.

Blaizot, J. P., Gogny, D., and Grammaticos, B. (1976). *Nuclear Physics*, **A265**, 315.

Blin, A. H., Brack, M., and Hiller, B. (1986). *Physics Letters*, **B182**, 239.

Blin, A. H., Hiller, B., Reinhard, P.-G., and Schuck, P. (1986). *Journal de Physique*, **C4**, 423.

Bogolubov, N. N. and Bogolubov, N. N. (1984). *Introduction in quantum-statistical mechanics* (Sov.ed.), Nauka, Moscow.

Bohigas, O., Lane, A. M., and Martorell. J. (1979). *Physics Reports*, **51**, 267.

Bohigas, O., Martorell. J., and Lane, A. M. (1976). *Physics Letters*, **B64**, 1.

Bohigas, O., Campi, X., Krivine, H., and Treiner, J. (1976). *Physics Letters*, **B64**, 381.

Bohr, A. (1952). *Kongelige Danske Videnskabenrnes Selskab. Matematics-Fisiske Meddelelser* , 26, 14.

Bohr, A. and Mottelson, R. B. (1969). *Nuclear structure*, Vol. *1*. Benjamin, New York, Amsterdam.

Bohr, A. and Mottelson, R. B. (1974). *Nuclear structure*, Vol. 2. Benjamin, New York, Amsterdam.

Bohr, N. and Kalckar, F. (1937). *Kongelige Danske Videnskabenrnes Selskab. Matematics-Fisiske Meddelelser* , **14**, 10.

Bohr, N. and Wheeler, J. A. (1939). *Physical Review*, **56**, 426.

Bonche, P., Grammaticos, B., and Koonin, S. E. (1978). *Physical Review*, **C17**, 1700.

Bonche, P., Koonin, S. E., and Negele, J. W. (1976). *Physical Review*, **C13**, 1226.

Bonche, P., Levit, S., and Vautherin, D. (1984). *Nuclear Physics*, **A427**, 278.

Bonche, P., Levit, S., and Vautherin, D. (1985). *Nuclear Physics*, **A436**, 265.

Bonche, P. and Quentin, P. (1978). *Physical Review*, **C18**, 1891.

Bonche, P. and Vautherin, D. (1981). *Nuclear Physics*, **A372**, 496.

Bonche, P. and Vautherin, D. (1982). *Astronomy and Astrophysics*, **112**, 168.

Bondorf, J. P., Donangelo, R., Mishustin, I. N., Pethick, C. I., Schulz, H., and Sneppen, K. (1985a). *Nuclear Physics*, **A443**, 321.

Bondorf, J. P., Donangelo, R., Mishustin, I. N., and Schulz, H., (1985*b*). *Nuclear Physics*, **A444**, 460.

Bortignon, D. F. and Broglia, R. A. (1981). *Nuclear Physics*, **A371**, 405.

Bowman, J.D. *et al.* (1983). *Physical Review Letters*, **50**, 1195.

Bowman, J.D. *et al.* (1984). *Journal de Physique*, **52**, C4-351.

Brack, M. (1977). *Physics Letters*, **B71**, 239.

Brack, M. (1984*a*). *Physical Review Letters*, **53**, 119.

Brack, M. (1984*b*). *Journal de Physique*, **C6**, 15.

Brack, M. (1985*a*). *Physical Review Letters*, **54**, 851.

Brack, M. (1985*b*). *Helvetica Physica Acta*, **58**, 715.

Brack, M. Guet, C., and Håkansson, H.-B. (1985). *Physics Reports*, **123**, 275.

Brack, M. and Pauli, H. C. (1973). *Nuclear Physics*, **A207**, 401.

Brack, M. and Quentin, P. (1974*a*). *Physics Letters*, **B52**, 159.

Brack, M. and Quentin, P. (1974*b*). *Physica Scripta*, **A10**, 163.

Brack, M. and Quentin, P. (1975). *Physics Letters*, **B56**, 421.

Brack, M. and Quentin, P. (1981). *Nuclear Physics*, **A361**, 35.

Brack, M. and Stöcker, W. (1983). *Nuclear Physics*, **A406**, 413.

Brack, M., Damgaard, J., Jensen, A. S., Pauli, H. C., Strutinski, V. M., and Wong, C. Y. (1972). *Reviews of Modern Physics*, **44**, 320.

Brink, D. M. and Di Toro, M. (1981). *Nuclear Physics*, **A372**, 151.

Brink, D. M., Giannoni, M. J., and Vénéroni, M. (1976). *Nuclear Physics*, **A258**, 237.

Brink, D. M. and Leonardi, R. (1976). *Nuclear Physics*, **A258**, 285.

Brink, D. M. and Stancu, F. (1975). *Nuclear Physics*, **A243**, 175.

Brockmann, R. and Machleidt, R. (1984). *Physics Letters*, **B149**, 283.

Brown, G. E. (1970). *Facets of physics*, p. 141. Academic Press, New York.

Brown, G. E. and Jackson, A. D. (1976). *The nucleon-nucleon interaction*. North Holland, Amsterdam.

Brückner, K. A. (1955). *Physical Review*, **97**, 1353.

Brückner, K. A., Buchler, J. R., and Kelly, M. M. (1968). *Physical Review*, **173**, 944.

Brückner, K. A., Chirico, J. H., and Meldner, H. W. (1971). *Physical Review*, **C4**, 732.

Brückner, K. A., Coon, S., and Dabrowski, J. (1968). *Physical Review*, **168**, 1184.

Brückner, K. A., Lin, Wing-fai, and Lombard, R. J. (1969). *Physical Review*, **181**, 1506.

Brückner, K. A., Buchler, J. R., Jorna, S., and Lombard, R. J. (1968). *Physical Review*, **171**, 1188.

Brückner, K. A., Buchler, J. R., Clark, R. C., and Lombard, R. J. (1969). *Physical Review*, **181**, 1543.

Brückner, K. A., Clark, R. C., Lin, Wing-fai, and Lombard, R. J. (1970). *Physical Review*, **C1**, 249.

Bryan, R. A. (1967). In *International conference on nuclear physics* (ed. R. Becker). Academic Press, New York.

Buenerd, M. (1984). *Journal de Physique*, **45**, C4-115.

Bunatian, G. G., Kolomietz, V. M., and Strutinski, V. M. (1972). *Nuclear Physics*, **A188**, 225.

Burov, V. V., Eldishev, Yu. N., and Lukyanov, V. K. (1974). *Communication of JINR*, **E4-8029**, Dubna.

Callaway, J. and March, N. H. (1984). Density functional methods: theory and applications. In *Solid state physics*, Vol.38. (ed. H. Ehrenreich and D. Turbull), p. 135. Academic Press, New York.

Campi, X. and Bouyssy, A. (1978). *Physics Letters*, **B73**, 263.

Campi, X. and Sprung, D.W.L. (1972). *Nuclear Physics*, **A419**, 401.

Campi, X. and Stringari, S. (1980). *Nuclear Physics*, **A337**, 313.

Campi, X. and Stringari, S. (1983). *Zeitschrift für Physik*, **A309**,239.

Cavinato, M., Marangoni, M., Ottaviani, P.L., and Saruis, A.M. (1982). *Nuclear Physics*, **A373**, 445.

Chin, S. A. (1977). *Annals of Physics*, **108**, 301.

Chu, Y. H., Jennings, B. K., and Brack, M. (1977). *Physics Letters*, **B68**, 407.

Civitarese, O., Broglia, R. A., and Dasso, C. H. (1984). *Annals of Physics*, **156**, 142.

Cloizeaux, J. des (1968). In *Many-body physics, Les houches 1967*, (ed. C. de Witt and R. Balian), p.5 Gordon and Breach, New York.

Coleman, A.J. (1963). *Reviews of Modern Physics*, **35**, 668.

Coleman, S. (1977). *Physical Review*, **D15**, 2929.

Cottingham, W. N., Lacombe, M., Loiseau, B., Richard, J. M., and Vinh Mau, R. (1973). *Physical Review*, **D8**, 800.

Cugnon, J., Lejeune, A., Baldo, M., and Lombardo, U. (1989). *Nuclear Physics*, **A492**, 173.

Curtin, M. W., Toki, H., and Scott, D. K. (1983). *Physics Letters*, **B123**, 289.

Cusson, R. Y. and Maruhn, J. A. (1976). *Physics Letters*, **B62**, 134.

Cusson, R. Y., Maruhn, J. A., and Meldner, H. W. (1978). *Physical Review*, **C18**, 2589.

Cusson, R. Y., Smith, R. K., and Maruhn, J. A. (1976). *Physical Review Letters*, **36**, 1166.

Cusson, R. Y., Reinhard, P.-G., Strayer, M. R., Maruhn, J. A., and Greiner, W. (1985). *Zeitschrift für Physik*, **A320**, 475.

Dahl, J.P. and Avery, J.(eds) (1984). *Local density approximations in quantum chemistry and solid state physics*. Plenum, New York.

Dalili, D., Németh, J., and Ngô, C. (1985). *Zeitschrift für Physik*, **A321**, 335.

Damgaard, J., Pauli, H. C., Pashkevich, V. V., and Strutinski, V. M. (1969). *Nuclear Physics*, **A135**, 432.

Day, B. D. (1967). *Reviews of Modern Physics*, **39**, 719.

Day, B. D. (1981*a*). *Physical Review Letters*, **47**, 226.

Day, B. D. (1981*b*). *Physical Review*, **C24**, 1203.

Day, B. D. (1983). *Communications in Nuclear and Particle Physics*, **11**, 115.

Day, B. D. and Wiringa, R. B. (1985). *Physical Review*, **C32**, 1057.

Dean, D.R. and Mosel, U. (1985). *Zeitschrift für Physik*, **A332**, 647.

Dean, T. J. and Fallieros, S. (1973). *Physical Review*, **C7**, 1709.

Deb, B. M. and Ghosh, S.K. (1987). The hydrodynamic formulation of time-dependent Kohn-Sham orbital density functional theory. In *The single-particle density in physics and chemistry* (ed. N. H. March and B. M. Deb), p. 219. Academic Press, New York.

Dehesa, J. S., Krewald, S., Speth, J., and Faessler, A. (1977). *Physical Review*, **C15**, 1858.

Delchev, Ya., Marchev, E., and Petkov, I.Zh. (1979). *Zeitschrift für Physik*, **A290**, 273.

Delocle, I., Bencheikh, K., Bennour, L., Brack, M., Libert, J., Meyer, J., and Quentin, P. (1989). In *10th session d'études biennale de physique nucleaire, Aussois-France (March 1989), LYCEN 89 report*, p. S.18.1.

De Shalit, A. and Feshbach, H. (1974). *Theoretical nuclear physics*, Vol. 1, Wiley, New York.

De Tar, C. (1978*a*). *Physical Review*, **D17**, 302.

De Tar, C. (1978*b*). *Physical Review*, **D17**, 323.

Dimitrova, S. S., Petkov, I. Zh., and Stoitsov, M. V. (1986*a*). *Communication of JINR*, **E4-86-658**, Dubna.

Dimitrova, S. S., Petkov, I. Zh., and Stoitsov, M. V. (1986*b*). *Zeitschrift für Physik*, **A325**, 15.

Dimitrova, S. S., Petkov, I. Zh., and Stoitsov, M. V. (1988*a*). *Nuclear Physics*, **A485**, 233.

Dimitrova, S. S., Petkov, I. Zh., and Stoitsov, M. V. (1988*b*). *Bulgarian Journal of Physics*, **15**, 203.

Dirac, P. A. M. (1930). *Proceedings of the Cambridge Philosophical Society*, **26**, 376.

Di Toro, M., Lombardo, U., and Russo, G. (1985). *Nuclear Physics*, **A435**, 173.

Donnelly, I. J. (1968). *Physics Letters*, **28**, 161.

Donnelly, I. J. (1969). *Nuclear Physics*, **A128**, 129.

Draper, J. E. *et al.* (1982). *Physical Review Letters*, **49**, 434.

Dreizler, R. M. and Providencia, J. Da (eds) (1985). *Density functional methods in physics, Nato ASI Series B*, Vol. 123, Plenum, New York.

Durand, M., Brack, M., and Schuck, P. (1978). *Zeitschrift für Physik*, **A286**, 381.

Durand, M., Brack, M., and Schuck, P. (1980). *Zeitschrift für Physik*, **A296**, 87.

Durso, J. W., Saarela, M., Brown, G. E., and Jackson, A. D. (1977). *Nuclear Physics*, **A278**, 445

Dutta, A. K., Arcoragi, J.-P., Pearson, J. M., Behrman, R. H., and Farine, M. (1986*a*). *Nuclear Physics*, **A454**, 374.

Dutta, A. K., Arcoragi, J.-P., Pearson, J. M., Behrman, R. H., and Tondeur,F. (1986*b*). *Nuclear Physics*, **A458**, 77.

Dworzecka, M. and Moszkowski,S. (1975). *Physical Review*, **C12**, 619.

Eckart, G. and Holzwarth, G. (1977). *Zeitschrift für Physik*, **A281**, 385.

Eckart, G., Holzwarth, G., and Schwesinger, B. (1980). *Physics Letters*, **B94**, 453.

Eisenberg, J. M. and Greiner, W. (1972). *Nuclear theory*, Vol. 3, North Holland, Amsterdam.

Engel, Y. M., Brink, D. M., Goeke, K., Krieger, S. J., and Vautherin, D. (1975). *Nuclear Physics*, **A249**, 215.

English, H. and English, R. (1984*a*). *Physics Status Solidy*, **B123**, 711.

English, H. and English, R. (1984*b*). *Physics Status Solidy*, **B124**, 373.

Epstein, S. T. (1974). *The variational method in quantum chemistry*, Academic Press, New York.

Epstein, S.T. and Hirschfelder, J.O. (1961). *Physical Review*, **123**, 1495.

Erdahl, R. and Smith, V. H. (eds) (1987). *Density matrices and density functionals*, Reidel, Dordrecht.

Erell, A., Alster, J., Lichtenstadt, J., Moinester, M.A. *et al.* (1984). *Physical Review Letters*, **52**, 2134.

Erell, A., Alster, J., Lichtenstadt, J., Moinester, M.A. *et al.* (1986). *Physical Review*, **C34**, 1822.

Euler, H. (1937). *Zeitschrift für Physik*, **105**, 553.

Faber, M. E., Egido, J. L., and Ring, P. (1983). *Physics Letters*, **B127**, 5.

Fabris, D. *et al.* (1987). *Physics Letters*, **B196**, 429.

Faessler, A., Ismail, M., Ohtsuka, N., Rashdan, M., and Wadia, W. (1987). *Zeitschrift für Physik*, **A326**, 501.

Fallieros, S. (1972). In *Proceedings of the international conference on nuclear structure studies using electron scattering and photoreactions*, Sendai, Japan, p. 185.

Fermi, E. (1928). *Zeitschrift für Physik*, **48**, 73.

Fernández Velicia, F. J. (1984). *Physical Review*, **A30**, 1194.

Ferrell, R. A. (1957). *Physical Review*, **107**, 1631.

Fleckner, J. and Mosel, U. (1977). *Nuclear Physics*, **A277**, 170.

Flieshbach, T. (1971). *Zeitschrift für Physik*, **A247**, 117.

Flocard, H., Koonin, S. E., and Weiss, M. S. (1978). *Physical Review*, **C17**, 1682.

Flocard, H. and Weiss, M. S. (1978). *Physical Review*, **C18**, 573.

Flocard, H., Quentin, P., Kermann, A. K., and Vautherin, D. (1973). *Nuclear Physics*, **A203**, 433.

Fock, V. (1930). *Zeitschrift für Physik*, **61**, 126.

Foldy, L. L. (1962). *Journal of Mathematical Physics*, **3**, 531.

Friedrich, J. and Reinhard, P. -G. (1986). *Physical Review*, **C33**, 335.

Gaardhoje, J. J., Ellegaard, C., Herskind, B., and Steadmon, S. G. (1984). *Physical Review Letters*, 53, 148.

Gazquer, J. L. and Ludeña, E. V. (1981). *Chemical Physics Letters*, **83**, 145.

Ghosh, S. K. and Deb, B. M. (1982). *Physics Reports*, **92**, 1.

Giannoni, M. J., Vautherin, D., Vénéroni, M., and Brink, D. M. (1976). *Physics Letters*, **B63**, 8.

Gilbert, T. L. (1975). *Physical Review*, **B12**, 2111.

Gleissl, P., Brack, M., Meyer, J., and Quentin, P. (1988). *Preprint of Nils Bohr Institute*, NBI-88-62, Copenhagen.

Goeke, K. (1976). *Nuclear Physics*, **A265**, 301.

Goeke, K. (1977). *Physical Review Letters*, **38**, 212.

Goeke, K. and Castel, B. (1979). *Physical Review*, **C19**, 201.

Goeke, K., Castel, B., and Reinhard, P.-G. (1980). *Nuclear Physics*, **A339**, 377.

Goeke, K., Gruemmer, F., and Reinhard, P.-G. (1983). *Annals of Physics*, **150**, 504.

Goeke, K., Lane, A. M., and Martorell. J. (1978). *Nuclear Physics*, **A296**, 109.

Goeke, K. and Reinhard, P.-G. (1978). *Annals of Physics*, **112**, 328.

Goeke, K. and Reinhard, P.-G. (1980). *Annals of Physics*, **124**, 249.

Goeke, K., Reinhard, P.-G., and Rowe, D. J. (1981). *Nuclear Physics*, **A359**, 408.

Goeke, K. and Speth, J. (1982). *Annual Review of Nuclear and Particle Sciences*, 32, 65.

Goldhaber, M. and Teller, E. (1948). *Physical Review*, **74**, 1046.

Goldstein, H. (1959). *Classical Mechanics*, Addison-Wesley, London.

Gombas, P. (1948). *Die statische Theorie des Atoms und ihre Anwendungen*, Springer Verlag, Vienna.

Göritz, G. -H. and Mosel, U.(1976). *Zeitschrift für Physik*, **A277**, 243.

Grammaticos, B. and Voros, A. (1979). *Annals of Physics*, **123**, 359.

Grammaticos, B. and Voros, A. (1980). *Annals of Physics*, **129**, 153.

Groenewald, H. H. (1946). *Physica*, 12, 405.

Guet, G. and Brack, M. (1980). *Zeitschrift für Physik*, **A297**, 247.

Guet, G., Håkansson, H.-B., and Brack, M. (1980). *Physics Letters*, **B97**, 7.

Guet, G., Strumberger, E., and Brack, M. (1988). *Physics Letters*, **B205**, 427.

Gunnarson, O. and Lundqvist, B. J. (1976). *Physical Review*, **B13**, 4274.

Gupta, U. and Rajagopal, A. K. (1982). *Physics Reports*, **87**, 260.

Hamada, T. and Johnston, I. D. (1962). *Nuclear Physics*, **34**, 382.

Hamamoto, I. and Siemens, P. (1976). *Nuclear Physics*, **A269**, 199.

Hara, Y. (1960). *Progress of Theoretical Physics*, **24**, 1179.

Harriman, J. E. (1981). *Physical Review*, **A24**, 680.

Hartree, D. (1927). *Proceedings of the Cambridge Philosophical Society*, **24**, 111.

Henderson, G. A. (1981). *Physical Review*, **A23**, 19.

Hodges, C. H. (1973). *Canadian Journal of Physics*, **51**, 1428.

Hodgson, P. E. (1975). *Reports on Progress of Physics*, **38**, 847.

Hohenberg, P. and Kohn, W. (1964). *Physical Review*, 136, B864.

Holzwarth, G. and Eckart, G. (1977). *Zeitschrift für Physik*, **A283**, 219.

Holzwarth, G. and Eckart, G. (1978). *Zeitschrift für Physik*, **A284**, 291.

Holzwarth, G. and Eckart, G. (1979). *Nuclear Physics*, **A325**, 1.

Horowiz, C.J. and Serot, B.D. (1984). *Physics Letters*, **B137**, 287.

Hoshino, T. and Arima, A. (1976). *Physical Review Letters*, **37**, 266.

Huang, K. (1963). *Statistical mechanics*, Wiley, New York.

Irom, F., Bowman J. D., Baer, H. W., Bolme, O. G. *et al.* (1986). *Physical Review*, **C34**, 2231.

Ismail, M., Rashdan, M., Faessler, A., Trefz, M., and Mansour, H. M. M. (1986). *Zeitschrift für Physik*, **A323**. 399.

Izumoto, T., Krewald, S., and Faessler, A. (1980*a*). *Nuclear Physics*, **A341**, 319.

Izumoto, T., Krewald, S., and Faessler, A. (1980*b*). *Nuclear Physics*, **A357**, 417.

Jackson, A. D. (1983). In *Annual review of nuclear science*, Vol. 33, p. 105, Palo Alto, California.

Jaeger, C., de Vries, H., and de Vries, C. (1974). *Atomic Data and Nuclear Data Tables*, 5-6, 479.

James, A. N., Andrews, P. T., Kirkby, P., and Lowe, B. G. (1969). *Nuclear Physics*, **A138**, 145.

Jensen, A. S. and Larsen, S. M. (1981). *Physica Scripta*, **24**, 534.

Jennings, B. K. (1973). *Nuclear Physics*, **A207**, 538.

Jennings, B. K. and Bhaduri, R. K. (1975). *Physical Review Letters*, **34**, 228.

Jennings, B. K., Bhaduri, R. K., and Brack, M. (1975). *Nuclear Physics*, **A253**, 29.

Jennings, B.K. and Jackson, A.D. (1980). *Nuclear Physics*, **A342**, 23.

Kadanoff, L. P. and Baym, G. (1962). *Quantum statistical mechanics*, p. 71. Benjamin, New York.

Kan, K. K., Lichtner, P. C., Dworzecka, M., and Griffin, J. J. (1980). *Physical Review*, **C21**, 1098.

Katriel, J., Appellof, C. J., and Davidson, E. R. (1981). *International Journal of Quantum Chemistry*, **19**, 293.

Kerman, A.K. and Koonin, S.E. (1976). *Annals of Physics*, **100**, 332.

Khadkikar, S. B., Rikus, L., Faessler, A., and Sartor, R. (1981). *Nuclear Physics*, **A369**, 495.

Kirkwood, J. G. (1933). *Physical Review*, **44**, 31.

Kirzhnits, D. A. (1957). *Soviet Physics*, JETP, **5**, 64.

Kirzhnits, D. A. (1967). *Field theoretical methods in many-body systems*, Pergamon, Oxford.

Knüpfer, W. and Huber, M. G. (1976*a*). *Physical Review*, **C14**, 2254.

Knüpfer, W. and Huber, M. G. (1976*b*). *Zeitschrift für Physik*, **A276**, 99.

Ko, C. M., Pauli, H. C., Brack, M., and Brown, G. E. (1974). *Nuclear Physics*, **A236**, 269.

Köhler, H. S. (1976). *Nuclear Physics*, **A258**, 301.

Kohn, W. (1966). In *Proceedings of the Midwest conference on theoretical physics*, Indiana University, Bloomington, Indiana.

Kohn, W. and Sham, L. J. (1965). *Physical Review*, **140**, A1133.

Kohn, W. and Vashishta, P. (1983). General density functional theory. In *Theory of the inhomogeneous electron gas* (ed. S. Lundqvist and N. H. March), p. 79. Plenum, New York.

Kohno, M. and Andō, K. (1979). *Progress of Theoretical Physics*, **61**, 1065.

Koonin, S. K. (1976). *Physics Letters*, **B61**, 227.

Koonin, S. K., Davies, K. T. R., Maruhn-Rezwani, V., Feldmeier, H., Krieger, S.J., and Negele, J.W. (1977). *Physical Review*, **C15**, 1359.

Koopmans, T. (1933). *Physica*, **1**, 104.

Krappe, H. J., Nix, J. R., and Sierk, A. J. (1972). *Physical Review*, **C20**, 992.

Krewald, S., Birkholz, J., Faessler, A., and Speth, J. (1974). *Physical Review Letters*, **33**, 1386.

Krewald, S., Klemt, V., Speth, J., and Faessler, A. (1977). *Nuclear Physics*, **A281**, 166.

Krieger, S. J. (1977). *Nuclear Physics*, **A276**, 12.

Krivchenkov, G. (1964). In *Collective problems in quantum mechanics* (ed. D. Ter Haar). Infosearch, London.

Krivine, H. (1984). *Journal de Physique*, **45**, C6.

Krivine, H., Schmit, C., and Treiner, J. (1982). *Physics Letters*, **B112**, 281.

Krivine, H. and Treiner, J. (1979). *Physics Letters*, **B88**, 212.

Krivine, H., Treiner, J., and Bohigas, O. (1980). *Nuclear Physics*, **A336**, 155.

Kryachko, E. S., Petkov, I. Zh., and Stoitsov, M. V. (1987*a*). *International Journal of Quantum Chemistry*, **32**, 473.

Kryachko, E. S., Petkov, I. Zh., and Stoitsov, M. V. (1987*b*). *International Journal of Quantum Chemistry*, **32**, 467.

Kryachko, E. S., Petkov, I. Zh., and Stoitsov, M. V. (1987*c*). *Ukrainian Journal of Physics*, **32**, 1748.

Kryachko, E. S., Petkov, I. Zh., and Stoitsov, M. V. (1987*d*). *Ukrainian Journal of Physics*, **32**, 1868.

Kumar, K., Le Couteur, K. J., and Roy, M. K. (1963). *Nuclear Physics*, **42**, 529.

Küpper, W. A., Wegmann, G., and Hilf, E. (1974). *Annals of Physics*, **88**, 454.

Kyrchev, G., Malov, L. A., Nesterenko, V. O., and Soloviev, V. G. (1978). *Soviet Journal of Nuclear Physics*, **25**, 506.

Lamb, D. Q., Lattimer, J. M., Pethick, C. J., and Ravenhall, D. G. (1981). *Nuclear Physics*, **A360**, 459.

Landau, L. D. (1957*a*). *Soviet Physics*, JETP, **3**, 920.

Landau, L. D. (1957*b*). *Soviet Physics*, JETP, **5**, 101.

Landau, L. D. (1958). *Soviet Physics*, JETP, **8**, 70.

Larsson, S. E., Leander, G., and Ragnarsson, I. (1978). *Nuclear Physics*, **A307**, 189.

Lassila, K. E., Hull, M. H., Ruppel, H. M., McDonald, F. A., and Breit, G. (1962). *Physical Review*, **126**, 881.

Lattimer, J. M. (1981). *Annual Review of Nuclear and Particle Sciences*, **31**, 337.

Lejeune, A., Grange, P., Matzolff, M., and Cugnon, J. (1986). *Nuclear Physics*, **A453**, 189.

Levy, M. (1979). *Proceedings of the National Academy of Sciences of the USA*, **76**, 6062.

Levy, M. (1982). *Physical Review*, **A26**, 1200.

Liberman, D. A. (1977). *Il Nuovo Cimento*, **A16**, 1542.

Lieb, E. H. (1982). In *Physics as natural philosophy: Essays in honor of Laszlo Tisza on his 75th birthday* (ed. H. Feshbach and A. Shimony). MIT Press, Cambridge.

Lieb, E. H. (1983). *International Journal of Quantum Chemistry*, **24**, 224.

Lipparini, E., Orlandini, G., and Leonardi, R. (1976). *Physical Review Letters*, **C36**, 660.

Lipparini, E. and Stringari, S. (1989). *Physics Reports*, **175**, 103.

Lin, Y. C. (1970). *Nuclear Physics*, **A140**, 359.

Liu, K. F. and Brown, G. E. (1976). *Nuclear Physics*, **A265**, 385.

Liu, K. F. and Van Giai, N. (1976). *Physics Letters*, **B65**, 23.

Lombard, R. J. (1973). *Annals of Physics*, **77**, 380.

Lombard, R. J. and Mass, D. (1986). *Annals of Physics*, **167**, 2.

Löwdin, P.-O. (1955). *Physical Review*, **97**, 1474.

Ludeña, E. V. (1982). *Journal of Chemical Physics*, **76**, 3157.

Ludeña, E. V. (1983). *International Journal of Quantum Chemistry*, **23**, 127.

Ludeña, E. V. and Kryachko, E. S.(1989). *Energy density functional theory in quantum chemistry*, Kluwer, Dordrecht.

Lundqvist, S. and March, N. H. (eds) (1983). *Theory of the inhomogeneous electron gas*, Plenum, New York.

McDonald, A. H. (1979). *Journal of Physics*, **C12**, 2977.

McWeeny, R. and Sutcliffe, B. T. (1969). *Methods of molecular quantum mechanics*, Academic Press, New York.

Mädler, P. (1984). *Physics of Elementary Particles and Atomic Nuclei*, **15**, 418.

Malov, L. A., Nesterenko, V. O., and Soloviev, V. G. (1976). *Physics Letters*, **B64**, 247.

March, N. H. and Deb, B. M. (eds) (1987). *The single-particle density in physics and chemistry*, Academic Press, New York.

March, N. H. and Pucci, R. (1981). *Journal of Chemical Physics*, **75**, 496.

March, N. H. and Young, W. H. (1958). *Proceedings of the Philosophical Society*, **72**, 182.

Marinova, L. P., Petkov, I. Zh., and Stoitsov, M. V. (1985). *Scientific Contribution of Youth*, p.22, Bulgarian Academy of Sciences, Sofia.

Marinova, L. P., Petkov, I. Zh., and Stoitsov, M. V. (1986*a*). *Comptes Rendus de l'Académie Bulgare des Sciences*, **39**, 55.

Marinova, L. P., Petkov, I. Zh., and Stoitsov, M. V. (1986*b*). *Bulgarian Journal of Physics*, **13**, 3.

Marinova, L. P., Petkov, I. Zh., and Stoitsov, M. V. (1989). *Il Nuovo Cimento*, **A101**, 941.

Marshalek, E. R. and Providencia, J. Da (1973). *Physical Review*, **C7**, 2281.

Martorell, J., Bohigas, O., Fallieros, S., and Lane, A. M. (1976). *Physics Letters*, **B60**, 313.

Maruhn, J. A. and Cusson, R. Y. (1976). *Nuclear Physics*, **A270**, 471.

Maruhn-Rezwani, V., Davies, K. T. R., and Koonin, S. E. (1977). *Physics Letters*, **B67**, 134.

Matveev, B. A. and Sorba, P. (1978). *Il Nuovo Cimento*, **45**, 357.

Meitner, L. and Frisch, O. R. (1939). *Nature*, **143**, 239.

Mermin, N. D. (1963). *Annals of Physics*, **21**, 99.

Mermin, N. D. (1965). *Physical Review*, **137**, A1441.

Messiah, A. (1966). *Quantum mechanics*, Vol. 2,p. 670, North Holland, Amsterdam.

Meyer, J., Quentin, P., and Brack, M. (1983). *Physics Letters*, **B133**, 279.

Migdal, A.B. (1967). *Theory of finite Fermi systems*, Wiley, New York.

Mikolas, V. and Tomasek, M. (1977). *Physics Letters*, **A64**, 109.

Möller, P. and Nix, J. R. (1981). *Nuclear Physics*, **A361**, 117.

Möller, P., Myers, W. D., Swiatecki, W. J., and Treiner, J. (1985). In *Proceedings of the 7th international conference on atomic masses and fundamental constants, AMOC-7*, (ed. O. Klepper), p. 457, TDH, Darmstadt.

Morrell, M. M., Parr, R. G., and Levy, M. (1975). *Journal of Chemical Physics*, **62**, 549.

Morsch, H. P. and Decowski, P. (1970). *Physics Letters*, **B82**, 1.

Morsch, H. P., Sukosd, C., Rogge, M., Turek, P., and Mayer-Boericke, C. (1980). *Physical Review*, **C22**, 489.

Morsch, H. P., Decowski, P., Rogge, M., and Turek, P. (1983). *Physical Review*, **C28**, 1947.

Mosel, U., Zint, P., and Passler, K. N. (1974). *Nuclear Physics*, **A236**, 252.

Müther, H., Waghmore, R. Y., and Faessler, A. (1976). *Physics Letters*, **B64**, 125.

Myers, W. D. (1976). *Atomic Data and Nuclear Data Tables*, **17**, 411.

Myers, W. D. and Swiatecki, W. J. (1966). *Nuclear Physics*, **81**, 1.

Myers, W. D. and Swiatecki, W. J. (1969). *Annals of Physics*, **55**, 395.

Myers, W. D. and Swiatecki, W. J. (1974). *Annals of Physics*, **84**, 186.

Myers, W. D. and Swiatecki, W. J. (1982). *Annual Review of Nuclear and Particle Sciences*, **32**, 309.

Naimark, M.A. (1976), *Theory of Group Representations*, Nauka, Moscow.

Negele, J. W. (1970). *Physical Review*, **C1**, 1260.

Negele, J. W. (1982). *Reviews of Modern Physics*, **54**, 913.

Negele, J. W. and Vautherin, D. (1972). *Physical Review*, **C5**, 1472.

Negele, J. W. and Vautherin, D. (1975). *Physical Review*, **C11**, 1031.

Negele, J. W., Koonon, S. E., Möller, P., Nix, J. R., and Sierk, A. J. (1978). *Physical Review*, **C17**, 1098.

Németh, J. and Bethe, H. A. (1968). *Nuclear Physics*, **A116**, 241.

Németh, J., Dalili, D., and Ngô, C. (1985). *Physics Letters*, **B145**, 11.

Neumann, J. Von (1932). *Mathematische Grundlagen der Quantenmechanik*. Springer Verlag, Berlin.

Newton, J. O., Herskind, B., Diamond, R.M., Dines, E.L., Draper, J.E., *et al.* (1981). *Physical Review Letters*, **46**, 1383.

Ngô, H. and Ngô, C., (1980). *Nuclear Physics*, **A384**, 140.

Ngô, C., Tamain, B., Beiner, M., Lombard, R. J., Mass, D., and Deubler, H. H. (1975a). *Nuclear Physics*, **A252**, 237.

Ngô, C., Tamain, B., Galin, J., and Beiner, (1975b). *Nuclear Physics*, **A240**, 353.

Nilsson, S. G. (1955). *Kongelige Danske Videnskabenrnes Selskab. Matematics-Fisiske Meddelelser* , **29**, 1.

Nix, J. R. and Sierk, A. J. (1980). *Physical Review*, **C21**, 396.

Nix, J. R. and Sierk, A. J. (1982). *Physical Review*, **C25**, 1068.

Nix, J. R. and Swiatecki, W. J. (1965). *Nuclear Physics*, **71**, 1.

Nogami, Y. and Warke, C. S. (1978). *Physical Review*, **C17**, 1905.

Ohtsuka, N., Linden, R., Faessler, A., and Malik, F. B. (1987). *Nuclear Physics*, **A465**, 550.

Oulne, M., Petkov, I. Zh., and Stoitsov, M. V. (1989). *Bulgarian Journal of Physics*, 16, 20.

Panagiotou, A. D., Curtin, M. W., Toki, H., Scott, D. K., and Siemens, P. J. (1984). *Physical Review Letters*, **52**, 496.

Pandaripande, V. R. (1969a). *Nuclear Physics*, **A135**, 419.

Pandaripande, V. R. (1969b). *Physics Letters*, **B29**, 1.

Pandaripande, V. R. and Wiringa, R. B. (1979). *Reviews of Modern Physics*, **51**, 821.

Pant, M. M. and Rajagopal, A. K. (1972). *Solid State Communications*, **10**, 1189.

Parr, R. G. (1983). *Annual Review of Physical Chemistry*, **34**, 631.

Parr, R.G., and Yang, W. (1989). *Density-Functional Theory and Molecules*, Oxford University Press, Oxford.

Parr, R. G., Donnelly, R. A., Levy, M., and Palke, W. E. (1978). *Journal of Chemical Physics*, 68, 3801.

Pearson, J. M. (1982). *Nuclear Physics*, **A376**, 501.

Perrot, F. (1979). *Physical Review*, **A20**, 586.

Pethick, C. J., Ravenhall, D. G., and Lattimer, J. M. (1983). *Physics Letters*, **B128**, 137.

Petkov, I. Zh. (1983). *Proceedings of the 6th international school on nuclear physics, neutron physics and nuclear energy, September 12-21*, Varna, Bulgaria.

Petkov, I. Zh. (1988). In *Proceedings of the 8th international school on nuclear physics, neutron physics and nuclear energy, September 19-28, Varna, Bulgaria* (ed. V.Andrejtscheff, Chr. Christov, and D. Elenkov), p. 54, World Scientific, Singapore.

Petkov, I. Zh. and Stoitsov, M. V. (1980a). *Comptes Rendus de l'Académie Bulgare des Sciences*, 33, 1623.

Petkov, I. Zh. and Stoitsov, M. V. (1980b). *Preprint ICTP*, **IC/80/34**, Trieste.

Petkov, I. Zh. and Stoitsov, M. V. (1981). *Comptes Rendus de l'Académie Bulgare des Sciences*, **34**, 1651.

Petkov, I. Zh. and Stoitsov, M. V. (1983a). *Theoretical and Mathematical Physics*, **55**, 584.

Petkov, I. Zh. and Stoitsov, M. V. (1983b). *Soviet Journal of Nuclear Physics*, **37**, 692.

Petkov, I. Zh. and Stoitsov, M. V. (1983c). *Communication of JINR*, **E4-83-282**, Dubna.

Petkov, I. Zh. and Stoitsov, M. V. (1985). *Communication of JINR*, **P4-85-785**, Dubna.

Petkov, I. Zh., Stoitsov, M. V., and Kryachko, E. S. (1986a). *International Journal of Quantum Chemistry*, **29**, 149.

Petkov, I. Zh., Stoitsov, M. V., and Kryachko, E. S. (1986b). *Ukrainian Journal of Physics*, **31**, 777.

Pi, M., Viñas, X., Garcias, F., and Barranco, M. (1988). *Physics Letters*, **B215**, 5.

Providencia, J. P. Da and Holzwarth, G. (1983). *Nuclear Physics*, **A398**, 59.

Raimes, S. (1972). *Many-electron theory*. North Holland, Amsterdam.

Rajagopal, A. K. (1978). *Journal of Physics*, **C11**, 2948.

Rajagopal, A. K. (1980). In *Advances in Chemical Physics*, Vol. 41 (ed. I. Prigogine and S. A. Rice), p. 59. Wiley, New York.

Rajagopal, A. K. (1985). In *Density functional methods in physics*, Nato Asi Series, Vol. 123 (ed. R. M. Dreizler and J. Da Providencia). Plenum, New York.

Rajagopal, A.K. and Callaway, J. (1973). *Physical Review*, **B7**, 1912.

Ramos, A., Polls, A., and Dickhoff, W. H. (1988). Saturation in nuclear matter: a new perspective. In *Condensed Matter Theories*, Vol. 3 (ed. J. S. Arponen, R. F. Bishop, and M. Manninen), p. 319. Plenum, New York.

Rashdan, M., Faessler, A., Ismail, M., and Wadia, W. (1987). *Nuclear Physics*, **A466**, 439.

Rashdan, M., Faessler, A., Ohtsuka, N., Linden, R., Wadia, W., and Ismail, M. (1988). *Zeitschrift für Physik*, **A330**, 417.

Ravenhall, D. G., Pethick, C. J., and Lattimer, J. M. (1983). *Nuclear Physics*, **A407**, 571.

Reichl, L. E. (1980). *A modern course in statistical physics*, p. 39, University of Texas Press, Austin.

Reid, P. V. (1968). *Annals of Physics*, 50, 414.

Reinhard, P.-G. and Friedrich, J. (1985). *Zeitschrift für Physik*, **A321**, 619.

Reinhard, P.-G. and Goeke, K. (1987). *Reports on Progress of Physics*, **50**, 1.

Reinhardt, H. (1981). *Nuclear Physics*, **A367**, 269.

Reinhardt, H. and Schulz, H. (1985). *Nuclear Physics*, **A432**, 630.

Ring, P. (1984). *Journal de Physique*, **C6**, 247.

Ring, P. and Schuck, P. (1980). *The nuclear many-body problem.* Springer Verlag, Berlin.

Ring, P. and Speth, J. (1974). *Nuclear Physics*, **A235**, 315.

Sagawa, H. and Bertsch, G. F. (1984). *Physics Letters*, **B146**, 138.

Sagawa, H. and Holzwarth, G. (1978). *Progress of Theoretical Physics*, **59**, 1213.

Sagawa, H. and Toki, H. (1986) *Progress of Theoretical Physics*, **76**, 433.

Sauer, G., Chandra, H., and Mosel, U. (1976). *Nuclear Physics*, **A264**, 221.

Schrödinger, E. (1926*a*). *Annalen der Physik*, **79**, 361.

Schrödinger, E. (1926*b*). *Annalen der Physik*, **79**, 489.

Schrödinger, E. (1926*c*). *Annalen der Physik*, **80**, 437.

Schrödinger, E. (1926*d*). *Annalen der Physik*, **81**, 109.

Seeger, P. A. (1967). In *Proceedings of the international symposium on nuclei far off the stability line, Lysekil 1966*, p. 495, Almqvist and Wiksell, Stockholm.

Serr, F. E., Bertsch, G. F., and Borysowicz, J. (1980). *Physics Letters*, **B92**, 241.

Seyler, R. G. and Blanchard, C. H. (1961). *Physical Review*, **124**, 227.

Seyler, R. G. and Blanchard, C. H. (1963). *Physical Review*, **131**, 355.

Shlomo, S. and Siemens, P. J. (1985). *Physical Review*, **C31**, 2291.

Sick, I., *et al.* (1979). *Physics Letters*, **B88**, 245.

Siemens, P. J. (1970a). *Physical Review*, **C1**, 98.

Siemens, P. J. (1970*b*). *Nuclear Physics*, **A141**, 225.

Simon, B. (1979). *Physics Letters*, **A71**, 211.

Skyrme, T. H. R. (1956). *Philosophical Magazine*, **1**, 1043.

Skyrme, T. H. R. (1957). *Proceedings of the Philosophical Society*, **A70**, 433.

Skyrme, T. H. R. (1959). *Nuclear Physics*, **9**, 615.

Slater, J. C. (1951). *Physical Review*, **81**, 385.

Slavov, B., Dimitrov, V., Goeke, K., and Reinhard, P.-G. (1986). *Nuclear Physics*, **A454**, 392.

Smith, V. H. (1984). Density functional theory and local potential approximations from momentum space considerations. In *Local density approximations in quantum chemistry and solid state.* (ed. J. P. Dahl and J. Avery), p. 1. Plenum, New York.

Snover, K. A. (1986). *Annual Review of Nuclear and Particle Sciences*, **36**, 545.

Soloviev, V. G. (1976). *Theory of complex nuclei*, Pergamon, Oxford.

Soloviev, V. G. (1978). *Nucleonika*, **23**, 1149.

Soloviev, V. G. and Shirikova, N. Yu. (1986). *Zeitschrift für Physik*, **A324**, 393.

Soloviev, V. G., Stoyanov, C., and Vdovin, A. I. (1977). *Nuclear Physics*, **A288**, 376.

Speth, J. and Van der Woode, A. (1981). *Reports on Progress of Physics*, **44**, 719.

Speth, J., Werner,E., and Wild,W. (1977).*Physics Reports*, **C33**, 127.

Stancu, Fl. and Brink, D. M. (1976). *Nuclear Physics*, **A270**, 236.

Starodubsky, V. E. (1985). *Yadernaja Fizika*, **42**, 78.

Steinwedel, H. and Jensen, J. H. D. (1950). *Zeitschrift für Naturforschung*, **A5**, 413.

Stock, H. (1975). *Nuclear Physics*, **A237**, 365.

Stock, H. and Scheid, W. (1975). *Nuclear Physics*, **A253**, 145.

Stöcker, H. and Greiner, W. (1986). *Physics Reports*, **137**, 277.

Stöcker, H., Cusson, R. Y., Maruhn, J. A., and Greiner, W. (1980). *Zeitschrift für Physik*, **A294**, 125.

Stocker, W. (1984). *Physics Letters*, **B142**, 319.

Stocker, W. and Burzlaff, J. (1975). *Nuclear Physics*, **A202**, 265.

Stocker, W., Gleissl, P., and Brack, M. (1987). *Nuclear Physics*, **A471**, 507.

Stocker, W., Süssmann, G., and Knaak, S. (1972). *Nuclear Physics*, **A187**, 38.

Stoddart, J. C. and March, N. H. (1971). *Annals of Physics*, **64**, 174.

Stoitsov, M. V. (1982). *Communication of JINR*, **P4-82-418**, Dubna.

Stoitsov, M. V. (1987). *Il Nuovo Cimento*, **A98**, 725.

Stoitsov, M. V. (1988). *Comptes Rendus de l'Académie Bulgare des Sciences*, **41**, 41.

Stoitsov, M. V. and Petkov, I. Zh. (1988). *Annals of Physics*, **148**,121.

Stoitsov, M. V., Petkov, I. Zh., and Dimitrova, S. S. (1986). *Izv. AN USSR*, ser. fiz., **50**, 2071.

Stoitsov, M. V., Petkov, I. Zh., and Kryachko, E. S. (1987). *Comptes Rendus de l'Académie Bulgare des Sciences*, **40**, 45.

Stoner, E. C. (1939). *Philosophical Magazine*, **28**, 257.

Stringari, S. (1977). *Nuclear Physics*, **A279**, 454.

Stringari, S. (1979). *Nuclear Physics*, **A325**, 199.

Stringari, S. (1983). *Annals of Physics*, **151**, 35.

Stringari, S. (1985). *Il Nuovo Cimento*, **A87**, 231.

Stringari, S. and Lipparini, E. (1982). *Physics Letters*, **B117**, 141.

Stringari, S. and Lipparini, E. (1986). *Physical Review*, **C34**, 370.

Stringari, S., Lipparini, E., Orlandini, G., Traini, M., and Leonardi, R. (1978*a*). *Nuclear Physics*, **A309**, 177.

Stringari, S., Lipparini, E., Orlandini, G., Traini, M., and Leonardi, R. (1978*b*). *Nuclear Physics*, **A309**, 189.

Strutinski, V. M. (1967). *Nuclear Physics*, **A95**, 420.

Strutinski, V. M. (1968). *Nuclear Physics*, **A122**, 1.

Strutinski, V. M. (1974). *Nuclear Physics*, **A218**, 169.

Strutinski, V. M. and Tyapin, A. S. (1964). *Soviet Physics*, JETP, **18**, 664.

Suraud, E. (1984). *Journal de Physique*, C6, 161.

Suraud, E., Barranco, M., and Treiner, J. (1987). *Journal de Physique*, **C2**, 11.

Suzuki, T. (1980). *Progress of Theoretical Physics*, **64**, 1627.

Suzuki, T. (1981). *Progress of Theoretical Physics*, **65**, 910.

Svenne, J. P. (1979). Fourteen years of selfconsistent field calculation: what have we learned. In *Advances in nuclear physics*, Vol. 11 (ed. J. W. Negele and E. Vogt), p. 179. Plenum, New York.

Tassie, L. J. (1956). *Australian Journal of Physics*, 9, 407.

Ter Haar, B. and Malfliet, R. (1987). *Physics Reports*, **C149**, 207.

Theophilou, A. L. (1979). *Journal of Physics*, **C12**, 5419.

Theophilou, A. L. (1987). Density functional theory for excited states. In *The single-particle density in physics and chemistry* (ed. N.H. March and B.M. Deb), p.201. Academic Press, New York.

Thomas, L. H. (1926). *Proceedings of the Cambridge Philosophical Society*, 23, 542.

Thouless, D. J. (1960). *Nuclear Physics*, **21**, 225.

Thouless, D. J. (1961). *Nuclear Physics*, **22**, 78.

Thouless, D. J. (1972). *The quantum mechanics of many-body systems*, (2nd edn). Academic Press, New York.

Tondeur, F. (1978*a*). *Nuclear Physics*, **A303**, 185.

Tondeur, F. (1978*b*). *Nuclear Physics*, **A311**, 51.

Tondeur, F. (1979). *Nuclear Physics*, **A315**, 353.

Tondeur, F. (1980). *Nuclear Physics*, **A338**, 77.

Tondeur, F. (1983). *Physics Letters*, **B123**, 139.

Tondeur, F., Pearson, J. M., and Farine, M. (1983). *Nuclear Physics*, **A394**, 462.

Tondeur, F., Brack, M., Farine, M., and Pearson, J. M. (1984). *Nuclear Physics*, **A420**, 297.

Tondeur, F., Dutta, J. M., Pearson, J., and Behrman, R. (1987). *Nuclear Physics*, **A470**, 93.

Trefz, M., Faessler, A., and Dickhoff, W. H. (1985). *Nuclear Physics*, **A443**, 499.

Treiner, J. and Krivine, H. (1986). *Annals of Physics*, **170**, 406.

Treiner, J., Krivine, H., Bohigas, O., and Martorell, J. (1981). *Nuclear Physics*, **A371**, 253.

Treiner, J., Myers, W. D., Swiatecki, W. J., and Weiss, M. S. (1986). *Nuclear Physics*, **A452**, 93.

Valone, S. M. (1980*a*). *Journal of Chemical Physics*, **73**, 1334.

Valone, S. M. (1980*b*). *Journal of Chemical Physics*, **73**, 4653.

Valone, S. M. and Capitani, J. F. (1981). *Physical Review*, **A23**, 2127.

Van Giai, N. and Sagawa, H. (1981). *Physics Letters*, **B196**, 379.

Vautherin, D. (1975). *Physics Letters*, **B57**, 425.

Vautherin, D. and Brink, D. M. (1970). *Physics Letters*, **B32**, 149.

Vautherin, D. and Brink, D. M. (1972). *Physical Review*, **C5**, 626.

Vautherin, D. and Vinh Mau, N. (1983). *Physics Letters*, **B120**, 261.

Vautherin, D. and Vinh Mau, N. (1984). *Nuclear Physics*, **A422**, 140.

Villars, F. (1977). *Nuclear Physics*, **A285**, 269.

Vinh Mau, R. (1979). The Paris nucleon-nucleon potential. In *Mesons in nuclei* (ed. M. Rho and D. Wilkinson). North Holland, Amsterdam.

Vlasov, A. (1954). *Journal of Physics (USSR)*, **9**, 25.

Voronov, V. V. and Soloviev, V. G. (1983). *Particles and Nuclei*, **14**, 1380.

Vries, H. de, de Jaeger, C., and de Vries, C. (1987). *Atomic Data and Nuclear Data Tables*, **36**, 495.

Walecka, J. D. (1974). *Annals of Physics*, **83**, 491.

Walecka, J. D. (1975). *Physics Letters*, **B59**, 109.

Wambach, J., Mishra, V., and Li, C. H. (1982). *Nuclear Physics*, **A380**, 285.

Waroquier, M., Heyde, K., and Wenes, G. (1983). *Nuclear Physics*, **A404**, 269.

Weizsäcker, C. F. Von (1935). *Zeitschrift für Physik*, **96**, 431.

Werntz, C. and Überall, H. (1966). *Physical Review*, **149**, 762.

Wigner, E. (1932). *Physical Review*, **40**, 749.

Wigner, E. (1934). *Physical Review*, **46**, 1002.

Wilets, L. (1956). *Reviews of Modern Physics*, **30**, 542.

Wilets, L. (1958). *Physical Review*, 101, 1805.

Wong, C. Y. and McDonald, J. A. (1977). *Physical Review*, C16. 1196.

Wong, C. Y., Maruhn, J. A., and Welton, T. A. (1975). *Nuclear Physics*, **A253**, 469.

Yoon, B. and Negele, J. W. (1977). *Physical Review*, **A16**, 1451.

Youngblood, D.H., Moss, J.M., Rozsa, C.M., Bronson, J.D., Bacher, A.D., and Brown, D. R. (1976). *Physical Review*, **C13**, 994.

Zint, P. (1977). *Zeitschrift für Physik*, **A281**, 373.

Author index

Abramowitz, M. 115
Acharya, P. K. 189
Aerts, A. Th. 26
Aleksandrov, L. 197
Alster, J. 245
Anastasio, M. R. 183
Andō, K. 259, 275
Andrews, P. T. 202
Antonov, A. N. 63 218–19
Appellof, C. J. 57
Arcoragi, J.-P. 131, 137–8, 174, 176, 216
Argyres, P. 116
Arima, A. 265
Arnold, M. 137
Auerbach, N. 267
Avery, J. 57

Bacher, A. D. 259
Bacher, F. 84, 132
Baer, H. W. 245
Balazs, N. L. 113, 120
Balbutsev, E. B. 250
Baldo, M. 295
Balescu, R. 280
Bar-Touv, J. 319
Baranger, M. 227, 229, 234–5, 253, 231–2
Bardeen, J. 211
Barette, J. 319
Barranco, M. 128, 152–3, 275, 277, 307, 310, 317, 319, 320–1
Bartel, J. 32, 46–7, 122–3, 130, 153, 159, 176, 290, 307, 311–14, 317
Barth, U. Von 61
Baym, G. 247
Behrman, R. H. 131, 137–8, 174, 176, 216
Beiner, M. 32, 44, 46–7, 109, 166, 168–9
Bell, L. S. 30
Bellicard, J. B. 130, 221
Bencheikh, K. 160
Bennour, L. 160
Berezin, F. A. 4
Berg, R. A. 85
Bertrand, F. E. 244, 271, 279
Bertsch, G. F. 26, 245, 250, 255, 257, 259, 264, 266–7, 319
Besold, W. 319
Bethe, H. A. 29, 42, 82, 84–5, 88–9, 90, 93, 132
Bhaduri, R. K. 113, 150–1
Birkholz, J. 267
Blaizot, J. P. 82, 264
Blanchard, C. H. 85, 134
Blin, A. H. 322
Bogolubov, N. N. 280–1
Bohigas, O. 32, 46, 103–5, 174–5, 259, 267, 275–9
Bohr, A. 23, 217–18, 257, 259
Bohr, N. 257
Bolme, O. G. 245
Bonche, P. 228, 234, 290, 294, 296–7, 300–2, 304, 320, 322
Bondorf, J. P. 315
Bortignon, D. F. 266–7
Borysowicz, J. 255
Bouyssy, A. 107
Bowman, J. D. 245
Brack, M. 32, 44, 46–7, 87–8, 113, 122, 124, 127–8, 130–3, 135–8, 141, 143–6, 150–3, 154, 156–9, 160, 170, 172, 173–4, 176, 184, 198, 205–6, 209, 211, 241–6, 277, 290, 294, 299, 300, 303, 307–8, 310, 311–14, 316–19, 322
Breit, G. 27
Brink, D. M. 31–2, 44, 46, 109, 197–9, 201–4, 207–9, 211, 233–8, 248, 251, 255, 272, 277
Brockmann, R. 83
Broglia, R. A. 266–7, 319
Bronson, J. D. 259
Brown, D. R. 259
Brown, G. E. 27, 170, 172, 198, 205, 264, 267, 270, 272
Brückner, K. A. 29, 91–5, 97, 99, 108–9, 165–7
Bryan, R. A. 27
Buchler, J. R. 91–5, 99, 108–9, 167, 307
Buenerd, M. 244, 271, 279
Bunatian, G. G. 146
Burov, V. V. 203
Burzlaff, J. 307

Callaway, J. 57–8, 61, 164
Campi, X. 47, 104–5, 107, 310, 317
Capitani, J. F. 62
Castel, B. 275–6
Cavinato, M. 367
Celenza, L. S. 183
Chandra, H. 294, 300, 317
Chin, S. A. 321
Chirico, J. H. 99
Chu, Y. H. 174
Civitarese, O. 319
Clark, R. C. 92–5, 99, 167
Cloizeaux, J. des 391, 393

Coleman, A. J. 55
Coleman, S. 321
Coon, S. 93–5, 97
Cooper, L. N. 211
Cottingham, W. N. 27
Cugnon, J. 295
Curtin, M. W. 321
Cusson, R. Y. 228

Dabrowski, J. 93–5, 97
Dahl, J. P. 57
Dalili, D. 252, 319
Damgaard, J. 137, 141, 146, 156, 158–9, 173
Dasso, C. H. 266, 319
Davidson, E. R. 57
Davies, K. T. R. 228
Day, B. D. 29, 82–3
De Shalit, A. 77
De Tar, C. 26
Dean, T. J. 272
Deb, B. M. 57, 63
Decowski, P. 245
Dehesa, J. S. 266
Delchev, Ya. 109
Delocle, I. 160
Deubler, H. H. 109
Di Toro, M. 248, 295
Diamond, R. M. 319
Dickhoff, W. H. 82, 110
Dimitrov, V. 235
Dimitrova, S. S. 198–9, 206–9, 238, 240, 242–6, 272, 280
Dines, E. L. 319
Dirac, P. A. M. 224
Donangelo, R. 315
Donnelly, I. J. 98
Donnelly, R. A. 57
Draper, J. E. 319
Dreizler, R. M. 57
Drenska, M. 197
Durand, M. 87–8, 122, 153, 290, 311–14, 317
Durso, J. W. 27
Dutta, A. K. 131, 137–8, 174, 176, 216
Dworzecka, M. 99, 228
Dymarz, R. 250

Eckart, G. 107, 255–9
Egido, J. L. 319
Eisenberg, J. M. 39, 171
Eldishev, Yu. N. 203
Ellegaard, C. 319
Engel, Y. M. 235–8, 255
English, H. 57
English, R. 57
Epstein, S. T. 35–6, 66–7
Erdahl, R. 57
Erell, A. 245
Euler, H. 84

Faber, M. E. 319
Fabris, D. 301
Faessler, A. 109, 110–12, 264–7
Farine, M. 32, 44, 46, 131, 137, 159, 174, 176, 211, 214–16, 138
Fallieros, S. 272, 275
Feldmeier, H. 228
Fermi, E. 84
Fernsndez Velicia, F. J. 311
Ferrell, R. A. 224
Feshbach, H. 77
Fleckner, J. 109
Flieshbach, T. 209, 211
Flocard, H. 32, 44, 46–7, 171, 228
Fock, V. 37
Foldy, L. L. 37
Friedrich, J. 46, 275
Frisch, O. R. 257

Gaardhoje, J. J. 319
Galin, J. 109
Garcias, F. 128, 152–3
Gazquer, J. L. 189
Ghosh, S. K. 57, 63
Giannoni, M. J. 233–4, 251
Giausiracusa, G. 295
Gilbert, T. L. 55
Gleissl, P. 241, 277
Goeke, K. 235–8, 255, 266, 270–1, 275–7
Gogny, D. 264
Goldhaber, M. 257
Goldstein, H. 118
Gombas, P. 57
Göritz, G.-H. 209, 211
Grammaticos, B. 113, 121, 124, 127, 228, 264
Grange, P. 295
Greiner, W. 39, 171, 228–9
Griffin, J. J. 228
Groenewald, H. H. 117
Gruemmer, F. 235
Guet, C. 32, 46–7, 124, 127–8, 130–3, 135–8, 145, 152, 154, 157–9, 160, 174, 176, 307–8, 313–14, 316–18
Gunnarson, O. 61
Gupta, U. 186

Håkansson, H.-B. 32, 46–7, 124, 128, 130–3, 135–8, 145, 154, 157–9, 160, 174, 176, 307–8, 313–14, 317–18

Hamada, T. 27
Hamamoto, I. 264
Hara, Y. 85
Harriman, J. E. 55, 162
Hartree, D. 37
Hedin, L. 61
Henderson, G. A. 63
Herskind, B. 319
Heyde, K. 32, 80
Hilf, E. 307
Hiller, B. 322
Hirschfelder, J. O. 66–7
Hodges, C. H. 113
Hodgson, P. E. 43, 63, 219
Hohenberg, P. 50–2, 56, 61
Holzwarth, G. 107, 251, 255–7, 259
Horowitz, C. J. 83
Hoshino, T. 265
Huang, K. 78
Huber, M. G. 265
Hull, M. H. 27

Irom, F. 245
Ismail, M. 110–11
Izumoto, T. 109

Jackson, A. D. 27, 82, 278
Jaeger, C. de 215–16, 222
James, A. N. 202
Jennings, B. K. 113, 120, 151, 174, 278
Jensen, A. S. 137, 141, 146, 156, 158–9, 173, 241, 257
Johnston, I. D. 27
Jorna, S. 92–3

Kadanoff, L. P. 247
Kalckar, F. 257
Kan, K. K. 228
Karadjov, D. 197
Katriel, J. 57
Kelly, M. M. 91, 108–9
Kerman, A. K. 12, 171, 274
Khadkikar, S. B. 109
Kirkby, P. 202
Kirkwood, J. G. 113, 120
Kirzhnits, D. A. 113, 120–1
Klein, A. 267
Klemt, V. 265–7
Knaak, S. 126
Knüpfer, W. 265
Ko, C. M. 170, 172, 198, 205
Köhler, H. S. 32, 46, 136
Kohn, W. 50–2, 56–8, 61, 161–2, 164
Kohno, M. 275
Kolomietz, V. M. 146
Koonin, S. E. 12, 274, 228
Koopmans, T. 39
Krappe, H. J. 137
Krewald, S. 109, 265–7
Krieger, S. J. 228, 235–8, 255
Krivchenkov, G. 149
Krivine, H. 32, 46, 82, 102–5, 105–6, 135, 138, 259, 227–9
Kryachko, E. S. 55, 57, 178, 181–3, 223
Kumar, K. 85
Kuo, T. T. S. 264
Küpper, W. A. 307
Kyrchev, G. 265

Lacombe, M. 27
Lamb, D. Q. 307
Landau, L. D. 247
Lane, A. M. 174–5, 267, 275–7
Larsen, S. M. 241
Larsson, S. E. 27
Lassila, K. E. 27
Lattimer, J. M. 290, 307, 311, 315
Le Couteur, K. J. 85
Leander, G. 27
Lejeune, A. 295
Leonardi, R. 272, 275–6, 277
Levit, S. 294, 296–7, 300–2, 304, 320, 322
Levy, M. 56–7, 76
Li, C. H. 265–7
Liberman, D. A. 26
Libert, J. 160
Lichtenstadt, J. 245
Lichtner, P. C. 228
Lieb, E. H. 55–6
Lin, Wing-fai 99, 165–6
Lin, Y. C. 90
Linden, R. 110–12
Lipparini, E. 138, 272, 275–7
Liu, K. F. 264, 267, 272
Loiseau, B. 27
Lombard, R. J. 92–6, 98–9, 100, 109, 165–9
Lombardo, U. 248, 295
Löwdin, P.-O. 8, 21, 37, 185
Lowe, B. G. 202
Ludeña, E. V. 55, 57, 189
Lukyanov, V. K. 203
Lundqvist, B. J. 61
Lundqvist, S. 57

McDonald, A. H. 62
McDonald, F. A. 27
McDonald, J. A. 259
McWeeny, R. 188
Machleidt, R. 83
Mädler, P. 229

Malfliet, R. 83
Malik, F. B. 111–12
Malov, L. A. 265
Mansour, H. M. M. 110
Marangoni, M. 267
March, N. H. 57, 61, 64, 76, 189
Marchev, E. 109
Marcos, S. 277, 319
Marinova, L. P. 49, 219, 220, 223
Marshalek, E. R. 270
Martorell, J. 174–5, 267, 275–9, 319
Maruhn, J. A. 228, 246
Maruhn-Rezwani, V. 228
Mass, D. 109, 169
Matveev, B. A. 340
Matzolff, M. 295
Mayer-Boericke, C. 245
Meitner, L. 257
Meldner, H. W. 99, 228
Mermin, N. D. 57–8, 281–3
Messiah, A. 10
Meyer, J. 160, 241, 277, 319
Migdal, A. B. 264
Mikhailov, I. N. 250
Mikolas, V. 62
Mishra, V. 265–7
Mishustin, I. N. 315
Moinester, M. A. 245
Möller, P. 137, 228
Morrell, M. M. 76
Morsch, H. P. 245
Mosel, U. 109, 209, 211, 249, 300, 317
Moss, J. M. 259
Moszkowski, S. 99
Mottelson, R. B. 23, 217–18, 259
Mulders, P. J. 26
Müther, H. 264
Myers, W. D. 132–8, 279

Naimark, M. A. 67
Navarro, J. 277, 319
Negele, J. W. 28, 31, 43, 47, 107, 147, 228–9, 295
Németh, J. 89, 90, 252, 319, 320
Nesterenko, V. O. 265
Neumann, J. Von 1, 20
Newton, J. O. 319
Ngô, C. 109, 252, 319, 320
Ngô, H. 109
Nilsson, S. G. 217
Nix, J. R. 137, 228, 250
Nogami, Y. 228, 89, 90, 252, 319

Ohtsuka, N. 110–12
Orlandini, G. 275–6
Ottaviani, P. L. 267
Oulne, M. 211, 215–16

Palke, W. E. 57
Panagiotou, A. D. 321
Pandaripande, V. R. 83, 98
Pant, M. M. 61
Parr, R. G. 57, 76
Pashkevich, V. V. 158, 173
Passler, K. N. 249, 300, 317
Pauli, H. C. 137, 141, 146, 150, 156, 158–9, 170, 172–3, 198, 205
Pearson, J. M. 32, 44, 46, 131, 137–8, 159, 176, 211, 214–16, 138
Petkov, I. Zh. 49, 63–7, 70–1, 106, 109, 178, 181–3, 193, 195, 197–9, 203–4, 206–9, 211, 215–19, 223, 283–5, 240, 242–6, 272, 288, 295, 304, 307, 311, 315
Pi, M. 128, 152–3, 310, 317, 319, 320
Piperova, I. 250
Polls, A. 82, 275, 277, 319
Pong, W. S. 183
Providencia, J. Da 57, 251, 270
Pucci, R. 76

Quentin, P. 32, 44, 46–7, 143–6, 130, 159, 160, 171, 176, 184, 206, 209, 234, 241, 277, 294, 299, 300, 303, 317, 319

Ragnarsson, I. 27
Raimes, S. 8
Rajagopal, A. K. 58, 60–1, 186
Ramos, A. 82
Rashdan, M. 110–11
Ravenhall, D. G. 307, 311, 315
Reichl, L. E. 280
Reid, P. V. 27
Reinhard, P.-G. 46, 228, 235, 270, 275–6, 307, 319, 321–2
Reinhardt, H. 307, 321
Richard, J. M. 27
Rikus, L. 109
Ring, P. 18, 148, 250, 267, 319
Rogge, M. 245
Ross, C. K. 150
Rowe, D. J. 235
Roy, M. K. 85
Rozsa, C. M. 259
Ruppel, H. M. 27
Russo, G. 248

Saarela, M. 27
Sagawa, H. 31, 255, 294, 297, 299, 319
Sandorfi, A. M. 319

Sartor, R. 109
Saruis, A. M. 267
Sauer, G. 294, 300, 317
Scheid, W. 99
Schmit, C. 227
Schrieffer, J. R. 211
Schrödinger, E. 1
Schuck, P. 18, 87–8, 122, 148, 250, 322
Schulz, H. 307, 315, 321
Schwesinger, B. 259
Scott, D. K. 321
Seeger, P. A. 137
Serot, B. D. 83
Serr, F. E. 255
Seyler, R. G. 85, 134
Shakin, C. M. 183
Sham, L. J. 161–2
Shirikova, N. Yu. 265
Shlomo, S. 241
Siciliano, E. R. 267
Sick, I. 221–2
Siemens, P. J. 89, 90–1, 241, 264, 321
Sierk, A. J. 137, 228, 250
Simon, B. 65
Skyrme, T. H. R. 29, 30
Slater, J. C. 85
Slavov, B. 235
Smith, R. K. 228
Smith, V. H. 57, 63
Sneppen, K. 315
Snover, K. A. 319
Soloviev, V. G. 264–5
Sorba, P. 340
Speth, J. 264, 265–7, 271
Sprung, D. W. L. 47
Stancu, F. 109, 209, 211
Starodubsky, V. E. 267
Steadmon, S. G. 319
Stegun, I. A. 115
Steinwedel, H. 257
Stock, H. 99
Stöcker, H. 228–9
Stocker, W. 126, 241, 307, 317
Stoddart, J. C. 61
Stoitsov, M. V. 49, 64–7, 70–1, 106, 178, 181–3, 193, 195, 197–9, 203–4, 206–9, 211–12, 215–19, 223, 238, 240, 242–6, 250, 272, 283–5, 288, 295–9, 301, 304, 319
Stoner, E. C. 306
Stoyanov, C. 265
Strayer, M. R. 228
Stricker, K. 255
Stringari, S. 105, 110, 138, 249, 259, 272, 275–6, 277, 317
Strumberger, E. 316, 318
Strutinski, V. M. 139, 141, 143, 146, 156, 158–9, 173
Sukosd, C. 245
Suraud, E. 307, 320–1
Süssmann, G. 126
Sutcliffe, B. T. 188
Suzuki, T. 255, 275
Svenne, J. P. 42
Swart, J. J. de 26
Swiatecki, W. J. 132–7, 137–8, 279

Tamain, B. 109
Tassie, L. J. 257
Teller, E. 257
Ter Haar, B. 83
Theophilou, A. L. 62
Thomas, L. H. 84
Thouless, D. J. 38, 41, 85, 225, 262, 270
Toepffer, C. 319
Toki, H. 294, 297, 299, 321
Tomasek, M. 62
Tomasi, E. 319, 320
Tondeur, F. 32, 44, 46–8, 138, 159, 169, 175–6, 211, 214–16
Traini, M. 275–6
Trefz, M. 110
Treiner, J. 32, 46, 82, 102–6, 137–8, 227, 259, 277–9, 307, 310, 319, 320–1
Tsai, S. F. 245, 264, 267
Turek, P. 245
Tyapin, A. S. 158

Überall, H. 257

Vaishvila, Z. 250
Vallieres, M. 123
Valone, S. M. 56, 62, 182
Van Giai, N. 31–2, 44, 46–7, 56, 267
Van der Woode, A. 266, 271
Vashishta, P. 58, 164
Vautherin, D. 31–2, 43–4, 46, 107, 147, 171, 197–9, 201–4, 207–8, 235–8, 290, 294–5, 251, 255, 296–7, 300–2, 304, 319, 320, 322, 355
Vdovin, A. I. 265
Vénéroni, M. 133–4
Villars, F. 325
Viñas, X. 128, 152–3, 310, 317
Vinh Mau, N. 319
Vinh Mau, R. 27
Vlasov, A. 274
Voronov, V. V. 265
Voros, A. 113, 121, 124, 127
Vries, C. de 215–16, 222
Vries, H. de 215–16, 222
Vwnwroni, M. 227, 229, 234–5, 253, 231–2
Vwnwroni, M. 251

Wadia, W. 110–11
Waghmore, R. Y. 264
Walecka, J. D. 321
Wambach, J. 265–7
Warke, C. S. 228
Waroquier, M. 32, 80
Wegmann, G. 307
Weiss, M. S. 138, 228
Weizsäcker, C. F. Von 74, 84, 132
Welton, T. A. 246
Wenes, G. 32, 80
Werner, E. 264
Werntz, C. 257
Wheeler, J. A. 257
Wigner, E. 113, 116, 120
Wild, W. 264
Wilets, L. 85
Wiringa, R. B. 83, 98
Wong, C. Y. 141, 146, 156, 158–9, 173, 246, 295

Yang, W. 57
Yoon, B. 228
Young, W. H. 189
Youngblood, D. H. 259

Zint, P. 211, 249, 300, 317

Subject index

adiabatic
  approximation 111–12
  time-dependent Hartree–Fock
    theory 229, 235, 253, 277
    quantized theory 235
    equations 230–3, 237–8, 262
annihiliation operator 14–16, 20, 59, 282, 291, 297
antisymmetrization 6, 40, 109
antisymmetrized
  matrix elements 17, 39
  two-body interaction 140
asymmetry 81, 82, 99, 135, 278–9, 316

^{138}Ba 216
bar nucleon force 27
Bloch
  density matrix 113–15, 118, 119, 149, 308–9
  equation 118, 149
Boltzmann
  constant 281
  equation 247
boson systems 4
bound states 299, 301
bubble formation 321–2

^{40}Ca 47–8, 99, 100, 112, 131–3, 145, 167, 198, 200, 203–4, 212, 216, 221–2, 246, 276, 302–3, 320
^{48}Ca 99, 133, 167, 216
^{140}Ce 99, 133
classical
  Hamiltonian 229, 234, 237, 240, 242
  turning point 87, 122, 152, 313
classically
  allowed region 87, 122
  forbidden region 123, 313
collective
  dynamics 256
  energy 251
    kinetic 232, 235, 237, 240–1, 251
    potential 232, 234–5, 237, 240–1, 251
  Hamiltonian 234
  mass tensor 234
  motion 227, 229, 230, 232–3, 240, 255, 260–2, 265, 274, 319
  parameters 233
  path 233, 135–6, 241
  strength 272
  variables 236, 238, 241
  vector 233, 242
  velocity 255
  wave functions 234
colliding nuclei 108–9, 111
collision
  term 247
  time 111
compound nucleus
  excitation energy 296, 315
  formation 228, 280
  multi-fragmentation 315
  stability 301
constrained Hartree–Fock
  calculations 160, 171–2, 275
  problem 171, 173, 236–7
  method 156, 170, 235, 253
  procedure 171, 174
  potential 173
  energy 276
Coulomb
  energy 94, 102, 302, 316
  force 45
  instability 302
cranking model 227
creation operator 14–16, 20, 59, 263, 282, 287, 291
current density 60–1, 239

deformation
  Fermi sphere 257, 259
  density 157, 171
  energy 137, 148, 160, 169, 171, 173, 175
  hexadecapole 155
  parameters 139, 155–6, 158, 173, 313
  quadrupole 103, 155, 171
deformed nuclei 110, 155, 159
density
  distribution 22, 50, 97, 240, 241, 246, 255
  functional 50, 58, 64, 72, 76, 84, 87–8, 101, 240, 253, 255, 259, 276
  matrix 19, 20, 23, 25–6, 55, 58, 107
    decomposition 230, 233
    expansion 107
    isospin 23
    one-body 20–1, 23, 63, 73, 78, 85, 88, 90, 113, 115, 117, 119, 140, 224–5, 229, 253, 255–6, 261
    operator 20, 58, 280–4, 286, 290–1, 294–5
    operator space 281, 284
    time-dependent 226, 229

density (*contd.*)
matrix (*contd.*)
time-even 229, 254
two-body 20, 24–5, 225
density-dependent forces 42–3
density-independent forces 42
dielectric theorem 269, 271
drip-line 137, 155

effective mass 102, 125, 128, 170, 311
eigen
class 284–5
functions 2–3, 6, 11, 33, 51–2, 114, 263, 268
orbit 68, 70
solution 75
values 2–3, 9, 10, 33, 36, 39, 62, 68, 75, 90, 97, 114, 164
energy
averaged 150, 155
kinetic 89, 92, 101, 103, 105, 109, 110, 121, 123, 125, 134, 153, 161
oscillating part 140, 142, 144, 146, 150–1, 154, 156, 173
potential 92, 110, 111, 119, 129, 160, 165
smooth part 139, 140, 144, 146, 150, 152, 154–6
surface 94, 107, 134, 227
symmetry 94–5, 111, 134
total 87, 96, 101, 105, 108–9, 111, 122, 134, 144, 164, 166, 171, 227, 232, 316
Energy Density Formalism
densities 166
calculations 98, 165
EDF 91, 93, 99, 109, 166
set of parameters 94
Energy Density Method 166, 168–9, 170
entropy 286, 288–9, 292, 296, 299, 313, 319
Euler–Lagrange equation 54, 56, 72, 75, 79, 86, 90, 96, 98, 104, 127, 152, 154, 162, 290, 296
Euler-like equation 238
evaporation time 290
excited
nuclei 280, 317
states 2, 12, 36–7, 61–3, 261, 264, 267, 319
density 62
Expectation Value Method 170, 172
Extended Thomas–Fermi model 131, 149, 155–6, 174, 227, 311, 315
external field 269, 237

^{56}Fe 216
Fermi
energy 141
integrals 306, 311
level 115, 297
liquid 257, 259
momentum 84, 86, 256
sphere 110, 240, 259
surface 139, 166, 168, 258, 259, 264
Fermi–Dirac
distribution 293
occupation numbers 308
Fermi-type local density 97, 136, 241, 257, 278, 296
fermion systems 4
fission
barrier 155–6, 159, 160, 317–18
barrier height 156, 160
mode 257
processes 317
fluid dynamic
continuum 259
equations 250, 252, 255
model 224, 245, 250, 257, 259, 277
ordinary fluid 257

G-matrix 29, 30–1, 42–3, 91, 110
gas of nucleons 290–1, 301, 320, 321
gas phase 299
generalized
occupation numbers 142
scaling
assumption 259, 274
model 259
property 237–8
Generator Coordinate Method 227
giant monopole resonances
excitation energies 243, 278, 279
experimental data 243, 279
DWBA analysis 245
isovector 240, 242–3, 320
EWSR 242, 268
isoscalar 240, 242–3, 279, 320
Gibbs free energy 280–2, 284, 289, 290–2, 295, 305
gradient corrections 93, 130
grand potential 57–8, 61, 280, 282, 290, 301
ground-state
density 52, 73, 75, 294
energy 11–12, 33, 35, 51, 53, 60–1, 68, 241, 255, 258, 274
minimum 227
problem 10, 57, 73, 294
solution 51
wave function 5, 52–3, 68, 70, 268, 274

Hartree–Fock
Hamiltonian 39, 123
calculations 42–3, 99, 155

energy density 124, 128
equations 39, 40, 43–4, 140, 164, 262
ground-state 39, 140, 155, 263, 270
iterations 41, 155, 174
kinetic energy 103, 105
local density 105, 129
potential 40, 42, 45, 85, 116, 228
single-particle spectrum 264–5
theory 37, 39, 41–2, 44, 85, 88, 91, 101, 104, 140, 143, 170, 174, 226–7, 232, 253
wave functions 39
^{4}He 73–5, 193, 212
Helmholtz free energy 286, 301, 311, 315
hydrodynamics limit 259
models 258–9
nuclear 249
ordinary 249, 258
Hohenberg–Kohn theorem 50–1, 57, 61–4, 113, 124, 224, 252
Hohenberg–Kohn–Mermin theorem 57–8, 60, 283, 285

idempotent condition 85, 226
imposed collective motion 253, 255–6
Improved Thomas–Fermi model 104, 257
independent-particle model 6, 40, 85, 104
inhomogeneity term 93, 105
isospin 2–3, 7, 23, 27, 77, 228, 242

kinetic equation 247
Kohn–Sham
energies 164–6
equations 163, 166
orbital 161, 163–4
theory 161, 163–6, 168, 170

Lagrange multipliers 12, 37, 44, 54, 95, 98, 162, 171, 236, 289, 293
Lagrangian 13, 15–17, 19, 37, 51
Laplace transform 308
leptodermous expansion 134–5, 137, 278, 315
Levinson theorem 302
Levy–Lieb approach 56, 58
Liquid Drop(let) Model 132–3, 138
at temperatures 315–16
difficulties 137
parameters 134–5, 315–16
mass formula 135, 137
barriers 318
improvements 137
liquid-vapour phase 307
local-scale transformation
transformations 64, 66–8, 71, 283–4, 287
function 64, 66, 71
group 64
method 70, 71–3
version of ATDHF 240
long-range interaction 89, 90, 93
low-lying resonances 264
low-temperature expansion 310, 317

major shell 141, 296–7
many particle
systems 33, 68, 90, 280
wave function 1, 3, 13, 22, 33, 52, 72, 252
many-body problem 1, 13, 19, 20, 26, 32, 42, 73, 76, 84, 113, 248, 292
mean field 4, 264, 266, 291–2, 301, 321
method of moments 248, 250
Modified Thomas–Fermi Model 107
monopole
scaling 104
strength function 278
vibrations 241
multipole resonances
dipole 254, 257–8, 272
on excited states 319
quadrupole 254, 257–9, 272

natural
occupation numbers 21
orbital 21, 76
neutron excess 93
^{56}Ni 99, 133, 145, 246
^{58}Ni 48, 132
non-collective states 264
non-local parameter 239
non-normal parity modes 260
N-representability 55–6
nuclear
force 27, 30, 91
incompressibility 277
matter 77–8, 80–4, 86, 89, 90, 106, 109, 110, 128, 135, 137, 168, 278, 306
at temperature 307, 321
density 79, 90
incompressibility 81, 129, 270
kinetic energy 79
saturation curve 79, 92–3
polarization 109
nucleon–nucleon forces 26, 27, 31, 42, 46, 107, 134, 136, 160, 168

^{16}O 47, 99, 133, 145, 198, 200, 203–4, 212, 216, 246, 266, 276
occupation
numbers 9, 13, 14, 21, 140–2, 168, 292, 294, 308, 319
representation 13, 21
probabilities 291, 293, 298

open channels 109
operator
  constraint 172, 235, 253, 271, 275
  density matrix 20, 58, 280–4, 286, 290–1, 294–5
  isoscalar 272
  isovector 272
  monopole 63
  multipole 271
  projection 36
  spin-exchange 30, 273
  time-odd 229, 253
  time-reversal 227
  transition 263
orbits 67–9, 70–1, 241, 284

Pauli principle 2, 40, 110
^{202}Pb 167
^{208}Pb 47–8, 99, 100, 131–3, 145, 167–8, 175, 198, 201, 203–4, 206, 212, 216, 246, 276, 297–9, 300, 304
^{226}Pb 216
phase
  equilibrium 289, 290
  space 248
poisson brackets 118
polarizability 269, 270
potential
  OPEP 27
  Woods–Saxon 85, 87, 128, 147, 153, 166
  chemical 54, 58, 61, 75, 87, 90, 162, 280–1, 289, 293–5, 310
  exchange 40, 45
  external 51–2, 54, 63
  four vector 59
  harmonic oscillator 85, 149, 150
  optical 82, 108
  rearrangement 43
  scattering 108–9, 110–11
  single-particle 114, 163, 165
  soft-core 27
  spin-orbit 45, 166
  vector 61
pressure 280, 295
^{240}Pu 145, 155, 160, 318

quadrupole moment 146, 148
quantization 235, 242
quasibound states 299, 301

Random Phase Approximation
  at finite temperature 319
  1p-1h model 263, 264–5, 267
  2p-2h model 265
  equations 262, 264
  ground-state 263
  RPA 261, 264, 277, 319
  shell model 264–5
  strength distribution 270
Ritz variational method 35
rms radii
  charge 94, 98, 130, 168
  mass 278, 296
  neutron 105
  proton 105
r-process 137

^{32}S 216
Schrödinger equation 1–9, 10–13, 25–6, 32–3, 36, 45, 51–4, 68, 77, 114, 224
second-quantization 15–17, 19, 37, 51
semiclassical
  approximation 102, 114, 259
  averaging 150
  densities 313
  expansion 113, 119, 120, 122, 127, 149, 151, 250, 309
    re-summation 122–3
  functional 303
  limit 149, 247
shape isomeric state 317
shell
  correction method 137, 139, 140–1, 146, 149, 154, 158, 173, 176
  density oscillations 166, 174
  effects 143, 149, 152–4, 156, 175, 303, 318
  model 150, 146–7, 265, 296, 300, 314
    calculations 139, 172
    level density 116, 141
short-range interaction 89, 90
single-particle
  basis 21, 22, 270
  energies 39, 41, 143, 161, 293, 305, 308
  Hamiltonian 39, 85, 114, 118, 123, 140, 142–4
  levels 115, 264
  operator 2, 21, 115, 116, 171
  orbital 21, 168
  problem 173
  propagator 114
  spectrum 146, 168, 264–5, 296
  states 6, 8, 39, 40, 77–8, 140, 251, 264, 291, 294, 296–7, 304
  structure 166
  wave-functions 6, 39, 40, 44, 85, 103, 140, 142, 152, 156, 161–4, 168, 171
Skyrme–Orce forces 46
Skyrme-type effective forces
  SI 32, 46, 81, 198, 204, 208, 212
  SII 32, 46, 81, 266, 300

SIII 32, 46–7, 81, 136, 145, 147–8, 160, 206, 209, 246, 276
SIV 32, 46, 81
SV 32, 46, 81, 246
SVI 32, 46, 81, 175
Ska 32, 46, 81, 136, 160, 246
Skb 32, 46, 81, 136
SkE2 32, 46, 80–1
SkM 32, 46, 48, 81, 145, 160
SkM* 32, 46–7, 81, 131–3, 136, 160, 243–4, 246, 298, 302–3, 316, 317
T 32, 46, 81
T1 32, 81, 214–15
T1* 214–15, 245
Slater
approximation 85, 88
determinant 8–9, 13, 37–8, 43, 77, 85, 164, 170–1, 173, 225–6, 229, 241, 250
time-even 227, 230–3, 235, 237, 240, 253
^{114}Sn 133, 145
^{116}Sn 132, 167
^{123}Sn 133
^{124}Sn 132, 167
^{132}Sn 216, 246
space
configuration 50, 116, 173, 227, 296, 300
coordinate 6, 30, 40, 46, 227, 235–6, 267
Fock 13, 291
Hilbert 1, 4, 6, 8–9, 36, 65–8, 226, 284
momentum 30, 63, 78, 110, 250, 257–8
R^3 50, 54, 64, 122, 152, 314
trial 37, 97
variational 37, 97, 130, 156
velocity 249
spin 2, 7, 23, 61, 77, 228
-exchange operator 30
-density 43, 61, 125, 128, 309, 310
-orbital splitting 166
-saturated system 31
spinless 23, 51
subspace 4–6, 12–13, 36–7, 97
sudden approximation 111–12
sum rules
RPA 270, 271, 275, 276
approach 261, 271, 275–7, 319, 320
moments 268
symmetry
energy 94–5, 111, 134, 136, 316
operators 3, 227
term 165

temperature
critical 299
effects 317
extreme 280
inverse 281
zero 288, 294, 308–9, 314–15
tensor force 89
Thomas–Fermi
approximation 84–5, 87, 91, 99, 110, 256, 277, 304, 307, 310, 317, 322
Bloch matrix 119
density matrix 85, 256
distribution 258
equation 87, 90, 252
functional 88, 98, 240, 277–8, 318
kinetic energy 86, 92, 103, 109, 125, 134, 306, 317
local density 87
spin-orbit potential 166
theory (failure) 90
Thomas–Fermi-like theories 85
Temperature-Dependent HF
calculations 294, 299, 301, 313, 315
density matrix 282, 308
operator 294
equations 293, 296, 301
Hamiltonian 293, 300
local density 296
subtraction procedures 299, 302
theory 282, 292, 294, 297, 319
truncation procedures 299, 302, 322
vapour solution 301'
Time-Dependent HF
calculations 228
equations 226–8
Hamiltonian 226
theory 224, 229
trajectories 227
transition density 244
trial
class 33, 285–7
densities 56, 135, 152, 156–7
density matrix operator 282, 283, 291–2
family 233
space 37, 97
variational set 33–7, 68, 97
wave functions 34–5, 37, 53, 68–9, 71, 250
twist mode 260

^{235}U 99

vapour
nucleons 299, 302, 304
phase 301, 302
variational
method 33–7, 68, 97–8, 130, 282
principle 9, 10–12, 33, 53, 71, 94, 131, 154–5, 174, 259, 281–2, 284–5, 292
time-dependent 12, 62

velocity field 238, 240, 242, 260
Vlasov equation 248
v-representability 55–6

Weizsäker term 75–6, 93, 95–7, 101, 104, 107, 111–12, 129, 288, 306–7
Weyl quantization rule 117
Wigner
  distribution function 117, 247
  representation 116
  transform 116–19, 120, 247, 254–6
Wigner–Kirkwood expansion 122–4, 309

$^{298}X_{114}$ 207–9
$^{354}X_{126}$ 216

^{168}Yb 145, 147–8

zero-point energy 235
zero-range force 30
^{80}Zr 99, 133
^{90}Xr 47–8, 99, 100, 145, 198, 203–4, 212, 216, 246, 276, 321